Perspectives and Challenges of Hair Analysis

Issues in Toxicology

Editor-in-chief:
Diana Anderson, *University of Bradford, UK*

Series editors:
Michael D. Waters, *Michael Waters Consulting, USA*
Timothy C. Marrs, *Edentox Associates, UK*
Alok Dhawan, *CSIR-Indian Institute of Toxicology Research, India*

Titles in the Series:

1: Hair in Toxicology: An Important Bio-Monitor
2: Male-mediated Developmental Toxicity
3: Cytochrome P450: Role in the Metabolism and Toxicity of Drugs and other Xenobiotics
4: Bile Acids: Toxicology and Bioactivity
5: The Comet Assay in Toxicology
6: Silver in Healthcare
7: *In Silico* Toxicology: Principles and Applications
8: Environmental Cardiology
9: Biomarkers and Human Biomonitoring, Volume 1: Ongoing Programs and Exposures
10: Biomarkers and Human Biomonitoring, Volume 2: Selected Biomarkers of Current Interest
11: Hormone-Disruptive Chemical Contaminants in Food
12: Mammalian Toxicology of Insecticides
13: The Cellular Response to the Genotoxic Insult: The Question of Threshold for Genotoxic Carcinogens
14: Toxicological Effects of Veterinary Medicinal Products in Humans: Volume 1
15: Toxicological Effects of Veterinary Medicinal Products in Humans: Volume 2
16: Aging and Vulnerability to Environmental Chemicals: Age-related Disorders and their Origins in Environmental Exposures
17: Chemical Toxicity Prediction: Category Formation and Read-Across
18: The Carcinogenicity of Metals: Human Risk Through Occupational and Environmental Exposure
19: Reducing, Refining and Replacing the Use of Animals in Toxicity Testing
20: Advances in Dermatological Sciences
21: Metabolic Profiling: Disease and Xenobiotics
22: Manganese in Health and Disease
23: Toxicology, Survival and Health Hazards of Combustion Products
24: Masked Mycotoxins in Food: Formation, Occurrence and Toxicological Relevance
25: Aerobiology: The Toxicology of Airborne Pathogens and Toxins

26: Chemical Warfare Toxicology, Volume 1: Fundamental Aspects
27: Chemical Warfare Toxicology, Volume 2: Management of Poisoning
28: Toxicogenomics in Predictive Carcinogenicity
29: Human Stem Cell Toxicology
30: The Comet Assay in Toxicology, 2nd edition
31: Computational Systems Pharmacology and Toxicology
32: Ecotoxicology and Genotoxicology: Non-traditional Terrestrial Models
33: Ecotoxicology and Genotoxicology: Non-traditional Aquatic Models
34: Oxidative Stress and Redox Signalling in Parkinson's Disease
35: Nanotoxicology: Experimental and Computational Perspectives
36: Regulatory Toxicology in the European Union
37: Formaldehyde: Exposure, Toxicity and Health Effects
38: Chemical Health Threats: Assessing and Alerting
39: The Micronucleus Assay in Toxicology
40: Nanoparticle–Protein Corona: Biophysics to Biology
41: Big Data in Predictive Toxicology
42: Challenges in Endocrine Disruptor Toxicology and Risk Assessment
43: Bisphenol A: A Multi-modal Endocrine Disruptor
44: Marsupial and Placental Mammal Species in Environmental Risk Assessment Strategies
45: Bird and Reptile Species in Environmental Risk Assessment Strategies
46: Amphibian Species in Environmental Risk Assessment Strategies
47: Perspectives and Challenges of Hair Analysis

How to obtain future titles on publication:
A standing order plan is available for this series. A standing order will bring delivery of each new volume immediately on publication.

For further information please contact:
Book Sales Department, Royal Society of Chemistry, Thomas Graham House, Science Park, Milton Road, Cambridge, CB4 0WF, UK
Telephone: +44 (0)1223 420066, Fax: +44 (0)1223 420247
Email: booksales@rsc.org
Visit our website at books.rsc.org

Perspectives and Challenges of Hair Analysis

Edited by

Pascal Kintz
X-Pertise Consulting, France
Email: pascal.kintz@wanadoo.fr

Alberto Salomone
University of Turin, Italy
Email: alberto.salomone@unito.it

and

Marco Vincenti
University of Turin, Italy
Email: marco.vincenti@unito.it

ROYAL SOCIETY OF CHEMISTRY

Issues in Toxicology No. 47

Print ISBN: 978-1-83916-726-3
PDF ISBN: 978-1-83767-194-6
EPUB ISBN: 978-1-83767-195-3
Print ISSN: 1757-7179
Electronic ISSN: 1757-7187

A catalogue record for this book is available from the British Library

The Royal Society of Chemistry is a charity, registered in England and Wales, Number 207890, and a company incorporated in England by Royal Charter (Registered No. RC000524), registered office: Burlington House, Piccadilly, London W1J 0BA, UK, Telephone: +44 (0)20 7437 8656.

For further information see our web site at www.rsc.org

Printed in the United Kingdom by CPI Group (UK) Ltd, Croydon, CR0 4YY, UK

Contents

Issues in Toxicology No. 47
Perspectives and Challenges of Hair Analysis
Edited by Pascal Kintz, Alberto Salomone and Marco Vincenti

Published by the Royal Society of Chemistry, www.rsc.org

CHAPTER 1

Anatomy and Biology of Hair at Different Ages

A. L. PELISSIER-ALICOT

Aix Marseille Univ, APHM, Hôpital La Timone, Service de Médecine Légale, Marseille, France
Email: apelissier@ap-hm.fr

1.1 Introduction

Hair is a common characteristic of all mammals. Beyond its aesthetic and social properties, it protects the skin from mechanical insults, participates in thermoregulation, and allows the perception of touch to be extended beyond the skin area. Hair is currently considered to be a highly complex functioning "mini-organ", that presents a continuous cycle of alternating growth and regression phases, in an environment requiring stem cells, surrounding niche components and various signaling pathways.[3] This regeneration also affects the pigment compartment that gives hair its colour. To build this complex structure, which is also unique in humans, sophisticated morphogenesis is required during embryonic development.[1–4] The aim of this chapter is to provide a comprehensive understanding of the anatomy and biology of the hair follicle, as well as the mechanisms that control its organogenesis.

1.2 Adult Hair

1.2.1 Classification of Human Hair

Mammalian skin is covered with body hair over almost the entire body surface, with the exception of the sole of the foot, the palm of the hand, the

Issues in Toxicology No. 47
Perspectives and Challenges of Hair Analysis
Edited by Pascal Kintz, Alberto Salomone and Marco Vincenti

Published by the Royal Society of Chemistry, www.rsc.org

oral side of the lip, and some areas of the external genitalia. These hairs are generally fine, with very little melanization, and grow very slowly. Others, such as head hair, eyelashes, and eyebrows, are longer, more pigmented, and more resistant.[1] From a macrostructural point of view, head hair varies in form (straight, wavy, curly), but also in length, diameter, color, and cross-sectional shape among the different ethnic groups and among singular individuals. While human hair is generally classified according to three conventional ethnic subgroups, namely Africans, Asians and Europeans, de la Mettrie *et al.* have demonstrated that, due to the great complexity of human biological diversity resulting from both multiple and mixed origins, hair is best classified into eight major types, based on three parameters: curl diameter, curl index, and number of waves.[5] This method allows a coherent classification applicable in dermatology as well as in forensic sciences and anthropology.

1.2.2 Composition of Human Hair

Hair is composed, depending on its moisture content, of approximately 65–95% proteins, 1–9% lipids, 0.1–5% pigments, 0.25–0.95% of trace elements and heavy metals, and water. 80% of hair proteins are made up of keratin, a fibrous protein that contains a high level of amino acids, mainly cysteine, which may be degraded and afterwards may be re-oxidized in a disulphide bonding form, and which gives the hair mechanical resistance, flexibility, durability, and functionality.[6–8]

1.2.3 Anatomy of Adult Hair

Hair is divided into two distinct parts: the hair shaft, visible on the surface of the scalp, and the root, which extends obliquely into the epidermal epithelium of the skin to a depth of about 3–4 mm.[6] The root includes the hair follicle and its appendages, the sweat and sebaceous glands and the arrector muscle, which form the pilosebaceous unit, as well as blood capillaries and nerve endings.[2] Surrounded by several layers, the root is embedded in the pilary canal.[1,2,6,9,10]

1.2.3.1 Hair Shaft

The hair shaft is composed of 95% dead and fully keratinized epithelial cells, and consists of three concentric layers from the periphery to the core (Figure 1.1). The outer layer, or cuticle, is composed of flattened rectangular cells which overlap the distal direction of the thread. The overlap is arranged so that only 1/6 of them are exposed. There are approximately 8–11 layers of cuticle, each layer consisting of a single cell. Each cell has a rectangular shape, a diameter of about 50–100 μm and a thickness of 0.5 μm.[8,11,12] It is made up, from the periphery to the center, of a thin outer membrane, the epicuticle, about 3 nm long, which is a protein envelope covered with a strong lipid structure; the A layer, mainly composed of cysteine (>30%), which gives the cuticle its resistance to physical and chemical attacks; layer B,

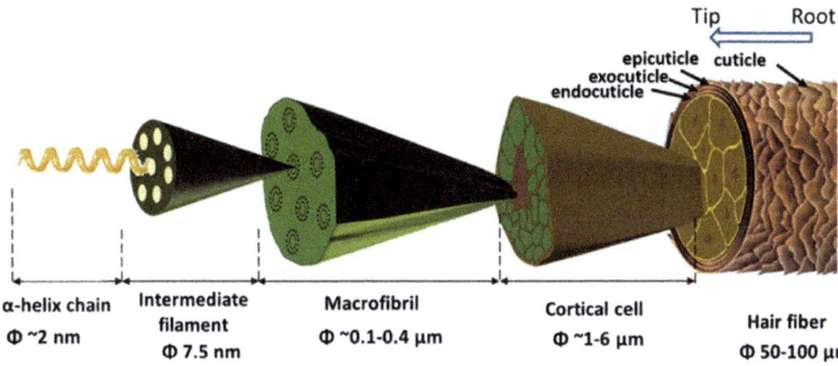

Figure 1.1 Schematic representation of the architectural organization of the human hair shaft. Reproduced from ref. 12 with permission from Elsevier, Copyright 2022.

or the exocuticle, also rich in cysteine (>15%) and physically rigid; and finally the endocuticle, less rich in cysteine and less rigid than the previous layers, which absorbs water when the hair is wet. The cuticle cells are cemented together by the cellular membrane complex (CMC), rich in essential fatty acids, in particular 18-methyl-eicosanoic acid (18-MEA), which is covalently bound to the proteins of the cuticle cells. 18-MEA is the main lipid in the hair composition contributing to the wet and dry combing properties.[8,13] Recently, Takayashi *et al.* identified another structure, located under the endocuticle of the innermost cell layer, which would ensure the anchoring of the cuticle to the cortex and protect the latter against external aggression, called CARB, for cuticle anchored resistant base.[14] CARB, rich in cystine and glycolipids, appears at the final stage of keratinization of the cuticular layers.[15]

The middle layer, or cortex, represents 90% of the total weight of the hair and forms the primary structure of the hair shaft. The elongated, interdigitated, spindle-shaped cortical cells, about 100 μm long and 1–6 μm thick, are merged and aligned or slightly inclined to the direction of the longitudinal axis of the fiber. They are composed of about ten macrofibrils, themselves composed of several hundred intermediate filaments or microfibrils, each microfibril comprising eight protofilaments. Each protofilament contains chains composed of four helices of α-keratin. The cohesion of the filaments is ensured by an interfilamentary matrix, rich in sulfur proteins, called keratin-associated proteins (KAPs). There are more than 100 KAPs grouped into three classes according to their composition (rich in cysteine, ultra-rich in cysteine and rich in glycine–tyrosine). These proteins establish disulphide bonds and hydrophobic interactions with keratin fibers.[16] The arrangement of the fibrils makes it possible to distinguish schematically two types of cortical cells: paracortical cells, the most abundant, present a stack of microfibrils in a compact hexagonal mode, the macrofibrils being fused and practically indistinct, and orthocortical cells, in which the macrofibrils are twisted along the longitudinal axis of the cell, the microfibrils being wound around the periphery of the macrofibrils. Some authors describe mesocortical macrofibrils, intermediate structures between the two previous ones.[8,12,17,18] This complex arrangement of fibrils, held together by KAPs, is essential for the elasticity and strength of the hair. Fibril arrangement varies according to the hair curvature. In straight hairs, all macrofibrils tend to adopt a straight conformation, as wavy hairs are characterized by interlaced orthocortical and mesocortical macrofibrils around paracortical macrofibrils. As the curl degree increases, the mesocortex disappears, leading to a majority of orthocortical cells.[19] The melanin granules, responsible for hair color, are located in the periphery of the cortex.[20]

The inner layer, or medulla, consists of a thin layer of pigmented keratinized cells, often disjointed by small air bubbles. It is abundant in beard hair, inconsistent in hair, and can present solutions of continuity. It does not seem to have any influence on the mechanical properties of the hair.[21,22]

1.2.3.2 Hair Follicle

The hair follicle (Figure 1.2) is composed of different clearly individualized compartments: the connective sheath and the dermal papilla, of dermal origin, the external epithelial sheath, the internal sheath, and the hair shaft of epithelial origin. Anatomically, the follicle comprises two main regions, delimited by the isthmus, a short portion located between the duct of the sebaceous gland and the protuberance of the arrector muscle.[10] The supra-isthmic region comprises the infundibulum, which is divided into an upper intra-epidermal part, the acro-infundibulum, and a lower intra-dermal part, the infra-infundibulum. The acro-infundibulum, in communication with the skin surface, is bordered by an epithelium in continuity with the epidermis, including a developed stratum corneum and a stratum granulosum.[2] The sub-isthmic region is formed by the hair root and its outer and inner epithelial sheaths. Between the sebaceous gland and the insertion of the arrector muscle, the bulge, considered by some authors as the distal zone of the bulb, is a specialized compartment of the outer root sheath which contains several contingents of totipotent stem cells (see below). One of them will give rise to the keratinocytes of the bulb and interfollicular space, as well as the sebocytes, and another contingent will provide melanocytes.[23–25] The permanent part of the hair follicle, between the surface of the skin and the lower end of the bulb region, does not undergo significant cyclical changes, unlike the lower part of the hair follicle.[2] The lower part of the follicle, which extends from the bulge to the base of the bulb, is surrounded by eight concentric sheaths from the periphery to the core:

- The conjunctive sheath, synthesized by fibroblasts, is an extracellular matrix formed of type I and III collagen and proteoglycans. It is in continuous with the dermal papilla and is crossed in its lower part by a capillary loop and by nerve endings.[1,2,10,26]

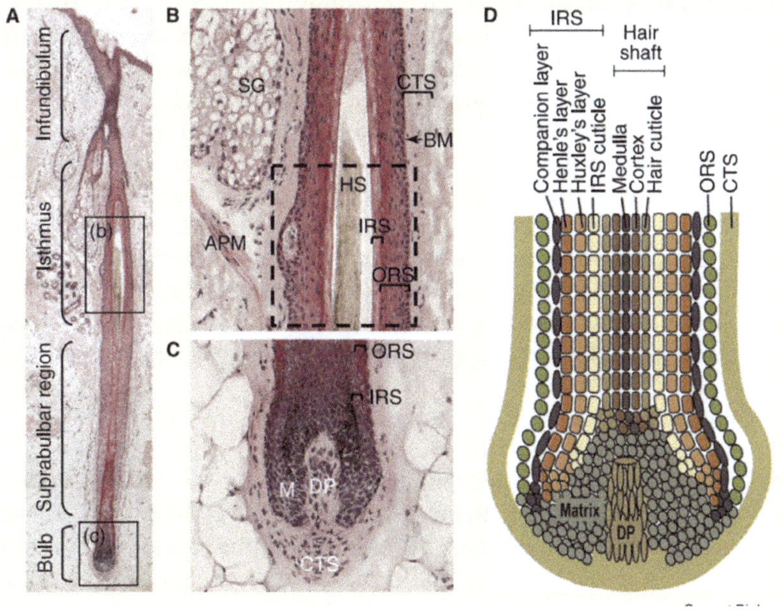

Figure 1.2 Histomorphology of the hair follicle. (A) Sagittal section through a human scalp hair follicle in the anagen phase, showing the permanent (infundibulum, isthmus) and anagen associated (suprabulbar and bulbar area) components of the hair follicle. (B) High magnification image of the isthmus. The dashed square indicates the approximate location of the bulge. (C) High magnification image of the bulb. (D) Schematic drawing illustrating the concentric layers of the bulb and the hair shaft. BM: basal membrane; CTS: connective tissue sheath; ORS: outer root sheath; IRS: inner root sheath, which is composed of the companion layer (CL), Henle's layer, Huxley's layer, and the inner root sheath of the cuticle; HS: hair shaft; SG: sebaceous gland; APM: arrector pili muscle; DP: dermal papilla; M: matrix. Reproduced from Ref. 10 with permission from Elsevier, Copyright 2022.

- The basal membrane, composed of type IV collagen, type 1 laminin, fibronectin, and proteoglycans, separates the dermal compartment from the epithelial compartment. The expression of some of its components such as laminin, is disrupted at the periphery of the bulb.[26,27]
- The outer root sheath, which extends from the matrix cells of the bulb to the orifice of the sebaceous duct, is keratinized from the isthmus. The cells of the outer root sheath express a wide variety of mediators, hormones, and receptors containing melanocytes, as well as Langerhans cells, dendritic cells involved in the immune response, and Merkel cells, rich in neuropeptides.[2,10,28]
- The inner root sheath, itself composed of three distinct layers: Henle's layer, a bed of cuboidal cells rich in trichohyalin granules, keratinizing early, Huxley's layer, made up of one or two layers of larger cells, whose

keratinization starts above the previous one, and finally the cuticle layer, consisting of a layer of flattened hyalinized cells from the isthmic region.[1] Some authors also recognize the existence of an additional layer, the so-called companion layer, which is single-celled and represents the transition between the outer and inner sheaths, over a length from the bulb to the isthmus.[10,29] According to Vogt *et al.*, it allows the inner root sheath to slide upward over the outer root sheath during hair growth.[2] The inner sheath narrows from bottom to top. From the middle of the isthmus, the different layers are merged into a single hyaline layer which disappears at the height of the isthmus. The cells of the inner sheath produce keratins and trichohyalin which serve as intracellular cement to support and shape the growing hair shaft.[1]

The follicle itself is divided into supra-bulbar and bulbar regions, the latter including the bulb and the dermal papilla.[2,10] The bulb is the portion of the follicle which actively produces the hair. It can be divided by the Auber's line into two regions. Below the Auber's line is the matrix, in contact with an expansion of the dermis, the dermal papilla, a mesenchymal invagination rich in densely packed fibroblasts, endothelial cells, hyaluronic acid, and chondroitin sulfate. It receives blood and lymphatic capillaries as well as nerve endings. Fibroblasts of the papilla secrete poorly differentiated matrix cells and anchor the follicle during the anagen phase.[1,10,25] They specifically express a large number of markers and growth factors (keratinocyte growth factor, bone morphogenetic protein, hepatocyte growth factor, insulin-like growth factor, stem cell factor, *etc.*), which generate complex signals for matrix cell proliferation, migration, and differentiation.[1,2,10] These signaling pathways also determine the size of the bulb, the dimension and curvature of the hair and the duration of the anagen phase (see below). In addition, the fibroblasts of the dermal papilla are thought to act as a reservoir of multipotent stem cells that differentiate into other cell lineages: chondrocytes, osteoblasts, and adipocytes.[30] The matrix zone, separated from the papilla by the basal membrane, is itself composed of several zones: the deep fertile zone, the melanocyte zone and the keratogenous zone. The deep fertile zone, in immediate contact with the papilla, is rich in epithelial progenitor cells which will multiply and divide at a very rapid rate during the anagen phase. The cells resulting from these divisions will accumulate keratin and melanosomes in the dedicated regions.[31] Then they move to the upper part of the bulb, which is divided into four parts. Above the Auber's line, in the wide portion of the bulb is the pre-elongation region in which the cells align themselves vertically and enlarge. Above the pre-elongation region is the cellular elongation region, narrower than the previous one, where the cells elongate. This zone ends about one third of the way from the tip of the papilla to the skin surface. At the end of this migration, the cells are enriched in keratin and pigments and have lost their nucleus.[1]

The follicle is, therefore, organized concentrically around a symmetry axis. The shape of the hair (straight, wavy, curly) is the result of an internal tension in the fiber: a follicle associated with a curly hair is characterized by an S-shaped curvature at the bulb, in relation to an asymmetry in the differentiation programs of the different compartments of the bulb, with the matrix cells proliferating more actively on the convex side than on the concave side of the retro-curvature, while the differentiation programs of the outer sheath, the inner sheath and, to a certain extent, the stalk (in particular the cuticle) start earlier on the concave than on the convex side.[32]

1.2.3.3 Melanogenesis

Melanocytes are oval or spindle-shaped dendritic cells, located at the apex on the dermal papilla, just below the precortical keratinocytes. The dendrites of the melanocytes insert between the cortical and medullary keratinocytes. Each melanocyte is connected to five keratinocytes to form the pigmentation unit. The dynamic process of hair pigmentation consists of three stages: the melanogenesis, the transfer of melanin granules to keratinocytes, and finally the formation of the pigmented hair shaft. Melanogenesis is characterized by a series of reactions, catalyzed by different enzymes, which result in the transformation of tyrosine into melanin pigments (Figure 1.3). This pigment production takes place during the anagen phase, under the control of neighboring cells, such as keratinocytes, fibroblasts and endothelial cells, which act by paracrine or autocrine mechanisms and can be modified by hormonal signals. This complex metabolism includes a common pathway, characterized by the conversion of L-tyrosine to L-DOPA (3,4-dihydroxyphenylalanine) and then of L-DOPA to DOPAquinone under the action of tyrosinase (TYR). The common pathway diverges at this stage. In the absence of cysteine, dopachrome is spontaneously and gradually converted to 5,6-dihydroxyindole (DHI) by decarboxylic rearrangement. This conversion can be accelerated by the tyrosinase-related protein-2 (TYRP2) (also called dopachrome tautomerase (DCT)), which catalyzes tautomerisation to 5,6-dihydroxyindole-2-carboxylic acid (DHICA). Polymerization of DHI and DHICA results in eumelanins. The role of TYR and TYRP1 (tyrosinase-related protein-1) in this last step is confirmed in mice but not yet in humans. The DHI eumelanins are responsible for the black or brown coloring, the DHICA eumelanins for the light brown coloring. In the presence of cysteine, dopaquinone is spontaneously oxidized to cysteinylDOPA to give 5-*S*- and 2-*S*-cysteinyldopa isomers with a 5 : 1 ratio. Oxidation of these cysteinyl dopa isomers leads to the production of the reddish-brown pigment pheomelanin through the intermediate compound 1,4-benzothiazinyl-alanine. Pheomelanins are responsible for the blond reddish coloring. These eu- and pheomelanins exist in reduced or oxidized form. In general, both pathways coexist and the hair colour is determined by the ratio eumelanin/pheomelanin which is genetically controlled.[31,33–38]

Melanosomes contain fibrillar or lamellar material with a characteristic periodicity. Mature melanosomes undergo a bidirectional migration from the perinuclear region to the tips of the dendrites. This intracellular movement

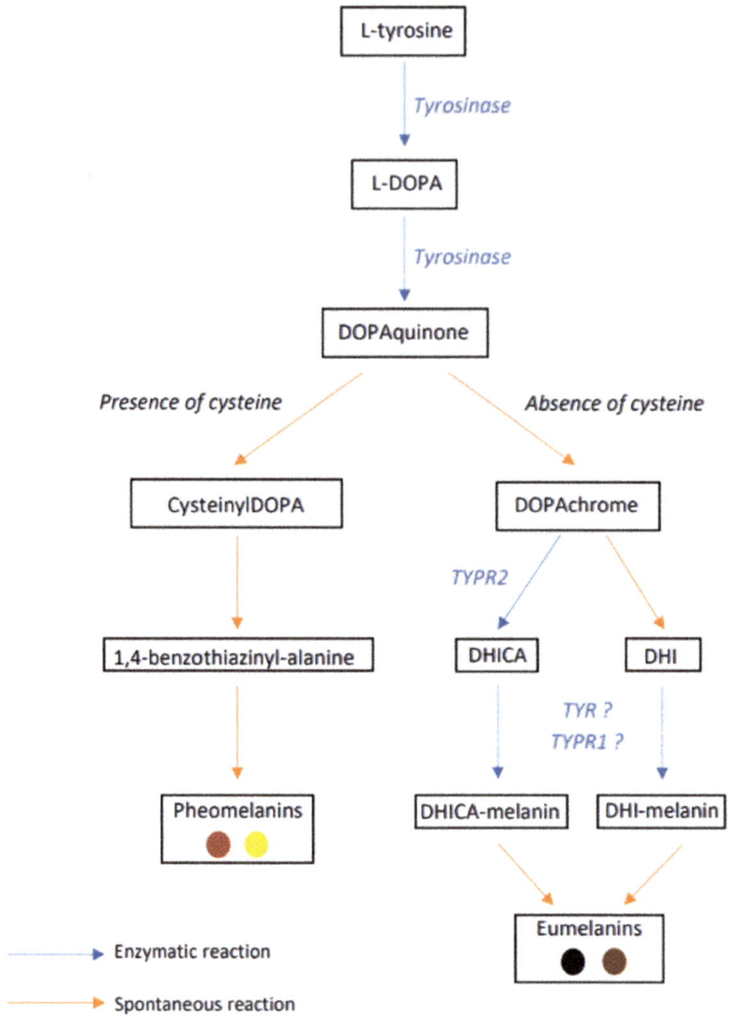

Figure 1.3 Simplified scheme of melanogenesis occurring in melanocytes during the anagen phase.

involves microtubules and various proteins. At the extremity of the dendrites, the melanosomes are transferred to the keratinocytes according to a mechanism that is still not well understood. Several hypotheses have been formulated: cytophagocytosis, release of melanosomes into the intercellular space followed by keratinocytes endocytosis keratinocytes, transfer of melanosomes by direct communication between the two cell types, process of cytophagocytosis of the dendritic ends by keratinocytes. The pigmentation unit is destroyed at the end of the anagen phase. During the neomorphogenesis, a contingent of dormant and inactive melanocytes located in the bulge is recruited to restock the bulb of the new follicle.[31,36]

The different stages of melanin production, distribution, and degradation depend on endogenous, genetic, and hormonal factors, as well as on exogenous factors such as the stimulation of photoprotection by UV.[1] During ageing, the first phenomenon is a decrease in tyrosinase activity. The migration of melasonomes will also be affected, as will the interactions between keratinocytes and melanocytes. This phenomenon, genetically programmed, may be accelerated by stress, smoking, or pathologies such as anaemia.[39]

1.2.3.4 Follicle Appendages

The follicle is associated to sweat and sebaceous glands, as well as to the arrector muscle, to form the pilosebaceous unit. Sebaceous glands are present on the entire surface of the body except the palms, soles, and dorsum of the foot. They are multilobed glands which end at the isthmus *via* their excretory canal. They are composed of several cell layers, from the periphery to the core: the basal membrane, the germinal layer, and then several sheets of undifferentiated cells, which express markers characteristic of stem cells and will give rise to the sebocytes. During this maturation period, which lasts about two weeks, the enzymatic equipment necessary for lipid synthesis is developed. The sebocytes then move to the core of the lobule and release the sebum into the infundibulum *via* the excretory duct. Sebum has a very complex composition. The "native" sebum, secreted by the sebocytes, is essentially composed of triglycerides (57%), mono- and diesters (26%), and squalene (12%). It is modified during transit in the infundibulum by the addition of lipids of epidermal origin such as sterols, sterol esters, and ceramides. The "final" sebum, excreted at the surface of the scalp, contributes to the formation of the surface hydrolipidic film which lubricates the hair and protects the epidermis.[40,41]

Eccrine sweat glands are widely distributed on the skin and scalp, with 2–5 billion in humans. These glands abut directly on the surface of the epidermis, unlike the apocrine sweat glands which are attached to the pilosebaceous follicles in the axillary, inguinal, and intergluteal folds. The eccrine sweat glands consist of a sweat glomerulus, a hairy acinus that produces sweat, located in the dermis or at the dermal-hypodermal junction, and which extends into the sweat duct that evacuates the sweat to the epidermis. The glomerulus itself comprises a secretory and an excretory part. The primitive sweat, produced by the secretory cells of the glomerulus, is isotonic to plasma. It is progressively modified by resorption phenomena in the excretory duct.[41,42]

The arrector muscle is a smooth muscle stretched between the dermal-epidermal region and the sub-isthmic region of the hair. Its contraction causes the hair to straighten, called horripilation.[1,2]

1.2.3.5 Vascularization and Innervation of Hair Follicles

The cutaneous vascularization is ensured by arterioles that ascend into the dermis, forming a plexus that supplies the cutaneous structures and in

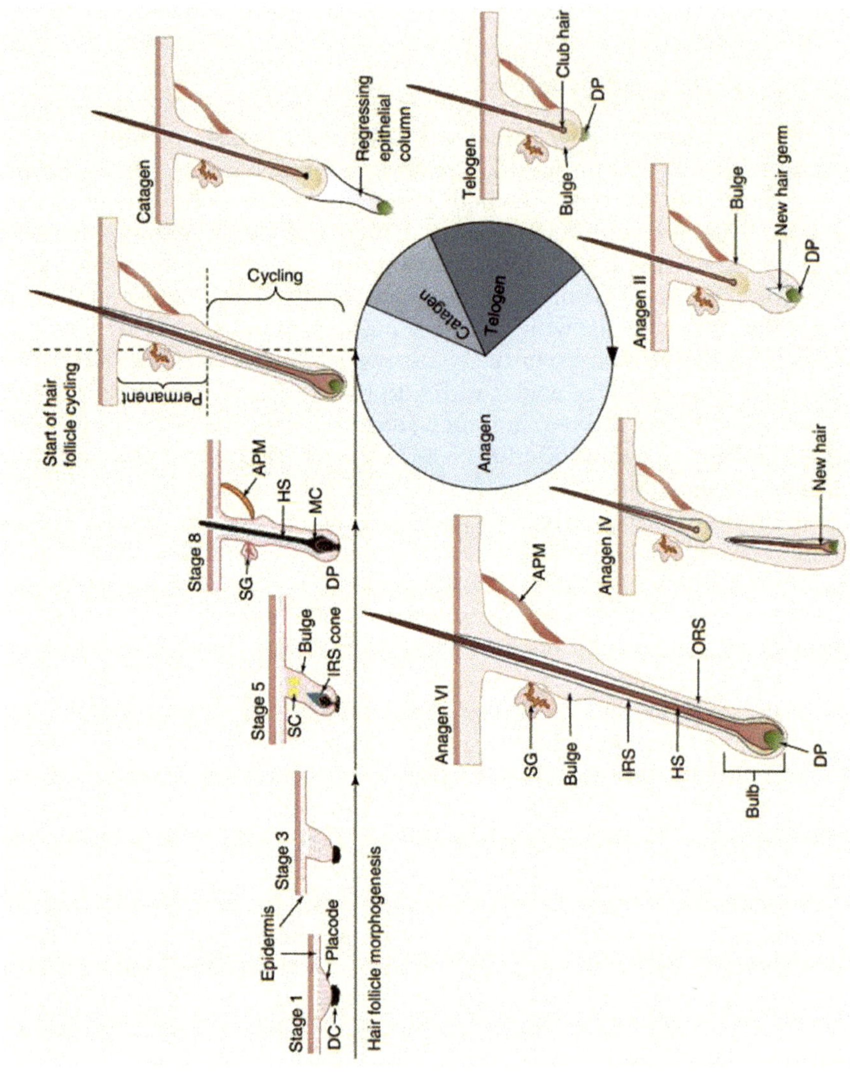
Stage 1
Epidermis
Placode
DC
Stage 3
Stage 5
SC
Bulge
IRS cone
Stage 8
SG
HS
APM
MC
DP
Hair follicle morphogenesis
Start of hair follicle cycling
Permanent
Cycling
Catagen
Regressing epithelial column
Telogen
Bulge
Club hair
DP
Anagen
Catagen
Telogen
Anagen II
Bulge
New hair germ
DP
Anagen IV
New hair
Anagen VI
APM
SG
Bulge
IRS
HS
ORS
Bulb
DP

particular the hair follicle. These arterioles are particularly concentrated in the lower part of the hair follicle and form a rich anastomotic network connected by cross shunts in contact with the papilla. A second capillary network is located in the sebaceous gland. It extends to the surface of the skin. These two networks are connected by a parallel system of larger vessels that descend along the sides of the follicle. The profound network, in contact with the papilla, is very developed during the anagen phase and involutes during the catagen and telogen phases, whereas the superficial networks do not show any modification.[1,2] These results are supported by a study by Ford *et al.*, which shows by the volumetric multispectral optoacoustic tomography method that the dermal papilla in the anagen phase has a high perfusion rate with an oxyhaemoglobin saturation of about 99%.[43] The venous circulation is parallel to the arterial circulation. Finally, each papilla has its own lymphatic system.

The nerves that supply the hair follicles are organized in the same way as the dermal nerve network and include sensory afferents and autonomic sympathetic nerves. They emerge in the dermis and ascend to innervate the hair follicle from the bulb to the epidermis. At the level of the sebaceous gland, a collar of nerves overlies certain follicles. Fibers may extend from this collar to innervate local structures or form a second dermal plexus. Nerve endings extend from this network to innervate the papillary dermis and the epidermis. [1,2]

1.2.4 Hair Cycling

The hair cycle consists of three phases: a growth (anagen) phase, a regression (catagen) phase, and a resting (telogen) phase (Figure 1.4). These phases are regulated by a very complex system involving cytokines, hormones, neurotransmitters, transcription factors, and enzymes. At each stage, the hair follicle epithelium, but also the mesenchyme, the extracellular matrix, and the vessels undergo dramatic changes.[2]

The anagen phase concerns about 80–90% of the hair mass and lasts two to three years in men and six to eight years in women; the growth rate, genetically determined, is about 0.22–0.52 mm per day or 0.6–1.4 cm per month. During this phase, the new hair appears in the hair follicle and

Figure 1.4 Key stages in hair development and hair cycling. Hair follicle morphogenesis (upper left) consists of eight stages. The first hair emerges from the skin and enters rapidly into the catagen phase. When the dermal papilla comes into contact with the bulge stem cells, signaling pathways are activated. As soon as a critical concentration of signals is reached, the anagen phase is initiated and a new hair follicle is formed. The first postnatal cycle is also initiated; this cycle will then be repeated during the individual's life. DC: dermal condensate; IRS: inner root sheath; SC: sebocytes; DP: dermal papilla; MC: melanocytes; HS: hair shaft; APM: arrector pili muscle; ORS: outer root sheath; SG: sebaceous gland. Reproduced from ref. 10 with permission from Elsevier, Copyright 2022.

grows through a mechanism of massive division of the cells above and around the dermal papilla. In fact, this process concerns all the epithelial compartment, with the highest activity observed in matrix cells.[2] The signal that triggers this division is emitted when the cells of the dermal papilla of the previous hair are in the telogen phase, which gradually rises to the surface and comes into contact with the stem cells of the bulge. Unlike epithelial progenitor cells, bulge cells are pluripotent and indefinitely divisible, and form a reservoir from which, at each telogen/anagen transition, a subpopulation is recruited to repopulate the bulge of the newly formed follicle and regenerate a pigmentation unit, while another subpopulation will remain in the bulge.[23,44]

The anagen phase can be divided into six stages (I–VI). During stages I to V (so-called proanagen stages), hair progenitor cells multiply very quickly and form an invagination that extends into the epidermis, in which all the components of a new follicle organize and differentiate. Some keratinocytes spread to the periphery of the hair follicle to form the outer and inner epithelial sheaths; others extend to form the hair shaft. They accumulate keratin and melanosomes. This is among the most intense and rapid mitotic activities in the body. This stage is controlled by several signaling pathways (BMP, EGF, EGF, and NOTCH, in particular) as well as by transcription factors, and is accompanied by a massive angiogenesis. Lymphatic vessels in the bulge could also be involved. During anagen VI (metanagen stage), the new hair fiber is produced, and the new hair shaft appears on the skin surface.[10,28,30,45–48]

Towards the end of the anagen phase, keratinocytes stop proliferating. Melanocyte dendrites retract, TYR and TYRP2 activity collapses, and the lower two-thirds of the hair follicle rapidly regress.[10,20] At the same time, blood vessels involute, growth factor production decreases, and epithelial cells of the lower telogen follicle no longer show significant DNA or RNA synthesis.[49] These different compartments actually undergo a process of terminal dedifferentiation and programmed apoptosis that leads to the rapid involution of the lower part of the follicle, while preserving a cluster of dermal fibroblasts that corresponds to the remnants of the dermal papilla. This short transition period of 2–4 weeks corresponds to the catagen phase.[2,10]

The follicle finally enters the telogen phase, which lasts about two to seven months and affects 10–20% of the hair mass. The hair follicle regresses to about half its original size.[2,10] A thin layer of epithelial cells covers a cluster of dermal fibroblasts which corresponds to the remains of the dermal papilla. At this stage, this is positioned directly under the bulge, which allows direct interactions between the bulge stem cells and the dermal papilla. Keratin synthesis is also stopped, with the exception of keratin 14 which continues in the epithelial sac to which the hair fiber is anchored. When the lower part of the follicle is completely detached from the papilla, the "club hair" is formed. The telogen hair shaft can be retained for months in this epithelial sac. Factors such as health status, nutritional status or season can influence the growth and duration of the telogen phase.[2,10]

Recently, the concept of an exogenous phase has emerged.[50] This phase corresponds to the period during which the process of release of the hair shaft is initiated and successfully executed. The exogen phase ends when the shaft is liberated. This hypothesis is based on the observation that a hair shaft plucked in the telogen phase shows major morphological differences with the root of a shed hair shaft. Whereas the telogen root still has compact nucleated cells surrounding a cornified nucleus, the shed hair root has very few cells separated at their outer edge by an intercellular cleavage. The assumption that newly anagenic hair regrows resting hair is challenged by the hypothesis of a highly controlled active process. Finally, the existence of an interval between exogen hair extrusion and emergence of new anagen hair, during which the follicle remains empty, the so called kenogen phase, has been recently identified.[2,51]

1.3 Hair Follicle Morphogenesis

The hair follicle presents three distinct origins: the infundibulum, isthmus, bulge, bulb, and keratinocytes are of ectodermal origin, the papilla is of mesodermal origin, and the melanocytes are of neuroectodermal origin.[52,53]

Hair development requires close collaboration between these structures during embryogenesis *via* growth factors, cytokines, neuropeptides, neurotransmitters, and hormones.[10] Three phases, induction, organogenesis, and cytodifferentiation, in turn subdivided into eight morphologically distinct stages, have been identified (Figure 1.4).[10,52–54] The induction phase begins around the eighth week of development and includes stage 0 and stage 1. Stage 0 is marked by the appearance of an interacting gradient of reciprocal interactions between the surface ectoderm, a superficial sheet that covers the entire body of the embryo, and the underlying mesoderm, thus creating an inductive field in the epidermis. In stage 1, the surface ectoderm gives rise to keratinocytes and the epithelial resulting structures: hair, sweat, and sebaceous glands. Some keratinocytes enlarge and elongate to form the ectodermal placodes.[10,47,54] These placodes will expand and form epithelial ingrowths that invaginate obliquely into the adjacent dermis, with the degree of angle determined by the location of the hair follicle. Unlike other mammals, the final number of follicles (between two and five million follicles, of which 100 000 to 150 000 are located on the scalp) is acquired at this stage. Placode formation subsequently expands in a wave caudally and ventrally over the skin of the fetus. The message initiating this phenomenon originates from a cluster of fibroblasts located in the underlying dermis, and the dialogue between these signaling pathways promotes the reciprocal proliferation of the ectoderm and mesoderm.[10,54] These include the growth factor TGF-β, bone morphogenetic protein (BMP) type 4, and lymphoid enhancer binding factor (LEF-1). Precise localization and growth of placodes is determined by signaling pathways including in particular the wingless protein (Wnt)/β-catenin for the hair follicle morphogenesis; in parallel, other signaling pathways inhibit placode fate in surrounding cells.[4,10,28]

The organogenesis phase, which corresponds to stages 2 to 5, begins around the sixteenth week. Stage 2 (the so-called germ stage) is marked by massive proliferation of germ cells of the ectoderm, leading to the formation of the dermal papilla. The resulting structure is called the hair germ. This phenomenon again requires an inductive interaction with the underlying mesoderm, which itself sends specific signals to the ectoderm cells, thus forming a reciprocal signaling pathway.[4,10] During stages 3 to 5, collectively known as the peg stage, the proximal keratinocytes continue to proliferate and surround the papilla. During stage 5, the keratinocytes form the inner root sheath, a process that will initiate terminal differentiation into the hair shaft.[4,10,28]

The cytodifferentiation phase, which begins between the eighteenth and twentieth week, corresponds to stages 6 to 8 of development. The lower end of the germ enlarges to form the bulb, which will rapidly deepen under the thrust of the papilla. The inner sheath completes its differentiation by forming a rigid structure in which the keratinocytes differentiate to form the hair shaft. At the same time the cylindrical layers of the outer sheath appear around the inner sheath. Other keratinocytes, located in the distal part of the follicle, will differentiate into sebocytes and apocrine gland precursors. Vasculo-nervous structures also appear at this stage. The sebaceous precursor develops in the obtuse angle formed by the epidermis and the future hair follicle. The first cells metabolize glycogen and secrete lipids. It becomes functional between the thirteenth and sixteenth week of life and secretes the vernix caseosa.[55]

Melanocytes are derived from undifferentiated, unpigmented cells, the melanoblasts, originating mainly from the neural crest cells, a transient population of cells arising from the dorsal part of the neural tube. Migration is modulated by the surrounding environment, consisting of environmental factors, extracellular matrices (ECM) and direct interactions with neighbouring cells encountered on their path. It begins early, between the sixth and seventh week of development, and reach the mesoderm, the ectoderm and the hair follicles between the eighth and fourteenth week. Some melanoblasts will proliferate and then differentiate into epidermal melanocytes, while others, called transit-amplification melanoblasts, leave the epidermis to distribute into the developing hair follicles as dopa-positive and dopa-negative cells.[31,33,56] At around E15.5, these melanoblasts concentrate near the placodes. They will then colonize the follicle in two distinct locations. A first contingent will colonize the bulge, where they form the melanocytic stem cell population, and serve as a reservoir with the hair follicle stem cells, responsible for the cyclic regeneration of hair follicle epithelial cells. A second contingent will colonize the hair bulb above the dermal papilla, where it will differentiate and proliferate to produce melanosomes. These melanoblasts will be incorporated into the first follicle to form the pigmentary unit at the apex of the dermal papilla. The first melanosomes will be produced around the twenty fourth week, and the first melanocyte dendrites around the twenty eighth week, at which time the first melanosomes will be incorporated into the keratinocytes. The bulge, for its part, is derived from a contingent of neural crest stem cells with a unique gene expression profile that differentiates them from embryonic neural crest cells and epidermal stem cells.[26,31,33,57]

The first hair emerges on the scalp around the sixteenth or seventeenth week of pregnancy. At the twentieth week, the scalp is entirely covered. From the twenty third week onwards, the hair has an oblique trajectory in relation to the skin, whose angle is determined by the degree of tension of the ectoderm caused by the concurrent development and maturation of the underlying brain. These first waves are made up of very fine (less than 30 μm in diameter), soft, short (about 2 mm long), non-pigmented and non-medullated hairs, which form the lanugo, visible in preterm neonates.

A second cycle of growth then occurs to form the downy or vellus hair, which is fine (diameter between 5 and 30 μm), short and poorly pigmented. The vellus hair penetrates the reticular dermis but not the subcutaneous fat layer. It is visible in the full-term neonate and continues to grow throughout life.[2,46,58] It should be noted that a significant proportion of hair remains in the vellus stage throughout life, even on the scalp, where it may represent up to 7–25% of the hair present. The intermediate hair is seen after birth on the scalp. It has a diameter of between 30 and 60 μm, a cuticle, sparse pigmentation and a fragmented (or absent) medulla. The intermediate hair will gradually develop into a pigmented terminal hair. This has a larger cross-section (>60 μm) as well as medulla, and penetrates deep into the underlying dermis, with a bulb embedded in the subcutaneous fat.[2,46] Its size and shape vary according to location and potential function. Initially, only the hairs on the eyebrows, eyelashes and scalp develop into terminal hairs. Progressively, depending on age, sex, and anatomical region, the transformation to terminal hair continues. At puberty, hair in certain areas of the body is replaced by longer, thicker, pigmented adult hair. This begins in the pubic and axillary areas in both sexes and continues with the eyelashes and eyebrows. Male facial hair begins to appear about two years after the onset of pubic hair growth.[2] The transformation of vellus hair to intermediate and then terminal hair is androgen dependent, as are the transformations observed at puberty in both sexes.[2,59,60] Growth hormone is also required for a full androgenic action.[61]

1.4 Variations in Synchronization and Hair Growth with Age

At the end of fetal life and the beginning of the neonatal period, all follicles are synchronized in the same phase of the cycle. A few weeks after birth, the follicles regress in two successive waves, starting in the frontal area and progressing towards the occipital area. As telogen hair predominates in the occipital area, hair loss often produces an area of temporary alopecia. Only a small area of hair above the occiput does not regress at this stage. The vellus hair shows a slow and synchronized growth. In subsequent cycles, intermediate hairs appear and show asynchronous cycles with a mosaic pattern. There is great inter-individual variability in the occurrence of this phenomenon. However, it is considered that between the twelfth and sixteenth month, the scalp is covered with terminal mosaic hairs.[2,58–60]

The rate of growth varies during the first months of life. During prepuberty, hair growth is most rapid at the vertex, whereas it slows down after puberty in both sexes compared to the growth rate measured at the occiput. The growth rate of terminal hair at the vertex and occiput is then relatively stable in adulthood, at about 0.35 mm per day. These regions present the more consistent linear growth rate compared to the other regions of the scalp.[61,62] Some variables may affect the growth rate, including seasons, pregnancy, and hormones.[2,46,63,64] To simplify the interpretation of hair concentrations, it is agreed that the average linear growth rate of human hair in the posterior vertex and occipital region is about 1 cm per month.[61,65,66]

1.5 Conclusion

As presented in this chapter, hair is a complex organ in constant interaction with its surrounding environment. During the first months of life, its anatomy and growth pattern diverge significantly from those of adult hair. The mechanisms of xenobiotic incorporation are closely related to hair cycling and melanogenesis, and the interpretation of the concentrations, often complex, must take into account these aspects.

References

1. B. Buffoli, F. Rinaldi, M. Labanca, E. Sorbellini, A. Trink, E. Guanziroli, R. Rezzana and L. F. Rodella, *Int. J. Dermatol.*, 2014, **53**, 331.
2. A. Vogt, K. J. McElwee and U. Blume-Peytavi, in *Hair Growth and Disorders*, ed. U. Blume-Peytavi, A. Tosti, D. A. Whiting and R. Trüeb, Springer, Berlin, 2008, ch. 1, p. 1.
3. K. S. Houschyar, M. R. Borrelli, C. Tapking, D. Popp, B. Puladi, M. Ooms, M. P. Chelliah, S. Rein, D. Pförringer, D. Thor, G. Reumuth, C. Wallner, L. K. Branski, F. Siemers, G. Grieb, M. Lehnhardt, A. S. Yazdi, Z. N. Maan and D. Duscher, *Dermatology*, 2020, **236**, 271.
4. S. Park, *Front. Cell. Dev. Biol.*, 2022, **10**, 933370.
5. R. de la Mettrie, R. D. Saint-Léger, G. Loussouarn, A. Garcel, C. Porter and A. Langaney, *Hum. Biol.*, 2007, **79**, 265.
6. M. R. Harkey, *Forensic Sci. Int.*, 1993, **63**, 9.
7. R. Kronstrand and C. Scott, in *Analytical and Practical Aspects of Drug Testing in Hair*, ed. P. Kintz, CRC Press, Boca Raton, 2006, ch. 1, p. 1.
8. M. V. R. Velasco, T. C. S. Dias, A. Z. Freitas, N. D. Vieira Júnior, C. A. S. O. Pinto, T. M. Kaneko and A. R. Baby, *Braz. J. Pharm. Sci.*, 2009, **45**, 153.
9. C. Popescu and H. Höcker, *Int. Rev. Cell. Mol. Biol.*, 2009, **277**, 137.

10. M. R. Schneider, R. Schmidt-Ullrich and R. Paus, *Curr. Biol.*, 2009, **19**, R132.
11. T. Takayashi and S. Yoshida, *Int. J. Cosmet. Sci.*, 2017, **39**, 327.
12. Y. Yu, W. Yang, B. Wang and M. A. Meyers, *Mater. Sci. Eng., C*, 2017, **73**, 152.
13. W. S. Lee, *J. Dermatol. Sci.*, 2011, **64**, 153.
14. T. Takayashi, *Int. J. Cosmet. Sci.*, 2019, **41**, 28.
15. T. Takayashi, *Int. J. Cosmet. Sci.*, 2021, **43**, 254.
16. M. A. Rogers, L. Langbein, S. Praetzel-Wunder, H. Winter and J. Schweizer, *Int. Rev. Cytol.*, 2006, **251**, 209.
17. D. P. Harland, R. J. Walls, J. A. Vernon, J. M. Dyer, J. L. Woods and F. Bell, *J. Struct. Biol.*, 2014, **185**, 397.
18. L. J. Wolfram, *J. Am. Acad. Dermatol.*, 2003, **48**, S106.
19. S. Thibaut, P. Barbarat, F. Leroy and B. A. Bernard, *Int. J. Dermatol.*, 2007, **46**, 7.
20. D. J. Tobin, *Int. J. Cosmet. Sci.*, 2008, **30**, 233.
21. L. Jones, *Clin. Dermatol.*, 2001, **19**, 95.
22. R. Wagner and I. Joekes, *J. Cosmet. Sci.*, 2007, **58**, 359.
23. S. Joulay-Veijouye, A. Yari, F. Heidari, N. Sajedi, F. G. Moghani and M. Nobakht, *Iran. Public Health*, 2017, **46**, 1167.
24. S. Tanimura, Y. Tadokoro, K. Inomata, N. Thanh Binh, W. Nishie, S. Yamazaki, H. Nakauchi, Y. Tanaka, J. R. McMillan, D. Sawamura, K. Yancey, H. Shimizu and E. K. Nishimura, *Cell Stem Cell*, 2011, **8**, 177.
25. J. L. Martel, J. H. Miao and T. Badri, *Anatomy, Hair Follicle*, StatPearls Publishing, Treasure Island, 2017.
26. S. Malgouries, S. Thibaut and B. A. Bernard, *Br. J. Dermatol.*, 2008, **158**, 234.
27. S. Commo and B. A. Bernard, *Br. J. Dermatol.*, 1997, **137**, 31.
28. R. Paus, E. A. Langan, S. Vidali, Y. Ramot and B. Andersen, *Trends Mol. Med.*, 2014, **20**, 559.
29. D. Krahl and K. Sellheyer, *Br. J. Dermatol.*, 2009, **161**, 678.
30. G. Cotsarelis, *J. Invest. Dermatol.*, 2006, **126**, 1459.
31. D. J. Tobin, in *Hair growth and disorders*, ed. U. Blume-Peytavi, A. Tosti, D. A. Whiting and R. Trüeb, Springer, Berlin, 2008, ch. 4, p. 51.
32. S. Thibaut, O. Gaillard, P. Bouhanna, D. W. Cannell and B. A. Bernard, *Br. J. Dermatol.*, 2005, **152**, 632.
33. J. L. Lin and D. E. Fisher, *Nature*, 2007, **445**, 843.
34. S. Ito and K. Wakamatsu, *J. Eur. Acad. Dermatol. Venereol.*, 2011, **25**, 1369.
35. G. Greco, L. Panzella, L. Verotta, M. d'Ischia and A. Napolitano, *J. Nat. Prod.*, 2011, **74**, 675.
36. M. Cichorek, M. Wachulska, A. Stasiewicz and A. Tymińska, *Adv. Dermatol. Alergol.*, 2013, **30**, 30.
37. L. Panzella, A. Ebato, A. Napolitano and K. Koike, *Int. J. Mol. Sci.*, 2018, **19**, 1753.
38. X. Lai, H. J. Wichers, M. Soler-Lopez and B. W. Dijkstra, *Chem. – Eur. J.*, 2018, **24**, 47.

39. D. J. Tobin and R. Paus, *Exp. Gerontol.*, 2001, **36**, 29.
40. E. Abell, in *Disorders of Hair Growth, Diagnosis and Treatment*, ed. E. A. Olsen, McGraw-Hill, New York, 1994, pp. 1–19.
41. O. Arda, N. Göksügür and Y. Tüzün, *Clin. Dermatol.*, 2014, **32**, 3.
42. C. Y. Cui and D. Schlessinger, *Exp. Dermatol.*, 2015, **24**, 644.
43. S. J. Ford, P. L. Bigliardi, T. C. P. Sardella, A. Urich, N. C. Burton, M. Kacprowicz, M. Bigliardi, M. Olivo and D. Razansky, *J. Invest. Dermatol.*, 2016, **136**, 753.
44. S. Commo and B. A. Bernard, *Pigm. Cell Res.*, 2000, **13**, 253.
45. G. Taylor, M. S. Lehrer, P. J. Jensen, T. T. Sun and R. M. Lavker, *Cell*, 2000, **102**, 451.
46. J. Gareri and G. Koren, *Forensic Sci. Int.*, 2010, **196**, 27.
47. S. E. Millar, *Dev. Cell*, 2015, **34**, 488.
48. S. Y. Yoon, L. C. Dieterich, S. Karaman, S. T. Proulx, S. B. Bachmann, C. Sciaroni and M. Detmar, *PLoS One*, 2019, **14**, e0220341.
49. T. Ellis, L. Gambardella, M. Horcher, S. Tschanz, J. Capol, P. Bertram, W. Jochum, Y. Barrandon and M. Busslinger, *Genes Dev.*, 2001, **15**, 2307.
50. K. Stenn, *J. Am. Acad. Dermatol.*, 2005, **52**, 374.
51. M. Guarrera and A. Rebora, *Skin Appendage Disord.*, 2017, **3**, 193.
52. R. Schmidt-Ullrich and R. Paus, *BioEssays*, 2005, **27**, 247.
53. N. Saxena, K. W. Mok and M. Rendl, *Exp. Dermatol.*, 2019, **28**, 332.
54. M. F. Forni, M. Trombetta-Lima and M. C. Sogayar, *Biol. Res.*, 2012, **45**, 215.
55. S. Liu, H. Zhang and E. Duan, *Int. J. Mol. Sci.*, 2013, **14**, 10869.
56. V. Petit and L. Larue, *Exp. Dermatol.*, 2016, **25**, 669.
57. Y. F. Hu, Z. J. Zhang and M. Sieber-Blum, *Stem Cells*, 2006, **24**, 2692.
58. P. Kintz, *Forensic Sci. Int.*, 2015, **249**, 314.
59. B. H. Miranda, D. J. Tobin, D. T. Sharpe and V. A. Randall, *Br. J. Dermatol.*, 2010, **163**, 287.
60. S. A. Furdon and D. A. Clark, *Adv. Neonatal Care*, 2003, **3**, 286.
61. R. Wennig, *Forensic Sci. Int.*, 2000, **107**, 5.
62. P. Kintz, V. Cirimele, C. Jamey and B. Ludes, *J. Forensic Sci.*, 2003, **48**, 195.
63. V. A. Randall, *Dermatol. Ther.*, 2008, **21**, 314.
64. V. A. Randall, *Semin. Cell Dev. Biol.*, 2007, **18**, 274.
65. P. Kintz, M. Villain and V. Cirimele, *Ther. Drug Monit.*, 2006, **28**, 442.
66. M. LeBeau, M. A. Montgomery and J. D. Brewer, *Forensic Sci. Int.*, 2011, **220**, 110.

CHAPTER 2

Interpretation: Incorporation of Xenobiotics Into Hair

ROBERT KRONSTRAND*[a] AND GAIL A. COOPER[b]

[a] National Board of Forensic Medicine, Artillerigatan 12, 58758 Linköping, Sweden; [b] Department of Forensic Toxicology, New York City Office of Chief Medical Examiner, New York City, New York 10016, USA
*Email: robert.kronstrand@rmv.se

2.1 Introduction

The incorporation of drugs into hair has been investigated for decades, still there are many questions unanswered. In this chapter, the aim is to put forward, evaluate, and discuss the research that builds the scientific foundation for answering the most common questions asked by the end point users of hair tests for drugs. These questions are:

(1) Was a drug administered?
(2) How much was administered?
(3) When was it administered?

They seem simple enough, but as you explore the science, it is clear that there are still aspects of incorporation that are not fully described.

Issues in Toxicology No. 47
Perspectives and Challenges of Hair Analysis
Edited by Pascal Kintz, Alberto Salomone and Marco Vincenti

Published by the Royal Society of Chemistry, www.rsc.org

2.2 Criteria for a Positive Report (Metabolites and Cut-offs)

2.2.1 Was a Drug Administered?

The question originates from the possibility of contamination from the environment since the hair is on the outside of the body, a positive result may not necessarily mean a detected drug was administered. Most researchers and practitioners agree that one cannot wash away external contamination entirely but instead rely on the simultaneous detection of drug metabolites to verify that an intake has occurred. The Society of Hair Testing (SoHT) has proposed some metabolites as useful markers for intake.[1] In Table 2.1, both recommended cut-offs as well as required metabolites are included. In this section we will discuss recommended cut-offs in relation to controlled studies and how metabolites can contribute to the conclusion that a person administered a drug rather than being exposed to it in some other way.

The risk of contamination from the environment depends on the formulation and route of administration. Drugs that mainly are medications and used as tablets like morphine, codeine, dihydrocodeine, oxycodone, buprenorphine,

Table 2.1 Drug cut-offs and required metabolites according to society of hair testing consensus 2021.

Drug group and analytes	Cut-off (pg mg^{-1})	Recommended metabolites to monitor
Morphine	200	
Codeine	200	
Dihydrocodeine	200	
Heroin	200	6-Acetylmorphine
Cocaine	500	Benzoylecgonine Norcocaine Cocaethylene Hydroxylated cocaine Hydroxylated benzoylecgonine Anhydroecgonine methyl ester
MDMA	200	
MDEA	200	
MDA	200	
Methylamphetamine	200	
Amphetamine	200	
THC	50	Carboxy-THC (0.2 pg mg^{-1})
CBD	50	
Tramadol	200	Desmethyltramadol
Oxycodone	100	
Methadone	200	EDDP
Buprenorphine	10	Norbuprenorphine
Ketamine	200	Norketamine

and even MDMA will be less likely to be associated with environmental contamination, whereas drugs such as heroin, cocaine, methamphetamine, and amphetamine that are provided in powdered form and those similar to cannabis that can be smoked are more likely to be associated with environmental contamination. The rationale for using metabolites formed in the body is of course that they are not present in the drug itself, powders, or plant material.

The cut-offs presented in Table 2.1 are, according to the SoHT, set so that a drug user can be identified. It is not more closely discussed what a drug user is but the consensus document also states that in, *e.g.*, drug facilitated crimes or single intake cases the cut-offs should not be considered but more sensitive detection limits are needed. Controlled studies are necessary to investigate the doses or dosing regiments needed to produce a positive result.

2.2.1.1 Cocaine Metabolites and Evaluation of Cut-off

Cocaine is a drug that presents special challenges when determining whether the source of cocaine present in hair is a result of active use or external contamination including passive exposure. The SoHT recommended cut-off is 500 $pg\,mg^{-1}$ to identify cocaine users with the additional caveat that consideration must be given to the presence of benzoylecgonine (BE), norcocaine, cocaethylene, hydroxy-cocaines or hydroxy-benzoylecgonine to confirm use. To confirm crack cocaine the presence of anhydroecgonine methyl ester (AEME) must be considered.

Cocaine has the longest list of metabolites to be considered when determining use within the SoHT Consensus on Drugs of Abuse Testing and this is a direct reflection of the extensive research that has been undertaken to identify a metabolite or metabolites that will confirm use.[1]

The presence of cocaine in hair above the cut-off in combination with the presence of BE does not provide sufficient certainty to confirm use, with many researchers reporting the detection of both in cases of passive exposure[2–4] and in intentionally cocaine-contaminated hair.[5,6]

The SoHT in their 1997 Statement[7] and the Substance Abuse and Mental Health Services Administration (SAMHSA) in the United States who provide Mandatory Guidelines for Federal Workplace Drug Testing programs,[8] both published the proposed use of a ratio between cocaine and BE, to determine use. The proposed ratio required the BE concentration in hair to be 5% or greater than the measured cocaine concentration. However, the publication of the SAMHSA guidelines for hair was 14 years ago and the Statement by the SoHT was 25 years ago. Subsequent publications by the Society of Hair Testing[1,9] and the current proposed SAMHSA guidelines for hair, last updated in 2020 (https://www.samhsa.gov/workplace/forms-guidelines), do not include the cocaine/BE ratio. Published studies have demonstrated that the use of this ratio does not distinguish drug use from passive exposure or contaminated drug-free hair.[4,5]

Alternative metabolites including norcocaine, produced following *N*-demethylation of cocaine, and cocaethylene, formed during the concomitant use of cocaine and ethanol, were initially identified as unique metabolites formed only through direct use of cocaine. Following reports that both metabolites could be found in cocaine powders and were detected in hair intentionally contaminated with cocaine, concerns were raised about the usefulness of hair as a direct determinant of cocaine use.[10] Hair remains the go to specimen for determining long-term drug exposure and with respect to all drugs including cocaine, hair test findings are to be considered as one part of the entire investigation and not as the only determinant of use.[11] The use of extensive hair wash protocols to improve removal of external contamination and the identification of additional unique metabolites provided greater certainty in identifying cocaine use.[12,13] However, it should also be noted that although possible, the circumstances required for norcocaine and coaethylene to result in a positive hair test from contamination are limited and more research is needed to evaluate the true extent of this route of incorporation.

Gambelunghe *et al.* reported norcocaine present in hair from chronic cocaine users utilizing an LOQ of 20 pg mg^{-1} and LOD of 10 pg mg^{-1} but the lengths of hair collected, or length of segment tested were not provided.[14] In their study were 90 Caucasian participants with predominantly black or brown hair, who had cocaine concentrations ranging from 1630 to 39 290 pg mg^{-1}, well above the SoHT cut-off of 500 pg mg^{-1}. BE was present in all specimens at concentrations ranging from 190 to 5770 pg mg^{-1}, while norcocaine was present in 58.9% of hair specimens at concentrations ranging from 220 to 3140 pg mg^{-1}. The presence of norcocaine in the hair specimens was proposed as a highly sensitive marker only for heavy cocaine use with 100% of the hair specimens with cocaine concentrations above 10 000 pg mg^{-1} also containing norcocaine.

Franz *et al.* reported norcocaine in hair collected from children of different age groups along with the adult in the family who was known to use drugs, to evaluate passive exposure.[2] Across all age groups ($N = 30$), the maximum concentration of norcocaine detected in hair was 400 pg mg^{-1}, compared with 11 700 pg mg^{-1} for the adults ($N = 624$). In comparison with the previous study by Gambelunghe *et al.*, the number of adults with norcocaine present in combination with cocaine, was higher at 71% and for the children this was as high as 50% for the youngest babies (<1 year) and was lower for older children. This is expected as younger children have greater contact with their parents or caregivers and therefore have more opportunity for passive exposure.

Both studies also measured cocaethylene present at a lower prevalence than norcocaine. Only 22.2% of chronic cocaine users had cocaethylene present at concentrations ranging from 220 to 1980 ng mg^{-1}.[14] The authors concluded that cocaethylene cannot be considered as a universal marker of active cocaine use due to the need for concomitant use of ethanol and cocaine.

The presence of hydroxy-cocaine metabolites was proposed in addition to an extensive wash protocol for determining cocaine use.[12,15] Musshoff *et al.* opted for a simpler criterion for determining cocaine use *versus* contamination.[16]

The developed method measured cocaine, BE, norcocaine, cocaethylene and the following "hydroxy-cocainics"; *o*-hydroxycocaine (*o*-OH-COC), *m*-hydroxycocaine (*m*-OH-COC), *p*-hydroxycocaine (*p*-OH-COC), *m*-hydroxybenzoylecgonine (*m*-OH-BE), *p*-hydroxybenzoylecgonine (*p*-OH-BE), *m*-hydroxynorcocaine (*m*-OH-NCOC), *p*-hydroxynorcocaine (*p*-OH-NCOC). The authors suggested the presence of at least one of the hydroxycocainics in addition to cocaine and a ratio involving norcocaine and cocaethylene in comparison to cocaine.

Franz *et al.* and Scholz *et al.* also suggested the presence of hydroxy-cocainics and additional ratios involving hydroxy-cocainics and/or ratios involving norcocaine and cocaethylene in comparison to cocaine.[17,18] Hart *et al.* most recently reported previously drug-free hair being positive for hydroxy-cocainics after intentional contamination with cocaine.[19] They also recognized the difficulty in developing hair contamination models that reflect real-life scenarios. Both Franz *et al.* and Scholz *et al.* reported detection of different hydroxy-cocainics in seized street cocaine and therefore their presence alone will not be sufficient to confirm use.

This is a developing area requiring more research and as Hart *et al.* highlighted, the ratios of the hydroxylated metabolites compared with cocaine may provide a solution to distinguishing use of cocaine *versus* contamination.[19]

When considering the SoHT cut-off for cocaine of 500 pg mg^{-1} to determine a drug user, there is a wealth of information on passive exposure where cocaine concentrations in children are in excess of the cut-off. Wang *et al.* summarized publications involving *in utero* exposure with cocaine ranging from 0 to 257 000 pg mg^{-1} (median 9800 pg mg^{-1}), cases of active exposure ranging from 6000 to 76 000 pg mg^{-1} (median 25 000 pg mg^{-1}) and cases of passive exposure ranging from 0 to 75 000 pg mg^{-1} (median 600 pg mg^{-1}).[4]

Controlled dose studies have been reported and Joseph *et al.* reported cocaine concentrations of 5300 $\text{pg mg}^{-1} \pm 1200$ at the start of a study involving five African-Americans.[20] Over the subsequent "wash-out" weeks, the concentrations decreased and on week four participants were administered three low doses of 75 mg/70 kg subcutaneously. Peak cocaine concentrations were reported between one to two weeks at 1500 $\text{pg mg}^{-1} \pm 400$ and 1800 $\text{pg mg}^{-1} \pm 200$. Further increasing doses administered over the course of the study, corresponded to higher cocaine concentrations in hair, suggesting a dose–response relationship under these controlled conditions.

Scheidweiler *et al.* had a similar study design as Joseph *et al.* but with ten volunteers and also reported dose-related concentrations of cocaine.[21] For the lowest dose administered the maximum hair cocaine concentrations was 2997 $\text{pg mg}^{-1} \pm 619$ but there may have been "baseline" contribution from the washout period before the study began. This was reported as 464 $\text{pg mg}^{-1} \pm 176$. Even with this subtracted from the low dose cocaine hair concentrations, they are well above the SoHT recommended cut-off.

Whether the cut-off recommended by the SoHT is appropriate can be debated but more important is the detection and quantification of the metabolites of cocaine and how their presence and prevalence in relation to each other can distinguish between cocaine use and contamination.

2.2.1.2 Amphetamine Metabolites and Evaluation of Cut-off

Five amphetamines are mentioned in the SoHT consensus, MDMA, MDEA, MDA, methamphetamine, and amphetamine. There are no metabolites required for the identification of drug use. On the other hand, MDA is a metabolite of both MDMA and MDEA and amphetamine is a metabolite of methamphetamine. However, since both MDA and amphetamine can be drugs on their own, their status as metabolites is not exclusive. There have been some efforts in finding specific metabolic analytes, *e.g.* *para*-hydroxylated amphetamine and methamphetamine. These are prominent metabolites in other species but minor in humans. Miki *et al.* (2003) successfully identified and quantified *para*-hydroxy-methamphetamine in hair from users.[22] The 15 subjects presented with concentrations of 0.87–56.4 ng mg^{-1} of methamphetamine, 0.05–3.5 ng mg^{-1} of amphetamine, and 0.02–1.4 ng mg^{-1} of *para*-hydroxy-methamphetamine with 0.67–11% *para*-hydroxy-methamphetamine. Since then, few publications have addressed specific metabolites for methamphetamine.

The recommended cut-off for methamphetamine is 200 pg mg^{-1} (0.2 ng mg^{-1}). Polettini and co-workers investigated the hair concentrations after controlled dosing of two doses (4×10 mg and 4×20 mg) of methamphetamine to 5–7 subjects.[23] Two weeks after administration all subjects had Cmax above the cut-off. The authors reason about the positivity of a real-life sample taken farther away from the time of administration and using a 3 cm segment for the analysis, reaching the conclusion that methamphetamine concentrations would be above or close to the cut-off regardless of hair pigmentation.

We failed to find any publications that studied the hair concentration of methamphetamine in patients receiving methamphetamine.

In a controlled dosing study with 50 mg of dextroamphetamine, Jakobsson and Kronstrand measured amphetamine concentrations in nine subjects where amphetamine was found in five of the subjects with peak concentrations of amphetamine in 5 mm segments between 50 and 100 pg mg^{-1}, all below the 200 pg mg^{-1} cut-off.[24] Another controlled dosing study with amphetamine was performed by Roberts *et al.* where seven doses of 10 mg dextroamphetamine was administered and hair concentrations in a 3 cm segment was determined after five weeks.[25] Most of the 40 subjects presented with concentrations lower than or close to 200 pg mg^{-1} but some had significantly higher concentrations, up to around 900 pg mg^{-1}. A study by Nyström *et al.* determined amphetamine in multiple hair samples from a few patients medicating with amphetamine.[26] At doses between 30 to 60 mg per day Nyström reported concentrations of amphetamine between 1.27 and 5.7 ng mg^{-1} in compliant patients.

Regarding MDMA, two controlled dosing studies were found. Martins *et al.* reported concentrations above the 200 pg mg^{-1} cut-off in proximal 2 cm segments from subjects receiving doses of 75 or 100 mg MDMA.[27] However, the hair segment also included time prior to the study when recreational use of MDMA may have occurred. In another controlled study, Madry *et al.* gave 2×125 mg of MDMA to 15 volunteers.[28] The average concentrations of MDMA from several segments ranged from 78–980 pg mg^{-1} with 10 subjects with average concentrations above 200 pg mg^{-1}.

In this paragraph, we have reviewed the scientific data from controlled studies and patients regarding methamphetamine, amphetamine, and MDMA in relation to the recommended cut-off of 200 pg mg^{-1} to identify drug use. For methamphetamine, controlled dosing data is limited but suggest that the cut-off can be exceeded after only a few doses over a limited time.

For amphetamine, the pool of data is somewhat larger, suggesting that single doses, or even multiple doses over a limited time may not be sufficient to produce a positive result whereas daily medication with therapeutic doses of amphetamine will result in a positive test.

For MDMA, the two identified studies suggest that a few recreational doses may produce positive results.

The use of metabolites to confirm use is not established for the amphetamines but a possible specific metabolite for methamphetamine may be the *para* hydroxylated metabolite.

In conclusion, the 200 pg mg^{-1} cut-off for identifying drug use seem appropriate but does not necessary implicate that the person is a chronic user.

2.2.1.3 Cannabis Metabolites and Evaluation of Cut-off

The identification of delta-9-tetrahydrocannabinol (THC) using a cut-off of 50 pg mg^{-1} in hair is recommended by the SoHT to determine use of cannabis and the SoHT consensus includes the statement that the presence of the metabolite,11-nor-9-carboxy-delta-9-tetrahydrocannabinol (THC-COOH), strongly supports cannabis use.[1] The required minimum limit of quantification (LOQ) for THC-COOH is 0.2 pg mg^{-1} and reflects the low rate of incorporation of this acidic metabolite but also presents a significant challenge for analytical method sensitivity.

The importance of not relying on the presence of THC only, due to contamination from THC containing smoke, has been well-documented when considering active use of cannabis.[29,30] Back in 2004, Uhl and Sachs described a gas chromatography tandem mass spectrometry (GC–MS–MS) method for the detection of THC-COOH with a lower limit of quantification (LLOQ) of 0.1 pg mg^{-1}.[31] Few laboratories at the time were able to afford a GC–MS–MS but other methodologies were developed to match the sensitivity. It is surprising then, that many of the published methods reviewed and summarized by Shah *et al.* over a 10 year period prior to 2019, did not include testing for THC-COOH in hair.[32] In addition, several of the studies that did include testing of THC-COOH, did not have sufficient sensitivity for their method LOQ. This significantly reduces the number of studies that can help provide support for the use of a THC-COOH cut-off of 0.2 pg mg^{-1}.

Huestis documented a controlled dose study using both smoking and oral administration protocols to known cannabis users and collected hair before and after administration.[33] The LOQs for THC and THC-COOH were 1 pg mg^{-1} and 0.1 pg mg^{-1} respectively.

The 38 male cannabis users were classified as daily users (one to several blunts per day) and non-daily users (one to five cannabis cigarettes per week), and the hair collected before the controlled dose was administered, had varied results. No THC or THC-COOH was detected in the hair of three participants who declared daily use of cannabis or in nine participants who declared non-daily use. Although self-reported, all participants had urine specimens collected and screened positive for cannabinoids before the study commenced. A further seven participants who smoked two 2.7% THC cigarettes or multiple oral doses of THC in hemp oil (116 mg) under the controlled study, had no THC or THC-COOH present in their hair.

Of the 53 hair samples collected in this study, two had THC only and 14 had THC-COOH only. If we apply the SoHT cut-offs, only seven (13%) hair samples had both THC and THC-COOH present at concentrations equal to or greater than 50 pg mg^{-1} for THC and 0.2 pg mg^{-1} for THC-COOH. All 53 samples were collected from known cannabis users but applying the SoHT cut-offs identified only 25% (5/20) of the daily cannabis users and 6.0% (2/33) of non-daily users.

Vaiano *et al.* utilized the SoHT cut-offs as their LLOQ for both THC and THC-COOH and found that of the 103 hair samples, where the participant declared use of cannabis, 49 (47.5%) had both THC and THC-COOH above the cut-off.[34] Less than half of the self-reported cannabis users were identified but this was three to four times more than those identified in the study by Huestis *et al.*, when applying the SoHT cut-offs. In addition, they did not identify any positive hair samples that had only THC or only THC-COOH present.

The authors did report positive hair findings for five cases where the participants claimed to have never used cannabis before and the THC concentrations ranged from 630 to 1020 pg mg^{-1} with a median of 750 pg mg^{-1}, and the THC-COOH concentrations ranged from 40 to 450 pg mg^{-1} with a median of 300 pg mg^{-1}. The self-report may be misreported in these cases as the concentrations are much higher than the cut-offs but it must also raise the question of whether this could be environmental contamination.

Moosman reported the identification of THC and THC-COOH in hair following the transfer of cannabinoids from a cannabis user to a non-user.[35,36] This transfer can occur through close contact due to sweat/sebum, contamination of hands or cannabis smoke and is of particular concern for children who are in close contact with adults who use cannabis. More data is needed to evaluate the extent of this mechanism for incorporation of THC and THC-COOH into hair and whether the apparent conservative cut-offs recommended by the SoHT are sufficient to identify drug use and not environmental contamination.

Alternative metabolites/precursors have also been investigated with the aim of providing greater certainty of cannabis use and include the metabolites, 11-hydroxy-delta-9-tetrahydrocannabinol (11-OH-THC),[37–39] and

11-nor-delta-9-tetrahydrocannabinol-9-carboxylic acid glucuronide (THC-COOH-gluc)[40] and the precursor, delta-9-tetrahydrocannabinolic acid A (THCA-A).[29,41] These metabolites/precursors have been proposed as alternatives or complimentary to the detection of THC and THC-COOH and with more research they should be considered as an additional requirement to identify and distinguish cannabis users from non-users.

Cannabidiol (CBD), is similar to THC in that it is present in the cannabis plant, in cannabis smoke and can be present in hair through use or from environmental contamination. In recent years, CBD-rich products have been approved for sale in different jurisdictions and may contain smaller quantities of THC. The SoHT recommends a cut-off of 50 $pg\,mg^{-1}$ for CBD to determine drug use and currently no metabolites are listed. Skopp *et al.* reported CBD concentrations ranging from 60 to 570 $pg\,mg^{-1}$ for known cannabis users.[42] This was a time when CBD-rich products were not available and they also reported three cases where CBD was not detected in the hair of cannabis users, while THC and cannabinol (CBN) were detected.

Five years later, Salomone reported CBD concentrations in hair from known cannabis users ranging from 18 to 1862 $pg\,mg^{-1}$.[43] The authors summarized 14 positive cases and six of these had higher CBD hair concentrations than the corresponding THC concentration. Of the same 14 cases, six of them have concentrations below the SoHT cut-off of 50 $pg\,mg^{-1}$ of CBD.

Gerace *et al.* measured CBD in the hair of a subject who previously had not used any cannabinoid-containing product, before smoking a joint each day for 26 consecutive days, that contained 250 mg of CBD-rich cannabis (6% CBD and 0.2% THC).[44] The head and pubic hair collected prior to the start of the study had no CBD, THC, or THC-COOH identified. One week after the study finished, the head hair contained 312 $pg\,mg^{-1}$ of CBD, while the pubic hair contained 178 $pg\,mg^{-1}$ of CBD. THC and THC-COOH were not detected.

Casati *et al.* proposed the use of the ratios of CBD:THC and 11-OH-THC:THC to reliably discriminate between illegal and "light" cannabis use, the latter describing CBD-rich cannabis products.[45] Lo Faro *et al.* developed and validated a method to measure THC and CBD and a number of their metabolites and applied their method to hair from four subjects.[46] 7-OH-cannabidiol was reported as being the most prevalent CBD metabolite in hair.

The findings in these preliminary studies show great promise and with more research will help inform future updates to the requirements for determining use of cannabis and CBD-rich products from hair testing.

2.2.1.4 Opioid Metabolites and Evaluation of Cut-off

For heroin and the other opiates, the recommended cut-off is 200 $pg\,mg^{-1}$ hair and the metabolite stated is 6-acetylmorphine. Morphine is also a metabolite to heroin as well as to the medication codeine. Codeine was the only drug for which we could find dosing studies with low total doses that could investigate the cut-off value. Rollins *et al.* found that after five doses of 90 mg codeine, the proximal 3 cm of the hair contained mean concentrations (ng mg^{-1}) of 1.429 in black hair, 0.208 in brown, 0.099 in blond, and 0.069 in red hair showing that at least black- and brown-haired subjects' codeine concentrations exceeded the cut-off.[47]

Regarding heroin, Kintz *et al.* investigated the dose concentration relationship in 20 subjects in a controlled heroin maintenance program.[48] Heroin doses ranged from 30 to 800 mg per day. 6-Acetylmorphine was the dominating analyte in 15 of the 20 subjects. Concentrations (ng mg^{-1}) ranged from 0.38 to 10.11 (6-acetylmorphine), and 0.71 to 5.20 for morphine with all subjects above the recommended cut-off.

Musshoff *et al.* performed a similar study in a cohort of 46 subjects in a heroin maintenance program.[49] Morphine concentrations varied between 0.03 and 8.37 ng mg^{-1} with 11% of the subjects being negative for morphine whereas 6-acetylmorphine was present in all subjects but in some cases below the recommended cut-off (range 0.05–4.24 ng mg^{-1}).

As a metabolite and marker for systemic use, 6-acetylmorphine is not perfect. Even though it is a specific *in vivo* metabolite from heroin, seizures of heroin are likely to contain 6-acetylmorphine as well. From the studies mentioned above, also heroin was detected in the hair samples and the possible degradation of deposited heroin is likely to contribute to the concentrations of 6-acetylmorphine in hair samples. However, the finding of 6-acetylmorphine in hair does prove exposure to heroin and cannot be explained by exposure to another opiate such as codeine or morphine.

For tramadol, the recommended cut-off is 200 pg mg^{-1} and the detection of desmethyltramadol is also required to prove tramadol use. The guidelines do not include which demethylated metabolite, *O*-desmethyltramadol, *N*-desmethyltramadol or di-desmethyltramadol, nor do they include a cut-off for the metabolite. Very few studies are available from living patients where doses were known, and hair was analyzed for tramadol and its metabolites. In an early study by Hadidi *et al.* they measured tramadol in 11 patients treated with various doses of tramadol (100–300 mg per day) and found concentrations between 0.176–16.3 ng mg^{-1} without any correlation between dose and hair concentration.[50] The duration of treatment varied from only a few days to seven months. Pinho *et al.* investigated tramadol and *O*-desmethyltramadol in six patients treated with 50–150 mg per day for at least 15 days prior to sampling.[51] Tramadol was found in concentrations between 0.71 and 9.05 ng mg^{-1} with parent/metabolite ratios between 0.021 and 0.451. Even with the small number of patients, the evidence shows that ratios vary twice as much as concentrations. A possible reason for this is the polymorphic metabolism of tramadol to *O*-desmethyltramadol through the CYP2D6 enzyme. Yu *et al.* performed a clinical study with repeated dosing of tramadol to 24 volunteers and analysed hair samples to investigate metabolic ratios.[52] They found differences between subjects with normal CYP2D6 activity presenting with an *O*-desmethyltramadol/tramadol ratio of 0.13 whereas those with reduced capacity had a ratio of 0.04, and the only poor metabolizer had an even lower ratio of 0.02. Johansen *et al.* (2020) confirmed the differences in the parent to metabolite ratios in a controlled dosing study and found that ratios were stable over time within a subject but varied greatly between subjects.[53] The 18 subjects were genotyped and in the predicted phenotype group of extensive metabolizers the mean *O*-desmethyltramadol/tramadol was 0.30 (range 0.037–0.70) and for intermediate metabolizers the mean ratio was 0.21 (range 0.08–0.48) with significant differences between groups.

In an effort to use metabolite ratios to distinguish external contamination from passive exposure, and use, Madry *et al.* designed study with employees from a pharmaceutical company.[54] Eight employees with varying exposure to tramadol in their daily work were included and compared with 75 patients. *N*-Desmethyltramadol/tramadol ratios in employees ranged from 0.04 to 0.14 and were not significantly different from those in patients. For *O*-desmethyltramadol/tramadol ratios were from 0.002 to 0.008 in employees which was significantly lower than those from patients ranging between 0.003 to 0.43. The authors conclude that a metabolite to tramadol concentration ratio above 0.02 may suggest that tramadol was actively ingested, confirming the lower end ratios in the study by Pinho *et al.*[51]

Tramadol concentrations in hair samples of patients were 120–29 000 pg mg^{-1} hair meaning that at least some patients would have presented with negative results using the 200 pg mg^{-1} cut-off for tramadol. On the other hand, Yu *et al.* (2018) showed that a month after receiving five doses of 100 mg the mean hair concentrations of tramadol were 589.6–713.6 pg mg^{-1} in subjects with varying metabolic capacity for CYP2D6.[52]

From the scientific data, the 200 pg mg^{-1} cut-off for identifying drug use seems appropriate but does not necessary implicate that the person is a chronic user.

The demethylated metabolites occur at lower concentrations than tramadol itself and verifying systemic exposure at the 200 pg mg^{-1} level will require limits of quantification considerably lower than for tramadol. In addition, CYP2D6 poor metabolizers may not produce enough *O*-desmethyltramadol to present with a positive result.

Methadone metabolizes primarily to 2-ethylidene-1,5-dimethyl-3,3-diphenylpyrrolidine (EDDP), while its secondary metabolite, 2-ethyl-5-methyl-3,3-diphenylpyraline (EMDP) is less frequently detected in biological samples, and when present is in very low concentrations. Five studies detailed methods that were developed to test for EMDP in hair in addition to methadone and EDDP, and all of them involved testing hair samples collected from subjects enrolled on a methadone maintenance treatment program (MMTP). Wilkins *et al.* tested hair from two subjects and EMDP was not detected.[55] Kelly *et al.* measured the enantiomers of methadone and its two metabolites in hair from 20 subjects and they also reported no cases where EMDP was detected.[56] Trace amounts of EMDP in hair was reported for one case out of 20 tested by Goldberger *et al.*,[57] while Sporkert and Pragst[58] reported EMDP present in two of the seven subjects in their study that were tested for EMDP, and the concentrations were 0.18 and 0.84 ng mg^{-1}. EMDP concentrations in hair collected from pregnant women ranged from 0.14 ng mg^{-1} to 1.07 ng mg^{-1} and was only present in a limited number of cases (Himes 2012).[59]

The sensitivity of each of the methods was a limiting factor in the detection of EMDP with varying lower limits of quantification (LLOQ) reported and the highest being 1.0 ng mg^{-1}, while the lowest limit of detection (LOD) reported was 0.05 ng mg^{-1}. The case with trace amounts of EMDP reported also had the highest methadone concentration reported in that study, at 15 ng mg^{-1} and trace amounts of EDDP (Goldberger 1998).[57] Similarly, the two cases reported with EMDP above the LLOQ, had the highest methadone and EDDP concentrations of all 26 subjects tested and was highlighted by the authors.[58] EMDP being present in hair with high methadone concentrations was also observed by Himes *et al.*[59]

Some of these studies were published over 20 years ago and the authors were achieving excellent sensitivity with the available technology. Now with improved sensitivity in the methods used in hair testing laboratories today, could EMDP provide insights into the extent of chronic use? There is unfortunately a lack of data to answer this question as the majority of methods published for testing methadone in hair only include the primary metabolite EDDP.

The SoHT recommends a methadone cut-off of 200 pg mg^{-1} in hair and the additional presence of EDDP is definitive proof of methadone use. It is important to re-emphasize that the SoHT Consensus on Drugs of Abuse Testing in Hair, excludes the use of cut-offs when interpreting drugs detected in hair collected from children. Methadone measured above the 200 pg mg^{-1} cut-off, along with the presence of EDDP is not necessarily definitive proof of use or administration of methadone in children. At the very least, this does confirm exposure and is discussed in more detail in Chapter 3.

No cut-off value is currently proposed for EDDP and instead laboratories will apply their LLOQ and LOD for identification. Methadone is widely prescribed and there are a number of publications providing concentration ranges measured in patients enrolled in MMTPs. Musshoff *et al.* conducted a study that involved 41 Caucasian subjects who had been enrolled on a supervised MMTP for at least one year and prescribed doses ranging from 15 to 260 mg per day.[60] Hair was washed prior to analysis and up to 3 cm were segmented from the proximal end closest to the scalp. Methadone and EDDP were present in all segments and methadone was always in higher concentration than EDDP. Methadone ranged from 0.25 to 13.29 ng mg^{-1} and for EDDP from 0.05 to 2.17 ng mg^{-1}. Interestingly the ratio of methadone to EDDP varied and was statistically significant depending on the region of the hair. The proximal segments had a methadone:EDDP ratio of 7.5 ± 5.7 while the distal segments decreased to 4.8 ± 1.4 ng mg^{-1}. In this study, all hair segments were positive for methadone at concentrations greater than the SoHT cut-off of 200 pg mg^{-1} including those on a lower daily dose of 15 mg per day. Paterson measured methadone and EDDP concentrations in 3 cm long hair segments from 50 MMTP patients on a daily oral dose ranging from 20 to 165 mg.[61] The lowest methadone concentration measured was 2.6 ng mg^{-1} and the authors acknowledged there was overlap in the concentrations measured for groups separated by low, medium, and high oral dose. The medium group who were prescribed a dose ranging from 60 to 100 mg daily, had hair methadone concentrations from 2.6 to 63.2 ng mg^{-1} and this fully encompassed the lowest concentration measured from the low dose group (4.2 ng mg^{-1}) and the highest concentration measured from the high dose group (43.9 ng mg^{-1}). EDDP concentrations measured or the prevalence of EDDP in the hair specimens tested was not disclosed by the authors, although the method did have a high LLOQ of 0.4 ng mg^{-1} for EDDP.

Sporkert and Pragst's method had an LOD of 0.05 ng mg^{-1} and an LLOQ of 0.16 ng mg^{-1} for EDDP and reported EDDP concentrations from 0.10 to 10.1 ng mg^{-1}.[58] Although there are many publications detailing the measurement of methadone and its primary metabolite EDDP in the literature, unfortunately many are limited by the number of hair specimens in the study with a measured concentration of EDDP due to insufficient method sensitivity.

The SoHT recommended cut-off for methadone of 200 ng mg^{-1} would appear to be appropriate to determine use of methadone but not to determine chronic use. A corresponding EDDP cut-off evaluation can only be determined when method sensitivity is 5- to 10-fold lower than methadone.

Like methadone, buprenorphine is used as a substitute drug when medically treating opiate addiction, but it is also widely used for its analgesic effects. Buprenorphine metabolizes to norbuprenorphine mainly by the CYP3A4 enzyme system. An early study on 14 maintenance patients on various doses detected buprenorphine concentrations between 0.02 (LLOQ) and 0.59 ng mg^{-1} and suggested a correlation between dose and hair concentration.[62] A study by Wilkins *et al.* extended the data from controlled administration of 8 mg per day between 40 and 180 days using LLOQs for buprenorphine and norbuprenorphine at 3 pg mg^{-1} and 5 pg mg^{-1}, respectively.[63] The 12 subjects presented with buprenorphine concentrations ranging from 3–124 pg mg^{-1} for buprenorphine and from 5–1518 pg mg^{-1} for norbuprenorphine. Norbuprenorphine was found in higher concentrations than the parent drug. It is difficult to interpret the quantitative data since hair sample segments represented both periods with active dosing and those where dosing was not active. In 2011, Skopp *et al.* examined the hair of 18 patients with doses in the range 2–12 mg per day and found a relationship between dose per kg bodyweight and the combined concentrations of buprenorphine and its metabolite.[64] All hair samples were positive and showed concentrations between 0.003–0.238 ng mg^{-1} and 0.043–0.961 ng mg^{-1} for buprenorphine and norbuprenorphine, respectively. Norbuprenorphine concentrations were always higher than those of buprenorphine.

In a study of the hair concentrations in 36 maintenance patients on doses between 2 to 18 mg per day Belivanis *et al.* reported 25% interquartile ranges of buprenorphine from 4.3 pg mg^{-1} to 11.4 pg mg^{-1} in the distal, middle, and proximal segments.[65] Norbuprenorphine concentrations were higher with 25% interquartile ranges from 20.4, 21.5, and 27.0 pg mg^{-1}, respectively.

The recommended cut-off for buprenorphine is 10 pg mg^{-1}. In the light of the existing literature, this may not always detect a chronic user. The metabolite norbuprenorphine seems to be present in higher concentrations than the parent drug.

Ketamine is a widely used anaesthetic in surgery but has also been an abused drug especially in Asian countries but also increasingly in Europe and the Americas. Therefore, it is not surprising that early papers on ketamine are from Asia.[66,67] In a paper with four subjects, Ping *et al.* gave a single dose of 10 mg ketamine and determined ketamine and norketamine concentrations in hair segments from samples obtained up to 16 weeks after dosing.[67] Ping *et al.* found average maximum hair concentrations of 19.0 pg mg^{-1} for ketamine and 18.7 pg mg^{-1} for norketamine. The highest ketamine concentration in a segment was 28.4 pg mg^{-1} in the sample taken eight weeks after dosing and the highest norketamine concentration was 38.2 pg mg^{-1} in the sample obtained one week after dosing. All concentrations were well below the recommended cut-off of 200 pg mg^{-1}. In another controlled study, Barreto *et al.* found ketamine concentrations in the range 61–111 pg mg^{-1} in three patients who received 1 mg ketamine per kg 6–7 weeks prior to sampling.[68] A case report where ketamine was given as treatment in a total dose of 171 mg over three days Larabi *et al.* found 58 pg mg^{-1} ketamine in a 2 cm segment corresponding to the treatment period.[69] We did not find any other controlled studies and the data suggests that single doses as well as repeated doses during a short time are not likely to results in concentration above the 200 pg mg^{-1} threshold. In 2015, Salomone and his co-workers proposed a cut-off of 500 pg mg^{-1} for repeated exposure to ketamine based on their own data as well as a number of previously published studies and suggested a cut-off for norketamine at 100 pg mg^{-1}.[70] A similar effort was made by Leung *et al.* where they investigated the ketamine and norketamine concentrations in hair from 977 ketamine abusers and proposed a cut-off of 400 pg mg^{-1} since that was the concentration where the metabolite norketamine was detected in more than 90% of the cases using a cut-off of 100 pg mg^{-1}.[71] An interesting study was published by Favretto *et al.* in 2016 where they found evidence that exposure from the work environment can produce ketamine positive hair samples.[72] Eleven veterinary physicians were recruited, and hair was analysed for ketamine and norketamine at various timepoints. All hair samples presented with positive results for ketamine in the range 10–840 ng mg^{-1} and norketamine was detected up to 80 pg mg^{-1}. This finding stresses the necessity to also detect the metabolite at a certain level to confirm "repeated use" rather than exposure from the handling of ketamine on a daily basis. In summary, the 200 pg mg^{-1} cut-off seem appropriate to detect repeated use or exposure and the simultaneous detection of norketamine at concentrations above 100 pg mg^{-1} will strengthen the conclusion that the subject actively administered ketamine.

In this section, we have critically evaluated the SoHT recommended cut-offs for cocaine, amphetamines, cannabis, opioids, and ketamine and discussed the identification of metabolites to confirm use rather than exposure. From the data it is noticeable that cut-offs for some drugs are low enough to detect a single or sporadic administration and a positive result does not mean that the person was a regular user or abuser. In addition, the detection of metabolites may not be unequivocal proof of ingestion.

2.3 How Much Drug Was Administered?

Dose concentration relationships exist and have been verified for some drugs but even so, results from studies are difficult to translate to case work to support interpretation of the amount of drug administered. First of all, even if there is a correlation on the group level, the variation is large and depends on many factors. A factor identified by many research groups is hair pigmentation.

2.3.1 Dose Concentration Relationships and Hair Pigmentation

We know from numerous *in vitro* and animal studies that hair pigments are involved in the binding of many drugs into hair, especially those that are weak bases and thus can interact with the negatively charged moiety of the melanin macromolecules. Studies on the effect of pigmentation in humans are few but have included prescribed drugs and drugs of abuse with papers describing both intra-subject differences in pigmented and non-pigmented hair as well as melanin/pigmentation relationships. These studies are summarized in Tables 2.2 and 2.3.

Prompted by their earlier studies, showing that haloperidol doses correlated with hair concentrations, but with inter-individual differences,

Table 2.2 Studies with more than five subjects that investigated drug concentrations in gray haired subjects.

	Drug	Subjects	Dose	Pigmented hair (mean ng mg^{-1})	Non-pigmented hair (mean ng mg^{-1})	Difference pig/white	Ref.
1	Haloperidol	Patients $N=7$	3–25 mg per day				73
2	Cocaine	Users $N=29$	—	COC 31.6 BE 5.1 CE 3.2	COC 14.0 BE 3.9 CE 0.5	Sign Sign Sign	74
3	Selegiline	Patients $N=16$	5–10 mg per day	MA 1.5 A 0.50	MA 0.44 A 0.17	Sign Sign	75
4	Clozapine	Patients $N=12$	8–62 mg per day	CLZ 1.7 dmCLZ 2.2	CLZ 0.8 dmCLZ 0.8	Sign Sign	76

Table 2.3 Studies with more than five subjects that studied the relationship between drug concentrations and hair colour or pigmentation.

Drug	Dose	Subjects	Drug results (mean or range) (ng mg^{-1})	Total melanin (mean or range) (μg mg^{-1})	Results summary	Ref.
Codeine	100 mg	Subjects $N=9$	0.024–0.566	2.12–13.58	Exponential relationship between the drug concentrations and melanin was observed ($R^2=0.95$)	77
Selegiline	15 mg	Subjects $N=10$	MA 0.0058–0.1179 A 0.002–0.022	0.8–20.7	Exponential relationship between the drug concentrations and melanin was observed ($R^2=0.93$ For MA and $R^2=0.1$ For A)	78
Carbamazepine	525 mg per day (mean)	Patients $N=15$ Patients $N=15$ Patients $N=3$ Patients $N=3$	Black 17.59 Brown 18.05 Blond 24.23 Gray 13.58	—	No significant differences between hair colour groups	82
Valproic acid	822 mg per day (mean)	Patients $N=17$ Patients $N=17$ Patients $N=3$ Patients $N=3$	Black 9.65 Brown 7.58 Blond 5.81 Gray 6.08	—	No significant differences between hair colour groups	82
Phenytoin	243 mg per day (mean)	Patients $N=24$ Patients $N=18$ Patients $N=8$ Patients $N=10$	Black 16.35 Brown 16.76 Blond 7.92 Gray 6.55	—	Significant differences between black/blond, black/gray, brown/blond, and brown/gray	82
Codeine	5 × 90 mg per day	Subjects $N=16$ Subjects $N=12$ Subjects $N=8$ Subjects $N=6$	Black 1.429 Brown 0.208 Blond 0.099 Red 0.069	7.11–14.8 3.57 2.33 2.62	An exponential relationship between the drug concentrations and melanin was observed ($R^2=0.73$)	47

Ofloxacin	800 mg	Subjects $N=5$ Subjects $N=13$ Subjects $N=8$ Subjects $N=6$	Black 30.64 Brown 6.03 Blond 3.51 Red 1.49	10.92 (eumelanin) 3.51 (eumelanin) 2.25 (eumelanin) 1.68 (eumelanin)	A linear relationship between drug concentration and melanin was observed ($R^2=0.728$)	47
Codeine	3 × 60 mg per day 3 × 120 mg per day	Subjects $N=10$ Subjects $N=10$	1.291 2725	11.9–53	A linear relationship between drug concentration and melanin ($R^2=0.43$ (low dose) and 0.62 (high dose)	21
Cocaine	3 × 75 mg per day 3 × 150 mg per day	Subjects $N=10$ Subjects $N=10$	COC 2.997 BE 0.314 COC 6.419 BE 0.708	11.9–53	A linear relationship between drug concentration and melanin ($R^2=0.45$ (low dose) and 0.80 (high dose)	21
Methamphetamine	4 × 10 mg 4 × 20 mg	Subjects $N=7$ Subjects $N=5$	MA 1.78 A 0.20 MA 3.0 A 0.35	41.8 (22.9–51.7)	Dose-concentration relationship within subjects but not between. Significant correlations were observed between Cmax and melanin.	23
THC	—	Users $N=1038$ Users $N=2376$ Users $N=144$ Users $N=8$ Users $N=17$	Black 6.92 (pg per 10 mg) Brown 7.28 (pg per 10 mg) Blond 6.59 (pg per 10 mg) Gray 12.51 (pg per 10 mg) Red 3.66 (pg per 10 mg)		No significant differences between hair colour groups	80

Uematsu *et al.* investigated the impact on melanin in the incorporation of haloperidol in grey-haired patients and showed that white hair contained less than 10% of the haloperidol in pigmented hair.[73] In 1996, Reid *et al.* investigated the distribution of cocaine, cocaethylene, and benzoylecgonine in pigmented and white hair from 29 gray haired users.[74] For cocaine, the measured concentration in pigmented hair was always higher than in white hair, however for some subjects the difference was small. The mean ratio between pigmented and white hair was 3.8 confirming the preference for pigmented hair but the range was wide, 1.125–18.83, showing inter-individual differences. This can partly be explained by variation in pigmentation of the pigmented hair portion. For benzoylecgonine the results were also significantly different but with several of the subjects presenting with equal or higher benzoylecgonine concentrations in the white hair portion. The zwitter-ionic nature of benzoylecgonine may influence its ability to form ionic bonds with melanin and preferring other main binding sites in the hair structure.

Kronstrand *et al.* performed a similar study but in patients medicating with selegiline, confirming what Reid saw for cocaine.[75] Selegiline is converted to methamphetamine and amphetamine in the body and the metabolites can be readily detected in hair. Paired results from the gray-haired subjects showed that there were always higher amphetamine and methamphetamine concentrations in the pigmented portion of the hair, a finding significant for both analytes. The concentration ratios, however, ranged from 1.8–8.0 and 1.8–5.4 for methamphetamine and amphetamine, respectively again emphasizing that there is inter-individual variation. In Kronstrand's study the melanin in the pigmented hairs were determined and ranged between 6.24 and 17.50 $\mu g\, mg^{-1}$ hair but did not correlate with the observed analyte concentrations.

Later on, in 2007, Kronstrand *et al.* investigated the incorporation of the antipsychotic medication clozapine and its metabolite, desmethylclozapine, in patients on low dose clozapine.[76] The effect of melanin was evident also in this study with clozapine and desmethylclozapine found in significantly higher concentrations in pigmented hair than in non-pigmented at mean ratios of 2.52 and 3.04, respectively.

These four studies confirm a preference for binding to pigmented hair that can be attributed to a strong ionic interaction between the, at physiological pH, positively charged weak bases and the polyanionic melanin polymer. However, because drugs could be detected in white hair, pigmentation cannot be the only factor involved. Binding to hair proteins accounts for a significant part of the drug accumulation in hair and to be more accurate, hair colour is not black or white but a continuous scale from black, dark brown, brown, light brown, blond, and even red hair that mainly contains pheomelanin rather than eumelanin.

With that in mind, a series of studies in the early 2000s investigated the differences of drug concentrations in hair with different colours and some also correlated drug incorporation to the melanin content of the hairs.

Kronstrand *et al.* performed two controlled dosing studies in 1999 and 2001 where the authors found exponential relationships between total melanin and codeine, methamphetamine, and amphetamine in hair.[77,78] Subjects' hair colours were quite different, from light blond to black. Rollins *et al.* confirmed the exponential relationship between total melanin and codeine in a study where subjects were given multiple doses of codeine over five days.[47] Scheidweiler *et al.* followed up these results in subjects given multiple doses of codeine and also found a significant relationship between total melanin content and codeine incorporation but with a linear relationship, possibly explained by the limited melanin concentration range in the subjects' hair.[21]

The concentrations found in these controlled dosing studies sometimes exceeded the recommended cut-off for identifying drug use.[1] The length of the segment used for analysis is important and a 3 cm segment is recommended for screening purposes. Shorter segments analysed after single doses may present with high concentrations since less negative hair is diluting the sample. Rollins *et al.* argued that, based on their results from analysis of the proximal 3 cm of the hair using a cut-off for codeine at 200 $pg\,mg^{-1}$ all subjects with black hair and half of the subjects with brown hair would be reported as positive, whereas no subjects with blond or red hair would be reported as positive.[47]

Scheidweiler's study from 2005, with multiple doses of cocaine also suggests a hair colour relationship for cocaine incorporation into hair but also demonstrated dose-related concentrations of cocaine, and metabolites.[21] After controlled dosing of either 75 mg or 150 mg per day of cocaine during three days the average maximum concentrations of cocaine were 3 $ng\,mg^{-1}$ and 6.4 $ng\,mg^{-1}$, respectively, concentrations that exceed the recommended cut-off of 0.5 $ng\,mg^{-1}$ to identify a drug user. Unfortunately, the hair sampling did not reflect a typical hair testing scenario because the hair was shaved weekly rather than sampled and segmented. Nevertheless, it stresses the need for studies that can verify the relationship between dose and hair concentrations.

In summary, hair pigmentation influences the incorporation and at low doses, it will impact the positivity rate. However, when investigating single doses or sporadic doses, for example in drug facilitated crimes, the recommendation is to not use the cut-offs but use more sensitive methods. That said, practitioners still want an answer to the question how much was administered, *e.g.*, to understand if a person is a light, moderate, or heavy user of a drug.

2.3.2 Interpretation of Hair Concentrations in Case Work

The ultimate scenario would be that one could estimate the administered dose from the hair concentrations found. Such relationships can exist on a population level but are not entirely transferable to an individual case. For example, Kintz *et al.* investigated the dose concentration relationship in 20 subjects in a controlled heroin maintenance program.[48] Heroin doses were in the range 30–800 mg per day and the hair concentrations were compared to the total dose over 100 days. There was a dose concentration relationship towards morphine ($r = 0.64$) but not the other analytes, but the authors concluded that using concentrations in hair to determine the amount of drug ingested was inapplicable. Musshoff *et al.* performed a similar study on 46 subjects in a maintenance program and found significant relationships between average daily dose and concentrations of both 6-acetylmorphine ($r = 0.588\ 67$), morphine ($r = 0.685\ 81$), and total opiates in hair ($r = 0.658\ 67$).[49] Morphine, with the longest half-life, had the best correlation but total opiates was very close. Since heroin and 6-acetylmorphine are unstable, and may degrade to morphine over time before sampling, during storage and analysis, concentrations may change, and dose concentration relationships may be better described using the sum of analyte concentrations.

Most researchers and practitioners agree that even though dose concentration relationships exist it is not recommended to estimate a dose based on the hair concentration. Indeed, it is not advisable for any other matrix either. On the other hand, it is common to assign a concentration, *e.g.*, in plasma, to a group based on sub-therapeutic, therapeutic, and toxic ranges. Similarly, papers have been published on hair concentrations in different populations to enable categorization of light, moderate or heavy users.

Vignali *et al.* described the hair concentrations of cocaine and benzoylecgonine in Africans and Caucasians based on their self-reported drug use.[79] In Table 2.4 we have summarized the data from Vignali showing not only differences between ethnic groups but also between use patterns. In Caucasians with daily use cocaine mean concentration was 39.82 ng mg^{-1}, compared to 18.96 ng mg^{-1} in subjects using 2–4 times per week, and again lower in subjects using cocaine 1–2 times a month (3.99 ng mg^{-1}). In Africans, the respective concentrations were 60.74 ng mg^{-1}, 33.42 ng mg^{-1}, and 14.98 ng mg^{-1}. Data sets like this could be useful to categorize results and an opinion may read: the measured concentration of 17 ng cocaine per mg hair

Table 2.4 Cocaine and benzoylecgonine concentrations and self-reported cocaine use. Reproduced from ref. 79 with permission from Elsevier, Copyright 2011.

	Daily use mean (CV)	2–4 times per week mean (CV)	1–2 times per month mean (CV)
Africans $N = 51$			
Cocaine	60.74 (44%)	33.42 (60%)	14.98 (49%)
Bensoylecgonine	6.01 (41%)	4.05 (38%)	1.83 (47%)
Caucasians $N = 31$			
Cocaine	39.82 (49%)	18.96 (59%)	3.99 (44%)
Bensoylecgonine	5.69 (49%)	2.61 (46%)	0.56 (44%)

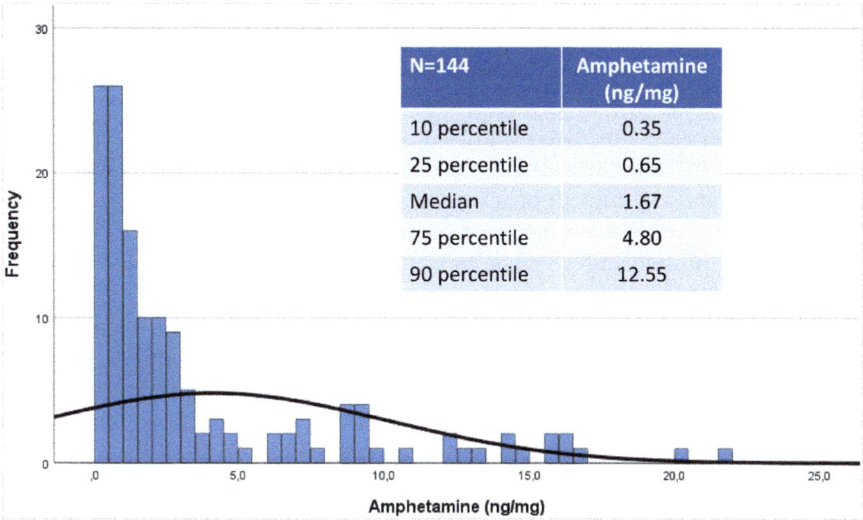

Figure 2.1 Frequency chart and descriptive statistics of amphetamine concentrations in case work. $N = 144$. Only autopsy case data from the inner segment S1 (0–5 mm) are included.

is in the same range as in Caucasians self-reporting cocaine use 2–4 times per week.

Rather than using ethnic origin, hair colour may be used to group individuals and enable a comparison of case work to literature data.[80–83]

Another approach would be to only state if a concentration is high or low depending on previous results from case work at the laboratory. Figure 2.1 shows the distribution of amphetamine concentrations in hair from autopsy cases that have been sent for analysis to the National Board of Forensic Medicine in Sweden during 2020 and 2021. All cases were negative for methamphetamine. The distribution is skewed toward the lower end. An opinion of a result may read: the measured concentration of 0.30 ng amphetamine per mg hair is below the 10th percentile and considered low.

2.4 When Was the Drug Administered?

When a drug was administered is commonly asked in drug facilitated crimes and those studies involving single intakes are the most helpful to the toxicologist when considering this question. Since drug administration is aimed at incapacitating the person, central nervous system depressants are of most interest. The scientific evidence is sparse since only a few subjects are included in many studies. In addition, study designs are different and direct comparisons are not always possible. The time between dosing and hair sampling as well as number and length of segments are important factors that will affect the results and so are the hair preparation procedures.

Nevertheless, we can draw some general conclusions before looking at individual drugs in more detail.

Drug concentrations are in the picogram per milligram range but even after a single dose, drugs can be detected several months after intake. However, the temporal pattern becomes more diffuse the longer the time between intake and sampling. In Table 2.5, controlled studies regarding single administration of central nervous system depressants are summarized showing studies for drugs where five or more subjects were included.[53,77,84–90]

One of the pioneering studies was published by Rollins *et al.* in 1996, perhaps by then, mass spectrometers were sensitive enough to measure picograms of analytes in small amounts of hair even though hair analysis as such was into its third decade.[84] Rollins' model drug was codeine, given in two doses. An observation was that in plucked hair, codeine could be detected in high concentrations just a few hours after administration. The codeine, probably situated in the germinative elements of the hair bulb peaked at 4 hours post dose and was 4.4 ng mg^{-1} after a 120 mg dose of codeine. These initial high concentrations were quickly declining because of falling plasma concentrations. The data from the distal segments up to 10 weeks after administered are difficult to interpret since the entire length of the hair was analysed but still a mean concentration of 30 pg mg^{-1} were seen in hair samples obtained at weeks three to eight. Following a similar design, but with cutting of the hair, Kronstrand *et al.* studied the role of hair pigmentation by analyzing the proximal 2 cm segment one month after an oral dose of 100 mg codeine to nine volunteers.[77] Proximal hair concentrations were in the range 27–566 pg mg^{-1} and were correlated to the hair melanin content. These studies prompted research aiming towards drug facilitated crimes using hair analysis to investigate flunitrazepam and clonazepam incorporation into hair following doses of 2 and 3 mg.[85,86] Like Kronstrand, samples were obtained up to 28 days and the proximal 2 cm were analysed. Metabolite concentrations were very low, less than 8 pg mg^{-1} for 7-aminoflunitrazepam and less than 23 pg mg^{-1} for 7-aminoclonazepam. Additional studies with estazolam[87] and zolpidem[88] further strengthened the scientific basis that after a single dose the drug is found in a segment corresponding to the time of administration. However, an important finding was that also adjacent segments, grown prior to dosing, were positive. In six of the 20 subjects receiving 10 mg zolpidem concentrations in the range 4.0–13.6 pg mg^{-1} were found in the distal segment 2–4 cm as compared to 135–555 pg mg^{-1} in the proximal 2 cm. This phenomenon had been described in numerous case reports before but since dosing regimens were unknown, no definitive conclusions can be drawn from those publications. Analyzing several segments and comparing the concentrations is commonly done and a concentration in a neighbouring segment less than three times that in the expected segment has been suggested to indicate "diffusion" and still compatible with a single intake.[91] But as we shall see later in the chapter, this depends on the drug administered, the length of the segment and perhaps also on the time between intake and sampling.

Table 2.5 Studies with more than five subjects that investigated the incorporation of drugs from single doses of sedatives.

	Drug	Subjects/dose	Sampling and segmentation	Decontamination	Sample preparation	Results	Ref.
1	Codeine	$N=7$ (60 mg) $N=12$ (120 mg)	Up to 10 weeks A 1 cm proximal segment and one distal segment remaining hair	None	NaOH, overnight, 37 °C	120 mg Distal segment Mean 30 pg mg^{-1} at 3–8 weeks	84
2	Codeine	$N=9$ Various hair colour 100 mg	Up to 28 days One 2 cm segment	None	NaOH, 20 min, 80 °C	0–2 cm at 28 days 24–566 pg mg^{-1}	77
3	Flunitrazepam	$N=10$ Various hair colours 2 mg	Up to 28 days A 1.5 cm segment	None	Pulverization, methanol sonication, acidic digestion overnight	0–2 cm at 28 days 7A-flu 1.4–8.0 pg mg^{-1} ($N=9$) Flunitrazepam sporadically detected	85
4	Clonazepam	$N=10$ 3 mg	Up to 28 days One 2 cm segment	None	Pulverization, methanol sonication, acidic digestion overnight	0–2 cm at 28 days 7A-clonazepam 1.8–23 pg mg^{-1} ($N=6$) Clonazepam not detected	86
5	Estazolam	$N=14$ Black hair Dose 1,2, 4, 5, or 6 mg	At 1 month 0–2 cm, 2–4 cm, 4–6 cm	2 × DCM	Pulverization, buffer pH 8.4 sonication 1 h	0–2 cm 0.56–2.60 pg mg^{-1} $N=13$ 2–4 cm 0.71–0.83 pg mg^{-1} $N=3$	87

Table 2.5 (*Continued*)

	Drug	Subjects/dose	Sampling and segmentation	Decontamination	Sample preparation	Results	Ref.
6	Zolpidem	$N=20$ Black hair 10 mg	At 1 month 2 cm segments	2 × DCM	NaOH, 1 h, 60 °C	0–2 cm Mean 250 pg mg^{-1} Range 135–555 pg mg^{-1} 2–4 cm Mean 7 pg mg^{-1} Range 4–14 pg mg^{-1}	88
7	GHB	$N=12$ 25 mg kg^{-1}	At 1 and 2 months 5 mm segments	Methylene chloride	Pulverization NaOH, 20 min, 80 °C	Before dosing (mean) 1.27 ng mg^{-1} 1 mo after dose 5.35 ng mg^{-1} 2 mo after dose 4.62 ng mg^{-1} Statistically different	89
8	Tramadol	$N=8$ (50 mg) $N=9$ (100 mg)	At 14, 30, 60, and 120 days 5 mm segments	Isopropanol, water, isopropanol	Pulverization, buffer incubation 18 h	50 mg (max conc) Tramadol 4–950 pg mg^{-1} 100 mg Tramadol 8–630 pg mg^{-1} *N*-Desmethyltramadol and *O*-desmethyltramadol also detected	53
9	Zopiclone	$N=8$ 5 mg $N=8$ 10 mg	At 14, 30, 60, and 120 days 5 mm segments	Isopropanol, water, isopropanol	Pulverization, buffer incubation 18 h	5 mg (max conc) 5–370 pg mg^{-1} 10 mg 17– 590 pg mg^{-1} Desmetylzopiclone also detected	90

A common prescription drug which is also widely misused is zopiclone. There are two studies that have investigated the incorporation of zopiclone.[90,92] Villain *et al.* administered 7.5 mg zopiclone and measured 5.4 and 9.0 $pg\,mg^{-1}$ in the proximal 2 cm segment cut one month after administration. Lund Hansen *et al.* gave 5 mg or 10 mg to 16 volunteers and measured zopiclone in 5 mm segments of hair samples obtained up to four months after administration.[90] One month after intake, the zopiclone concentration in the proximal 20 mm segment were 40 and 60 $pg\,mg^{-1}$, considerably higher than those in the study by Villain. A possible explanation could be that the hair extraction procedure was different with hair snippets compared to powdered hair that have been shown to increase the surface area and extraction recovery.

In a few subjects Lund Hansen also analyzed 10 mm and 20 mm segments to compare concentrations and temporal resolution with 5 mm segments. From Figure 2.2 we can learn a few things. First it is common to see several positive segments after a single intake, even if the hair segments are as long as 20 mm. A combination of factors contributes to this phenomenon. A physiological factor is the hair growth cycles where positive hair strands may enter the resting phase causing "fronting" with segments presumably representing hair grown after the intake being positive. Since only a smaller part of the hair strands are in the resting phase this contribution should be rather small. A major contributor is the shifting of hairs during sampling, transportation, and analysis. Variation in hair growth rate over time can also contribute as sampling times further away from the intake show increased spread. Shorter segments suffer more from parallel shift as can be seen in Figure 2.2.

After a single intake there is a dilution effect that can reduce the positivity rate when using long segments. In their experiment, Lund Hansen always saw higher drug concentrations in 5 mm segments than in longer segments, at some sampling times two to three times higher concentrations in the shorter segments. The explanation is that there is more drug free hair in a longer segment thus reducing the concentration.

Zopiclone did not show any wash-out effect in Lund Hansen's study and the explanation could be that zopiclone mainly incorporates through the blood stream and tightly binds to hair components. Zopiclone concentrations also increased over the first months indicating that incorporation *via* sweat was less important. In stark contrast to this is the incorporation of tramadol.

In an identical study, Johansen *et al.* studied the incorporation of tramadol in subjects given either 50 or 100 mg of tramadol orally.[53] A common side effect of tramadol is hyperhidrosis, excessive sweating, which contributes to initially high concentrations that are loosely bound to hair and successively washed out through normal hygiene. In Figure 2.3, we can see examples of the wash-out effect as well as how tramadol is distributed into several segments disturbing the temporal pattern. A decrease in both the maximum concentration and cumulated amount of tramadol were observed throughout the study suggesting that a substantial amount of tramadol was removed by normal hygiene.

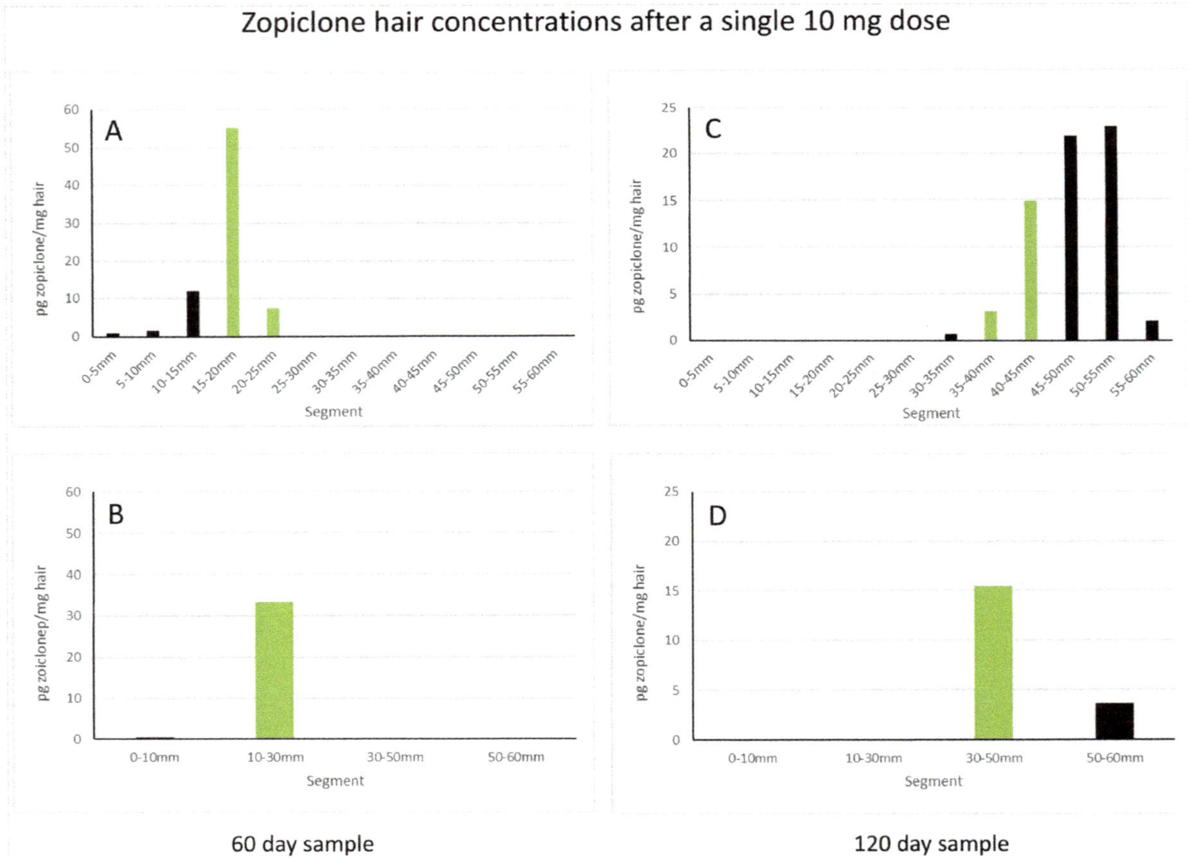

Figure 2.2 Zopiclone hair concentrations after a 10 mg dose. (A) Results from 5 mm segments 60 days after administration. (B) Results from 10–20 mm segments 60 days after administration. (C) Results from 5 mm segments 120 days after administration. (D) Results from 10–20 mm segments 120 days after administration. Green bars represent the segment per s where administration occurred. Reproduced from ref. 90 with permission from Elsevier, Copyright 2020.

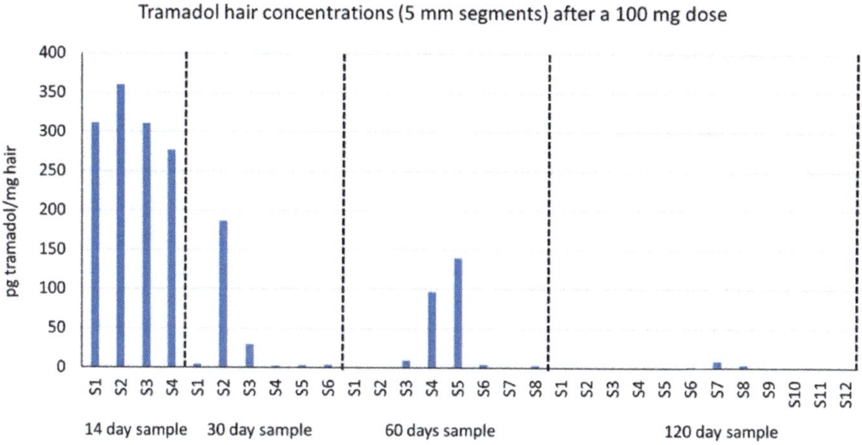

Figure 2.3 Tramadol hair concentrations after a 100 mg dose. Four samples taken at different times after dosing illustrates how tramadol incorporation from sweat dominates at early time points but is successively washed out through normal hygiene. Reproduced from ref. 53 with permission from Elsevier, Copyright 2020.

A drug of interest in drug facilitated crimes is gamma-hydroxybutyrate (GHB). Since GHB also is an endogenous compound additional measures must be taken to analyse hair and to interpret an intake. Bertol *et al.* provided some guidance in their well-designed study of baseline levels of GHB as well as the temporal pattern in short segments after a single intake of 25 $mg\,kg^{-1}$ body weight in 12 subjects.[89] Importantly, hair pigmentation, age, or gender did not affect baseline GHB concentrations. Segments of 5 mm each seemed to be short enough to distinguish increased concentrations of GHB at the timepoint of administration as compared to adjacent segments. One and two months after administration the mean concentrations were 5.35 ± 1.36 $ng\,mg^{-1}$ and 4.62 ± 1.69 $ng\,mg^{-1}$ approximately four and three times higher than those of adjacent segments. The study also showed that the GHB concentration decreased about 30% from one to two months indicating that sampling times further away from the incident may be less effective in distinguishing a single intake from baseline GHB concentrations.

Precise segmental hair analysis is necessary to interpret the temporal pattern and care must be taken during the sampling, transport, as well as during segmentation and analysis. Even under experimental conditions, the temporal pattern is sometimes difficult to discern.

2.5 Summary

In this chapter we have addressed three topics regarding the interpretation of drug incorporation into human hair. The most fundamental question is whether a drug was administered; the recommended cut-offs and the use of metabolites have been reviewed. Generally, the cut-offs recommended by the

SoHT are low enough to detect even sporadic use of drugs with the exception of buprenorphine where even daily use may be underreported using the 10 $pg\,mg^{-1}$ cut-off. Data from benzodiazepines and other sedatives are largely missing and research on these substances would be a welcome contribution. Metabolites, useful indicators as they are, still leave some questions since some are present in drug powders and may be the result of degradation of the parent compound. Using certain ratios between parent drug and metabolites to distinguish between intake and exposure has largely been abandoned in current guidelines. Each case finding must be interpreted in the light of the case history and other toxicological findings.

When the question 'Has a drug been administered?', is answered with yes, the next question is how much was ingested? It is accepted that even though dose concentration relationships exist, practitioners do not estimate the dose from a hair concentration. Many factors influence the incorporation of drugs into hair and hair pigmentation has been identified as a key factor. The binding of drug molecules to the hair pigment, melanin, is well described in the literature and several human studies have shown a positive relationship between melanin and drug concentration. Melanin is scarcely measured when performing hair analysis for drugs and a more pragmatic view is to record the hair colour and relate drug concentrations to hair colour groups, when data is available. Research describing population data stratified on use patterns and hair colour will provide useful data for practitioners.

The third question relates to the time of intake, important, *e.g.*, in drug facilitated crimes. Here, single dose studies provide a good scientific base for interpretation. However, drugs do incorporate differently based on their physiological effects and chemical properties and studies are needed for drugs not yet described. In addition to sensitive analytical methods, precise sampling and segmentation is crucial to provide reliable data for interpretation.

References

1. *2021 SoHT Consensus on Drugs of Abuse (DoA) Testing in Hair*, Society of Hair Testing, 2021.
2. T. Franz, G. Skopp and F. Musshoff, *Int. J. Leg. Med.*, 2021, **135**, 465–471.
3. S. Pichini, O. Garcia-Algar, A. T. Alvarez, M. Mercadal, C. Mortali, M. Gottardi, F. Svaizer and R. Pacifici, *Int. J. Environ. Res. Public Health*, 2014, **11**, 8267–8275.
4. X. Wang and O. H. Drummer, *Forensic Sci. Int.*, 2015, **257**, 458–472.
5. J. D. Ropero-Miller and P.R. Stout, *Analysis of Cocaine Analytes in Human Hair: Evaluation of Concentration Ratios in Different Hair Types, Cocaine Sources, Drug-User Populations, and Surface-Contaminated Specimens*, 2009.
6. P. R. Stout, J. D. Ropero-Miller, M. R. Baylor and J. M. Mitchell, *J. Anal. Toxicol.*, 2006, **30**, 490–500.
7. Society of Hair Testing, *Forensic Sci. Int.*, 1997, **84**, 3–6.
8. D. M. Bush, *Forensic Sci. Int.*, 2008, **174**, 111–119.

9. G. A. Cooper, R. Kronstrand and P. Kintz, Society of Hair Testing guidelines for drug testing in hair, *Forensic Sci. Int.*, 2012, **218**, 20–24.
10. M. A. LeBeau and M. A. Montgomery, *J. Anal. Toxicol.*, 2009, **33**, 343–344.
11. F. Pragst, H. Sachs and P. Kintz, *J. Anal. Toxicol.*, 2010, **34**, 354–355author reply 355-356.
12. M. Montgomery, M. LeBeau and C. Morris-Kukoski, *J. Anal. Toxicol.*, 2017, **41**, 161–162.
13. C. L. Morris-Kukoski, M. A. Montgomery and R. L. Hammer, *J. Anal. Toxicol.*, 2014, **38**, 628–636.
14. C. Gambelunghe, R. Rossi, K. Aroni, A. Gili, M. Bacci, V. Pascali and N. Fucci, *Drug Test. Anal.*, 2017, **9**, 161–167.
15. M. Schaffer, C. C. Cheng, O. Chao, V. Hill and P. Matsui, *Anal. Bioanal. Chem.*, 2016, **408**, 2043–2054.
16. F. Musshoff, H. Fels, A. Carli and D. Piombino-Mascali, *Forensic Sci. Int.*, 2017, **270**, 20–24.
17. T. Franz, F. Scheufler, K. Stein, M. Uhl, T. Dame, G. Schwarz, H. Sachs, G. Skopp and F. Musshoff, *Forensic Sci. Int.*, 2018, **288**, 223–226.
18. C. Scholz, B. B. Quednow, M. Herdener, T. Kraemer and M. R. Baumgartner, *J. Anal. Toxicol.*, 2019, **43**, 543–552.
19. E. D. Hart, S. Vikingsson, R. E. Winecker, A. L. Evans, E. J. Cone, J. M. Mitchell, E. D. Hayes and R. R. Flegel, *J. Anal. Toxicol.*, 2022, **46**(5), 573–576.
20. R. E. Joseph Jr., K. M. Höld, D. G. Wilkins, D. E. Rollins and E. J. Cone, *J. Anal. Toxicol.*, 1999, **23**, 396–408.
21. K. B. Scheidweiler, E. J. Cone, E. T. Moolchan and M. A. Huestis, *J. Pharmacol. Exp. Ther.*, 2005, **313**, 909–915.
22. A. Miki, M. Katagi and H. Tsuchihashi, *J. Anal. Toxicol.*, 2003, **27**, 95–102.
23. A. Polettini, E. J. Cone, D. A. Gorelick and M. A. Huestis, *Anal. Chim. Acta*, 2012, **726**, 35–43.
24. G. Jakobsson and R. Kronstrand, *Drug Test. Anal.*, 2014, **6**(Suppl 1), 22–29.
25. J. K. Roberts, S. F. Cook, C. Stockmann, D. E. Rollins, D. G. Wilkins and C. M. Sherwin, *Clin. Drug Invest.*, 2015, **35**, 633–643.
26. I. Nyström, T. Trygg, P. Woxler, J. Ahlner and R. Kronstrand, *J. Anal. Toxicol.*, 2005, **29**, 682–688.
27. L. F. Martins, M. Yegles, N. Samyn, J. G. Ramaekers and R. Wennig, *Forensic Sci. Int.*, 2007, **172**, 150–155.
28. M. M. Madry, A. E. Steuer, C. M. Hysek, M. E. Liechti, M. R. Baumgartner and T. Kraemer, *Anal. Bioanal. Chem.*, 2016, **408**, 545–556.
29. B. Moosmann, N. Roth and V. Auwarter, *Drug Test. Anal.*, 2014, **6**, 119–125.
30. J. Thorspecken, G. Skopp and L. Potsch, *Clin. Chem.*, 2004, **50**, 596–602.
31. M. Uhl and H. Sachs, *Forensic Sci. Int.*, 2004, **145**, 143–147.
32. I. Shah, B. Al-Dabbagh, A. E. Salem, S. A. A. Hamid, N. Muhammad and D. P. Naughton, *BMC Chem.*, 2019, **13**, 106.
33. M. A. Huestis, R. A. Gustafson, E. T. Moolchan, A. Barnes, J. A. Bourland, S. A. Sweeney, E. F. Hayes, P. M. Carpenter and M. L. Smith, *Forensic Sci. Int.*, 2007, **169**, 129–136.

34. F. Vaiano, L. Scuffi, A. Lachi, C. Trignano, A. Argo, F. Mari and E. Bertol, *J. Pharm. Biomed. Anal.*, 2023, **225**, 115237.
35. B. Moosmann, N. Roth and V. Auwarter, *Sci. Rep.*, 2015, **5**, 14906.
36. B. Moosmann, N. Roth, M. Hastedt, A. Jacobsen-Bauer, F. Pragst and V. Auwarter, *Drug Test. Anal.*, 2015, **7**, 349–357.
37. S. Casati, I. Angeli, A. Ravelli, M. Del Fabbro, M. Minoli and M. Orioli, *Forensic Sci. Int.*, 2019, **304**, 109951.
38. T. Franz, G. Skopp, G. Schwarz and F. Musshoff, *Drug Test. Anal.*, 2018, **10**, 1573–1578.
39. P. Hehet, T. Franz, N. Kunert and F. Musshoff, *Drug Test. Anal.*, 2022, **14**(9), 1614–1622.
40. S. Pichini, E. Marchei, S. Martello, M. Gottardi, M. Pellegrini, F. Svaizer, A. Lotti, M. Chiarotti and R. Pacifici, *Forensic Sci. Int.*, 2015, **249**, 47–51.
41. V. Auwarter, A. Wohlfarth, J. Traber, D. Thieme and W. Weinmann, *Forensic Sci. Int.*, 2010, **196**, 10–13.
42. G. Skopp, P. Strohbeck-Kuehner, K. Mann and D. Hermann, *Forensic Sci. Int.*, 2007, **170**, 46–50.
43. A. Salomone, E. Gerace, F. D'Urso, D. D. Corcia and M. Vincenti, *J. Mass Spectrom.*, 2012, **47**, 604–610.
44. E. Gerace, S. P. Bakanova, D. Di Corcia, A. Salomone and M. Vincenti, *Forensic Sci. Int.*, 2021, **318**, 110561.
45. S. Casati, I. Angeli, R. F. Bergamaschi, A. Ravelli, G. Colombo, G. Binelli, M. Minoli and M. Orioli, *Forensic Sci. Int.*, 2022, **333**, 111237.
46. A. F. Lo Faro, B. Venanzi, G. Pilli, U. Ripani, G. Basile, S. Pichini and F. P. Busardò, *J. Pharm. Biomed. Anal.*, 2022, **217**, 114841.
47. D. E. Rollins, D. G. Wilkins, G. G. Krueger, M. P. Augsburger, A. Mizuno, C. O'Neal, C. R. Borges and M. H. Slawson, *J. Anal. Toxicol.*, 2003, **27**, 545–551.
48. P. Kintz, P. Bundeli, R. Brenneisen and B. Ludes, *J. Anal. Toxicol.*, 1998, **22**, 231–236.
49. F. Musshoff, K. Lachenmeier, H. Wollersen, D. Lichtermann and B. Madea, *J. Anal. Toxicol.*, 2005, **29**, 345–352.
50. K. A. Hadidi, J. K. Almasad, T. Al-Nsour and S. Abu-Ragheib, *Forensic Sci. Int.*, 2003, **135**, 129–136.
51. S. Pinho, A. Oliveira, I. Costa, C. A. Gouveia, F. Carvalho, R. F. Moreira and R. J. Dinis-Oliveira, *Biomed. Chromatogr.*, 2013, **27**, 1003–1011.
52. H. Yu, M. Choi, J. H. Jang, B. Park, Y. H. Seo, C. H. Jeong, J. W. Bae and S. Lee, *Arch. Pharmacal Res.*, 2018, **41**, 554–563.
53. S. S. Johansen, L. Dang, M. K. K. Nielsen, P. Haage, F. C. Kugelberg and R. Kronstrand, *Forensic Sci. Int.*, 2020, **316**, 110546.
54. M. M. Madry, K. Y. Rust, R. Guglielmello, M. R. Baumgartner and T. Kraemer, *Forensic Sci. Int.*, 2012, **223**, 330–334.
55. D. G. Wilkins, P. R. Nagasawa, S. P. Gygi, R. L. Foltz and D. E. Rollins, *J. Anal. Toxicol.*, 1996, **20**, 355–361.
56. T. Kelly, P. Doble and M. Dawson, *J. Chromatogr. B: Anal. Technol. Biomed. Life Sci.*, 2005, **814**, 315–323.

57. B. A. Goldberger, A. G. Darraj, Y. H. Caplan and E. J. Cone, *J. Anal. Toxicol.*, 1998, **22**, 526–530.
58. F. Sporkert and F. Pragst, *J. Chromatogr. B: Biomed. Sci. Appl.*, 2000, **746**, 255–264.
59. S. K. Himes, R. S. Goodwin, C. M. Rock, H. E. Jones, R. E. Johnson, D. G. Wilkins and M. A. Huestis, *Ther. Drug Monit.*, 2012, **34**, 337–344.
60. F. Musshoff, K. Lachenmeier, D. W. Lachenmeier, H. Wollersen and B. Madea, *Forensic Sci., Med., Pathol.*, 2005, **1**, 97–103.
61. S. Paterson, R. Cordero, M. McPhillips and S. Carman, *J. Anal. Toxicol.*, 2003, **27**, 20–23.
62. P. Kintz, V. Cirimele, Y. Edel, C. Jamey and P. Mangin, *J. Forensic Sci.*, 1994, **39**, 1497–1503.
63. D. G. Wilkins, D. E. Rollins, A. S. Valdez, A. Mizuno, G. G. Krueger and E. J. Cone, *J. Anal. Toxicol.*, 1999, **23**, 409–415.
64. G. Skopp, A. Kniest, J. Haisser, K. Mann and D. Hermann, *Int. J. Leg. Med.*, 2011, **125**, 277–281.
65. S. Belivanis, M. N. Tzatzarakis, E. Vakonaki, L. Kovatsi, M. Mantsi, A. Alegakis, M. P. Kavvalakis, D. Vynias and A. M. Tsatsakis, *Drug Test. Anal.*, 2014, **6**(Suppl 1), 93–100.
66. P. Xiang, M. Shen and X. Zhuo, *Forensic Sci. Int.*, 2006, **162**, 131–134.
67. P. Xiang, Q. Sun, B. Shen and M. Shen, *Int. J. Leg. Med.*, 2011, **125**, 831–840.
68. A. S. Barreto, V. F. Brant, E. Spinelli and S. V. Rodrigues, *J. Chromatogr. B: Anal. Technol. Biomed. Life Sci.*, 2016, **1033–1034**, 200–209.
69. I. A. Larabi, I. Etting and J. C. Alvarez, *Drug Test. Anal.*, 2023, DOI: 10.1002/dta.3444.
70. A. Salomone, E. Gerace, P. Diana, M. Romeo, V. Malvaso, D. D. Corcia and M. Vincenti, *Forensic Sci. Int.*, 2015, **248**, 119–123.
71. K. W. Leung, Z. C. Wong, J. Y. Ho, A. W. Yip, J. S. Ng, S. P. Ip, W. Y. Ng, K. K. Ho, R. Duan, K. Y. Zhu and K. W. Tsim, *Forensic Sci. Int.*, 2016, **259**, 53–58.
72. D. Favretto, S. Vogliardi, M. Tucci, I. Simoncello, R. E. Mazloum and R. Snenghi, *Forensic Sci. Int.*, 2016, **265**, 193–199.
73. T. Uematsu, R. Sato, O. Fujimori and M. Nakashima, *Arch. Dermatol. Res.*, 1990, **282**, 120–125.
74. R. W. Reid, F. L. O'Connor, A. G. Deakin, D. M. Ivery and J. W. Crayton, *J. Toxicol., Clin. Toxicol.*, 1996, **34**, 685–690.
75. R. Kronstrand, J. Ahlner, N. Dizdar and G. Larson, *J. Anal. Toxicol.*, 2003, **27**, 135–141.
76. R. Kronstrand, M. Roman, M. Hedman, J. Ahlner and N. Dizdar, *Forensic Sci., Med., Pathol.*, 2007, **3**, 107–114.
77. R. Kronstrand, S. Forstberg-Peterson, B. Kagedal, J. Ahlner and G. Larson, *Clin. Chem.*, 1999, **45**, 1485–1494.
78. R. Kronstrand, M. C. Andersson, J. Ahlner and G. Larson, *J. Anal. Toxicol.*, 2001, **25**, 594–601.
79. C. Vignali, C. Stramesi, M. Vecchio and A. Groppi, *Forensic Sci. Int.*, 2012, **215**, 77–80.

80. T. Mieczkowski, *Life Sci.*, 2003, **74**, 463–469.
81. T. Mieczkowski and M. Kruger, *J. Forensic Leg. Med.*, 2007, **14**, 7–15.
82. T. Mieczkowski, A. M. Tsatsakis, M. Kruger and T. Psillakis, *BMC Clin. Pharmacol.*, 2001, **1**, 2.
83. F. Ursitti, J. Klein, E. Sellers and G. Koren, *J. Toxicol., Clin. Toxicol.*, 2001, **39**, 361–366.
84. D. E. Rollins, D. G. Wilkins and G. G. Krueger, *Eur. J. Clin. Pharmacol.*, 1996, **50**, 391–397.
85. A. Negrusz, C. M. Moore, K. B. Hinkel, T. L. Stockham, M. Verma, M. J. Strong and P. G. Janicak, *J. Forensic Sci.*, 2001, **46**, 1143–1151.
86. A. Negrusz, A. M. Bowen, C. M. Moore, S. M. Dowd, M. J. Strong and P. G. Janicak, *J. Anal. Toxicol.*, 2002, **26**, 471–478.
87. P. Xiang, Q. Sun, B. Shen, P. Chen, W. Liu and M. Shen, *Forensic Sci. Int.*, 2011, **204**, 19–26.
88. X. Cui, P. Xiang, J. Zhang, Y. Shi, B. Shen and M. Shen, *J. Anal. Toxicol.*, 2013, **37**, 369–375.
89. E. Bertol, F. Mari, F. Vaiano, G. Romano, S. Zaami, G. Baglìo and F. P. Busardò, *Drug Test. Anal.*, 2015, **7**, 376–384.
90. S. L. Hansen, S. S. Johansen, M. K. K. Nielsen, G. Nilsson and R. Kronstrand, *Forensic Sci. Int.*, 2020, **306**, 110074.
91. P. Kintz, *Ther. Drug Monit.*, 2013, **35**, 408–410.
92. M. Villain, M. Cheze, A. Tracqui, B. Ludes and P. Kintz, *Forensic Sci. Int.*, 2004, **145**, 117–121.

CHAPTER 3

Effects of External Contamination and Cosmetic Treatments on Hair Testing Can be Explained. . .Most of the Time!

LOLITA TSANACLIS* AND JOHN F. C. WICKS

Cansford Laboratories Limited, Cardiff, 1a Pentwyn Business Centre, Wharfedale Road, CF23 7HB, UK
*Email: lolita.tsanaclis@cansfordlabs.co.uk

3.1 Introduction

Hair serves as a repository for drugs and other substances delivered by the blood to the hair root. Drugs are deposited in the internal portion of the hair, known as the cortex, which is covered by a cuticle that makes the hair an extremely tough and durable natural fibre.[1] These characteristics provide a retrospective window of detection that can show trends in drug use over time and identify the drugs used.

However, there are factors that affect the quantities of drugs and related compounds detected in the hair. These include physicochemical properties of the compounds, inter-individual variations of drug incorporation rates, hair site and growth rate. The quantity of drugs detected in hair are also affected by external factors like the use of hair cosmetic treatments as well as issues regarding external contamination and the implications

Issues in Toxicology No. 47
Perspectives and Challenges of Hair Analysis
Edited by Pascal Kintz, Alberto Salomone and Marco Vincenti

Published by the Royal Society of Chemistry, www.rsc.org

of sample washing procedures applied to the sample in the testing laboratory.

External contamination of hair with drugs used to be regarded as one of the most questionable points regarding the validity of hair testing, and the risk of false positive as a result of external contamination could see a hair testing result dismissed.[2–4] For this reason, the search for the ideal wash protocol has been the Holy Grail of drug testing in hair.

There is a risk of a false positive result even when laboratory practice and policies attempt complete removal of any external contamination prior to analysis of the hair. Metabolites represent the biotransformation product of drugs when they are ingested, and their presence in a hair sample confirms drug use. When relevant metabolites are not detected, and if all drugs present as external contamination are not removed by the wash, we have the risk of a false positive result. The issue is that it is not possible to know that all external contamination has been removed, nor the amount of drug that is extracted by the wash procedure. While most external contamination is likely to be removed by most wash protocols prior to analysis, it is impossible to guarantee the complete removal of all external contamination. It is unlikely, therefore, that we will find a new revolutionising hair decontamination procedure approach, nor an analytical method that will avoid the effects of chemical cosmetic treatment.

This chapter discusses some of the approaches concerning external contamination and cosmetic treatment of hair described by the relevant scientific literature and provides further new data from our own practice. Our aim is to expand the possibilities and probabilities for interpretation of the results in such cases.

3.2 The Significance of External Contamination in Hair Drug Testing

3.2.1 Environmental Exposure and Passive Use: A Fact of Life

Drugs can be incorporated into the hair from voluntary ingestion or involuntary ingestion (passive use) and can be deposited on the hair from environmental exposure or from sweat and other secretions.[5,6]

Environmental exposure to drugs is an important issue to consider when testing hair for drugs, since it is possible that an individual in the company of drug users can experience external contamination of their hair. This is particularly significant for those common drug groups that are usually smoked or snorted, such as cannabis and cocaine, but can also occur with any drug. Environmental exposure can also occur in many other forms, for example, a person handling drugs in the form of tablets or powder and subsequently rubbing their hair with their hands. Environmental exposure can be the result of other people's use of drugs in the vicinity, but can also be from a person's own drug use. Environmental exposure to drugs can cause negligible involuntary ingestion when a small amount of drug present in the environment is inhaled or ingested, while causing simultaneous external contamination of the hair. In such cases, metabolites may be present, but because they are at low levels, the analytical method may not be able to detect them.

3.2.2 External Contamination from Sweat or Sebum

External contamination can also include the deposition of substances on the hair from sweat and other secretions, either from a person's own drug use or the transfer from another person's sweat.[6] If use has been made of a drug that can be deposited externally through sweat or sebum, hair tested for the drug can potentially give a positive result on a drug-free individual or show a different period of drug consumption for a drug user.

Consideration must be given to the interpretation of results in such cases, as the deposition of drugs onto an external portion of hair follicles can depend on several factors, such as the physicochemical of the drugs or the integrity of the follicle and is affected by inter-follicle differences.[7]

3.2.3 External Contamination from Own Use

Concomitant external contamination and systemic exposure is in fact most likely for drug users. Individuals are subjected to drug exposure from their own drug use which might result in the external contamination of their own hair. Indeed, we have shown that this can occur when we detected cocaine in the wash residue of 76% of the hair samples from confirmed crack cocaine users.[8] However, some publications recommend that results are reported as externally contaminated when the amount of cocaine in the wash residue is above certain levels, regardless of the amount in the hair.[9–11] This risky practice would have certainly caused underreporting of 76% in the cited study above,[8] by reporting the results of crack cocaine users as external contamination. The results of confirmed externally contaminated hair should therefore not be issued without consideration of the results found in the hair sample.

3.2.4 Ways to Distinguish Between External Contamination of Hair and Systemic Exposure

For several decades, researchers have advocated the elimination of possible external contamination by using decontamination processes, identification of metabolites at certain levels, and using body hair that is not exposed.[12] The confirmation of drug use in hair testing is achieved not solely by confidence in the analytical result and use of relevant cut-offs but by validated wash protocols with proven non-*in vitro* generation of the metabolites during analysis.[8,13] It is essential, therefore, in drug hair testing, particularly for those common drug groups that are usually smoked or snorted, that the parent drug is detected together with the relevant metabolites to confirm use.[14]

This means that when testing for certain drugs in hair, the detection of the parent drug in isolation does not absolutely prove drug use, particularly for those drug groups that are usually smoked or snorted, as smoke or powder may be in the environment. When relevant metabolites are not detected, however, the presence of the parent drug still indicates unequivocally that the individual has been associated with the drug in some way. This may happen when low doses are used, or levels reflect historical use, due to the use of cut-off or limitations of the analytical method.[15] Thus, the result of the hair test may become corroborative evidence, but results are not conclusive.

Hair testing laboratories sometimes face the challenge, when means for realistic and likely interpretation of results for differentiating between systemic exposure and external contamination are not met. For example, while the detection of the parent drugs in such cases (such as △9-tetrahydrocannabinolic (THC) or cocaine) in isolation indicates unequivocally that the individual has been associated with the drugs, it does not absolutely prove drug use. Whether the laboratory report should report the results as "not detected" (as suggested by Cairns and Hill *et al.*)[9,10] is debatable.

3.2.5 Alternative Ways of Explaining External Contamination Cases

Models have been advocated as alternative ways of explaining and considering cases of external contamination. Examples have been published of results considered indicative of external contamination when segmental analyses show uniform hair measurements or when the environment of cohabiting drug users has been evaluated.[16,17]

Kuwayama's group[18] studied a micro-segmental analysis and by examining peak shape patterns revealed differences in the distribution profiles of drugs in hair. As a result, they claimed that they could distinguish hair samples with and without external contamination.

The confirmation of cannabis external contamination using indirect markers for compounds that are not incorporated into hair through the bloodstream after oral intake has been suggested for THC.[19,20] The main question about this practice is that one must be categorically sure that drug users do not contaminate their own external hair.

Several alternative metabolites have been suggested to aid the distinction between external contamination and illicit drug usage. For example, for cocaine use the detection of alternative metabolites such as cocaine-*N*-oxide diastereomers, hydroxy-metabolites of cocaine, benzoylecgonine or norcocaine.[21–26] For heroin use, metabolite ratios of hydromorphone to morphine and hydrocodone to codeine have been used as indicators to distinguish external contamination.[27]

3.2.6 Challenge to Laboratories: In Search of the Holy Grail

3.2.6.1 The Four Requirements of Hair Decontamination

Currently, there is no question that hair samples should be decontaminated before testing to remove impurities that naturally occur as well as any external contamination. There is a consensus that, when testing hair samples for drug use, the sample must be decontaminated prior to analysis by the testing laboratory by a validated wash protocol.[28] The ideal decontamination protocol should:

(1) completely remove all external contaminants originating from the environment;
(2) completely remove all contamination deposited *via* sweat and sebum;
(3) avoid shifting contaminants into the hair matrix;
(4) avoid removal of incorporated drugs or metabolites.

3.2.6.2 Which Protocol to Use to Decontaminate Hair Samples

While many practical research studies have been published regarding different decontamination procedure protocols, they produce many dissimilar outcomes, and none perfectly achieve the four requirements above for the ideal decontamination protocol. Müller and Iwersen-Bergman[29] discussed the different ways in which hair decontamination protocols in hair EtG analysis are described in publications as "washed", "rinsed", "bathed", "submersed", "shaken", or even "ultrasonicated", and referred to these terms as "highly variable and scarcely described". The same comments could be applied to publications regarding hair testing for drugs, where different decontamination methods are described with different combinations of aqueous and/or organic wash solutions at different sequences, methods, and timings, including the sophisticated statistical design of experiments.[5,9,10,15,22,30–35]

3.2.6.3 Models of Contamination

Published methods also differ on how blank hair samples are contaminated *in vitro* (porous and non-porous hair samples have been used, contaminated by soaking or rubbing powders), and on how studies were carried out with hair samples from drug users. Negative human head hair samples soaked in a solution of drugs or rubbed with powders have been used to provide contaminated hair to evaluate protocols for differentiating between ingestion *versus* external contamination.[4,30,32,34,36,37]

Whilst using hair artificially contaminated with cocaine does not accurately represent reality, the investigative publications evaluating and describing wash experiments have been important and provide important data to the scientific community for consideration. They show that some protocols using specific solvents or solvent combinations provide different grades of efficiency. Yet, decisive evidence for a particular wash protocol able to remove all external contamination is not yet available from either real hair samples or from *in vitro* experiments.

While hair samples contaminated *via* sweat and sebum with zolpidem-soaked hair were washed out efficiently, even from genuine contamination,[34] other studies showed that a water-based wash could not completely remove tilidine, a synthetic opioid, deposited by the sweat and sebum.[38] Another interesting study requested volunteers to handle amphetamine as if they were drug users and showed detectable amphetamine in their hair samples for four to eight weeks.[39]

3.2.6.4 Decontaminate the Whole Lock or Just the Segment?

Another constraint of some of the studies that evaluate whether a wash was complete or not is that it is not always clear whether the experiments were based on washing the whole lock of the hair sample provided for analysis before segmentation, or washing the segment being tested.

This is significant because the technique used to decontaminate a long hair lock may produce different outcomes when the lock is short or curly or both. In contrast, when the decontamination technique is applied to the segment being tested, the wash efficiency is likely to be more consistent, with the added critical advantage of being able to keep the wash residue of the exact segment being tested.

3.2.6.5 Yes, We Do Need to Decontaminate, Whatever Method…

There is, however, agreement between the various publications on the subject that different decontamination procedures produce different results, and that a perfect decontamination protocol is difficult to achieve. This is because any wash procedure is affected by the choice and volume of solvent, drug-binding, temperature, time, and intactness of the hair, before a criterion can be established based on the drug contents of the wash and hair that indicates the hair may or may not be contaminated.[10]

We cannot assume, therefore, that an approach from one laboratory practice is applicable to all laboratories. It does not mean that a specific solvent or combination of solvents using certain volumes with a shaking protocol for a certain length of time used by one laboratory is suitable to remove all external contamination. For example, some studies showed that use of either water or methanol was considered a good washing solvent because it only removes cocaine on externally contaminated hair while preserving the incorporated compounds.[32,40,41] On the other hand, other studies established that methanol was their choice of solvent, but that prolonged or repeated decontamination procedures with methanol should be avoided, especially to highly damaged/porous hair, as it provokes drug loss from hair.[8,15,35,42]

There are two common understandings from the decontamination studies:

(1) results vary between the decontamination procedures investigated;
(2) it is difficult to ensure (and to prove) that the externally deposited drug has been eliminated.

Future studies will certainly be undertaken, but it is unlikely that a perfect decontamination procedure for forensic hair analysis will ever be established. Therefore, consensus needs to be achieved on the most effective way of explaining external contamination and interpreting results.

3.2.7 Any (Almost Ideal) Hair Decontamination Protocol is Helpful as Nothing is Absolute

The bulk of the relevant scientific literature converges on completely removing from the hair any potential external contamination (usually of drugs that are normally smoked or inhaled, but applicable to any context).

Publications investigating different wash procedures and evaluating their effectiveness are important to demonstrate individual laboratories' approaches, validate their own protocols, and disseminate their findings. They help other laboratories determine how best to develop their own methods, but their conclusions are applicable to their own specific procedures and cannot be extrapolated to all hair testing by all laboratories. The general extrapolation from individual studies cannot be applicable to all because we know there are a mixture of factors that affect the wash efficiency and the results.

3.2.7.1 Our Method of Decontamination

In our laboratory, hair samples are cut into segments according to client requests and each segment is washed separately, twice, each time with 1 mL of methanol, and shaken briskly each time for less than one minute for both washes in total before transferring the methanol residue to a clean test tube for storage. If analysis of the wash residue is required, the extract is spiked with deuterated internal standards, dried, and then reconstituted with the applicable solvent before analysis by LC–MS/MS.[15]

The wash procedure using methanol is restricted to one minute to maximise the wash efficiency and minimise the extraction of drugs from the hair cortex. It is not possible to know, however, whether all external contamination is indeed removed nor whether the short soaking time may start the extraction of compounds when present in the cortex. We also showed the presence in the wash residue of analytes that are not usually susceptible to external contamination, for example metabolites and drugs such as dihydrocodeine. This, therefore, confirms that the washing procedure can remove drugs from the hair shaft to some degree.[15]

Our aim, however, is not to completely remove all external contamination but to remove most of the external contamination. The short wash procedure causes minimal shift of contaminants into the hair and minimal removal of incorporated compounds. This enables us to compare the results of the wash residue analysis with that of the corresponding hair, when required. In our laboratory, sample wash residues are mainly analysed when the levels of relevant metabolites do not satisfy drug usage criteria, or systematically when testing children's hair.

We have proposed the following criteria for hair analysis interpretation without relevant metabolites:[15]

(1) when drugs are detected at less than 10% in wash residue in comparison to hair levels, it indicates drug use as opposed to environmental contamination;
(2) when levels in the wash residue are between 10% and 50% of hair levels, drug use is possible, but drug association is indicated;
(3) when the levels detected in the wash residue are greater than 50% of hair levels, the analysis is likely to indicate that the source of most of the drug in the wash residue is from external contamination and drug use is questionable, but drug association is indicated.

The wash protocol described above has been used in our laboratories for over 20 years and fully validated to fulfil ISO 17025 requirements. For the *in vitro* validation we spiked blank hair samples with cocaine and found that, while our washing protocol was evidently complete, no significant levels of relevant metabolites were generated during the washing nor extraction.[42] The implication is that the wash protocol used in our laboratory is efficient in removing external contamination from hair and that the protocol does not produce *in vitro* levels of the relevant metabolites that could mislead the final conclusions.

3.2.7.2 Evidence of Wash Efficiency Using Hair from Crack Cocaine Users

Most publications give valuable information about which solvents wash away artificial contamination from hair samples, and the wash efficiency is extrapolated to real samples. The fact is that external contamination is unlikely to be a homogeneous event, *i.e.*, it is not a uniform occurrence throughout the length of hair. People do not soak their hair in solutions containing drugs nor inside bags of drug powder. People's hair can be externally contaminated when they touch powdered drugs and inconsistently touch their hair. People living with people who smoke drugs can have uncontrolled amounts of the parent drug deposited externally. When they experience external contamination in these cases, the majority is washed out in routine hygiene. Perhaps not all of it will be washed out, but the wash residue to hair ratio will indicate this. Therefore, the evaluation of wash protocols using hair samples from drug users provide an authentic assessment of the scale of the issue.

Our paper evaluating the wash residue to hair ratios from hair collected from 138 crack cocaine users provided informative data.[8]

The study revealed key points:

(1) cocaine was detected in the wash residue in most of the cases (76%), indicating external contamination;
(2) the wash residue to hair ratios were below 0.07 in 96% of cases, including those cases where cocaine levels were as high as 130 $ng\,mg^{-1}$, confirming the efficacy of the relative comparison between wash residue and hair levels;
(3) in five cases (0.7%), the wash residue to hair ratios were above 0.1, although the hair results for metabolite confirmed use.

The results are comparable with those found from children removed from clandestine laboratories following police intervention, which showed evidence of methamphetamine consumption and where 79% of wash residues were positive.[43]

The data provided by these studies, where external contamination from individuals' own drug use, voluntarily and involuntarily, was expected and confirmed, provided evidence of the usefulness of the comparison between externally deposited cocaine and the concentrations in hair. Whilst it is not possible to differentiate between the drug extracted from the hair and the drug attached to the outside of the hair, it becomes possible to interpret the result conclusively by comparing levels of drug in the wash residue with levels detected in the corresponding segment of hair.

The assessments of wash efficacy made with real hair samples, as opposed to artificially contaminated samples, provide a more realistic interpretation They show how rarely external contamination affects the interpretation of results in real life. The conclusion is that when drugs are detected in hair, individuals have clearly been in an environment where drugs are present, but it is only on rare occasions that it is unclear whether the drugs present are the result of drug use or external contamination. In those cases, the test results need to be interpreted in the light of corroborating evidence from clinical data or social context.[8]

3.2.7.3 *Evidence of Wash Efficiency from General Routine*

The results presented in Table 3.1 are a compilation of data for cocaine and THC obtained from the routine work in our laboratory. They represent hair samples received for analysis from a wide variety of agencies including drug dependency clinics, hospitals, police investigations, sports clubs, social services, probation services, institutions, and the legal profession. All the samples were submitted to the wash protocol as described in 2.7.1 above.

The data showed that of the 103 643 samples tested for cannabinoids and the 100 683 hair samples tested for cocaine, only 1341 (1.3%) and 2729 (2.7%)

Table 3.1 Wash (W) to hair (H) mean ± SD ratios with number of cases and percentages when criteria for use or consumption for THC and for cocaine had not been met based on the detection of metabolites with corresponding wash residue testing.[a]

Analyte	Hair samples tested, n	Corresponding hair wash residues tested, n (%)	Interpretation based on ratio W/H[42]					
			0 and <0.1		>0.1 and ≤0.5		>0.5	
			Mean ± SD	n (%)	Mean ± SD	n (%)	Mean ± SD	n (%)
THC	103 643	1341 (1.3%)	0.01 ± 0.02	686 (0.7%)	0.31 ± 0.13	438 (0.4%)	1.54 ± 1.23	217 (0.2%)
Cocaine	100 682	2729 (2.7%)	0.01 ± 0.02	1815 (1.8%)	0.26 ± 0.12	616 (0.6%)	1.99 ± 3.29	298 (0.3%)
			Drug use		**Drug use likely**		**Drug use possible**	

[a] W: wash residue; H: hair

respectively required wash residues to be tested as the metabolite criteria were not met.

The data shows that:

(1) it was not possible to confirm drug use in 0.6% of the cases for cannabis use and 0.9% of the cases for cocaine use;
(2) data substantiated our previous published results that using hair sample data combined with decontamination data, the relative wash residue to hair ratio is a useful additional step for the routine assessment of drug use.[8,15,42]

Our laboratories and the laboratories associated with us currently or in the past still use the same protocol of washing the segment being tested. This procedure allows the possibility of analysing the wash residue and making comparisons between drug content in the hair and on the exterior of the hair. In our practice, external contamination can, therefore, be manageable in most cases, as confirmed by the data in Table 3.1.

In a separate study, we analysed the wash residues for 41 available hair samples where EtG was not detected but PEth was positive, with the aim of indicating alcohol use in the days prior to sample collection.[44] No EtG was detected in the hair wash residues, and the results confirmed the understanding that the incorporation of EtG from sweat and sebum in the hair is less relevant than the incorporation of EtG from blood.[44–46]

3.2.8 External Contamination: Considerations and Recommendations

3.2.8.1 On the Diversity of Published Wash Protocols

Whilst papers that deal with external drug decontamination of hair do not always clearly state the decontamination protocols used, much of the published research has helped laboratories to select their own procedures. They also largely helped improve interpretation and alternative explanations of test results for the differentiation between systemic exposure and external contamination. We have used our own protocol for over 20 years, but it is likely that most wash protocols utilised by different laboratories are suitable for the application of the concept of wash residue hair ratio.

The different published experiments show that drugs bound to the exterior of the hair behave differently in different conditions and that sometimes it is not possible to remove them completely, either by an aqueous solvent, an organic solvent, or a combination of the two. Evidently, some protocols use a combination of solvents, volumes and timings that are not universally suitable to all laboratory practices, and every laboratory needs to validate its own procedures. The perfect wash solution does not exist and, even if it did, it is unlikely that it would be a universal wash, applicable to all laboratories and hair types. However, whichever protocol is selected by a laboratory, it must be validated, and it needs to demonstrate that most, if not all, of the externally deposited drug is removed with minimal extraction of any inner drug.

3.2.8.2 On the Most Practical Approach

When applying any validated wash procedure to real hair samples, it is important to be aware that it is not possible to know with absolute certainty, using any washing procedure, whether the wash protocol is complete. As it is not possible to know that all external contamination has been removed by the wash procedure, nor that there has been no diffusion from inner hair during washing, it is not possible to determine whether a positive sample result is solely due to external contamination or to both external contamination and ingestion. It is not necessary to remove all traces of contamination if the sample result is to be interpreted correctly as possibly contaminated, when the amount in the wash is evaluated against the amount in the hair. This is in agreement with a recent publication by Hill and co-workers,[10] which confirms that it is not necessary to remove all traces of contamination if the sample result is to be interpreted correctly as possibly contaminated, and that the result of the washing is likely to be best evaluated by what Hill *et al.* refer to as "wash residue to hair criterion". But although it is important to understand the extent of solvent removal of externally deposited drugs, this does not clarify specific real cases on a routine basis, which is crucial if interpretations or deciding conclusions are to be based on these findings.

The difficulty some laboratories have in applying the wash residue to hair criterion is that many laboratories apply the decontaminating wash procedure to the whole lock of hair, and the washed hair lock is subsequently segmented over the period required. In such cases, it becomes challenging to relate the levels detected in the wash residue of the whole lock to levels detected in the segment being tested. Such laboratory practices aim to achieve the complete removal of all external contamination if present, even though this is difficult to prove on an individual basis. While we do need to distinguish between external hair contamination and systemic exposure or measure the extent of the external contamination in relation to the amount of drug in the hair, in our experience these are concurrent most of the time.

Our findings extend and confirm our earlier contributions[8,15,41] with the updated data displayed in Table 3.1 showing that it is possible to achieve consistent interpretive conclusions.

3.2.8.3 Recommendations

Measurements of both hair samples and corresponding wash residue gives a much broader insight into the possibilities of positive results being the consequence of external contamination. They may still not be conclusive in some cases, but they can be better indicators of the probability of external contamination. As technology improves with increasing method sensitivity, it is likely that more metabolites will be identified and routinely tested by laboratories involved in hair drug analysis and their routine detection for the diagnosis of drug use will improve. In the meantime, our recommendations when testing hair samples for drugs use are:

(1) hair samples must be washed or decontaminated before being submitted to extraction and analysis;
(2) individual laboratories may prefer a single solvent, aqueous or organic, a combination of a mixture or sequential, over limited periods of time;
(3) the best wash protocol must be validated by each laboratory and show that it can remove all or most external contamination;
(4) the hair segment being tested (not the whole lock of hair) must be washed;
(5) the corresponding hair segment wash residue must be kept if analysis is necessary;
(6) the detection of appropriate levels of metabolites is the main approach in hair testing and their presence in hair samples confirms drug use;
(7) the use of a cut-off guards against the risk of a false positive from accidental contamination or exposure to drugs;
(8) the analysis of the corresponding wash residue is a helpful tool to indicate external contamination or possible ingestion in the small number of cases where only the parent drug is detected and the relevant metabolites are not;
(9) when criteria for drug use are not met, the results of hair analysis would only be corroborative or suggestive evidence of drug use, which might require additional explanation.

3.3 The Implications of Strong Chemicals and Dreadlocks in Hair Drug Testing

3.3.1 Strong Chemical Cosmetic Treatments

It is well known that chemical hair treatments, such as hair dyes and bleaches, or environmental aggressions, such as ultraviolet radiation or thermal application, can damage hair so that it holds less drug than it would otherwise. As a result, single or low drug usage may not be detected when hair is tested for the presence of drugs.[13,15,47–51] While in regularly shampooed hair drugs are usually well detected for at least a year after intake, the levels decline following the use of certain products. Anything that affects the integrity of the hair shaft might affect the levels of drugs and analytes that are incorporated in the hair.[13]

A study published in 2017 identified over four million web sources in a Google search for "beat hair drug test", showing the popularity of attempts to influence hair test results with various "shampoos" and "cleansers", among other products.[52]

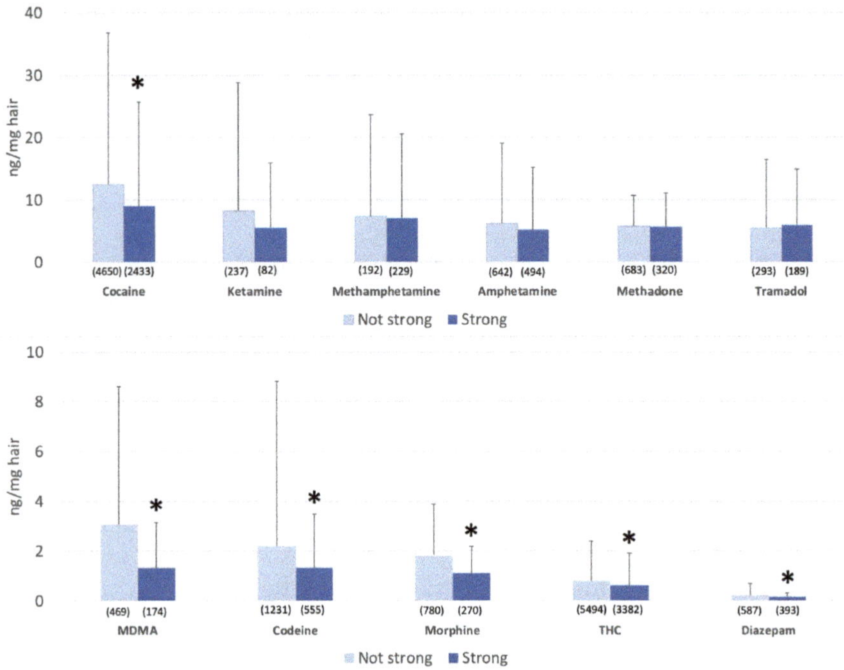

Figure 3.1 Mean and standard deviation levels (number of cases) of representative drugs in hair from individuals when the type of hair treatment was declared, grouped by normal hygiene or softer hair products (Not strong) or stronger chemical cosmetic (Strong). (*: $p<0.05$ Not strong *vs.* Strong by independent samples *t*-test).

In our experience, while levels can be reduced, it is unlikely that shampoos advertised on the Internet can clear hair of all drug residues when drugs are used regularly. Figure 3.1 displays the mean and standard deviation levels of a representative range of drugs tested in our laboratory on hair samples from individuals who declared the type of hair cosmetic treatment they had used. One group was coded as "Not strong" ($n=21\,161$; 65%) for samples when it was declared just normal hygiene or gentle hair products had been used. A second group was coded as "Strong" ($n=11\,571$; 35%) when stronger chemical or thermal treatment was declared. The data showed that the levels detected were significantly lower in the "Strong" group for some analytes but for other analytes did not appear to be significantly affected ($p>0.05$). The implication is that strong cosmetic chemical treatment may not affect the levels of compounds in hair in a uniform way. The analytes in the "Strong" group showing lower levels ($p<0.05$) were in concordance with the findings of an earlier study for matching analytes.[48]

Figure 3.2 displays the mean and standard deviation levels of hair samples that were tested for EtG and EtPa concurrently in our laboratory where the type of hair cosmetic treatment applied had been declared. Alcohol markers

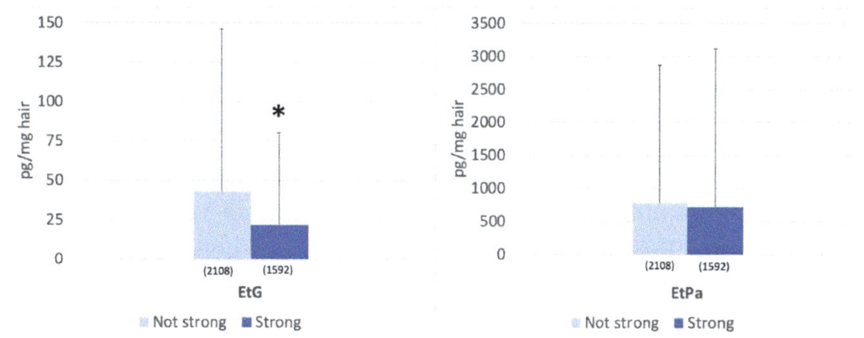

Figure 3.2 Mean and standard deviation levels (number of cases) of alcohol markers in hair from individuals when the type of hair treatment was declared, grouped by normal hygiene or softer hair products (Not strong) or stronger chemical cosmetic (Strong), (*: $p < 0.05$ Not strong *vs.* Strong by independent samples *t*-test).

results were (as in Figure 3.1) divided into two groups according to the type of cosmetic treatment: "Not strong" ($n = 2108$; 57%) and "Strong" ($n = 1592$; 43%). The lower results found of EtG in the "Strong" group ($p < 0.05$) confirm the previously published findings where the effects of cosmetic treatments on EtG levels in hair such as hair colouring or bleaching, and sunlight was shown.[53–56] Conversely, EtPa results in the "Strong" group were not significantly different ($p > 0.05$), which confirms the value of testing for EtPa, in combination with EtG.[44] Testing for alcohol markers using fingernails could be an alternative option to avoid underreporting EtG results in hair when strong cosmetic chemical treatments are known to be used.[57] We do recommend as normal practice that, when head hair is treated with strong chemical products such as hair colouring and bleach, the collection of body hair as an alternative to head hair, and/or testing for EtPa.

In a study evaluating drug levels and positivity rates in non-treated and cosmetically treated hair, the results showed that levels were significantly lower for drugs in treated hair when compared to non-treated hair, but no significant change was observed for EtG.[58] The same study showed no significant difference in the positivity rate observed between treated hair and non-treated hair for both drugs and EtG. This study shows that dyed or bleached hair is not necessarily of no use in detecting drugs and/or alcohol consumption, but there is always a possibility of a false negative result.

3.3.2 Natural Environment Conditions

Long-term effects of weather and sunshine, rain or wind may cause damage to the hair shaft that may impact on drug concentration.[13,51] Studies have shown that levels of EtG decreased when hair samples were incubated for up to 10 hours in chlorinated water,[59] but a different study using a solar simulator showed that EtG content in hair could either decrease or increase depending on the colour, integrity, and thickness of the hair.[50] The implication is that the interpretation of hair drug levels may by problematic in certain forensic cases, for example, when people are found dead and exposed to environmental conditions in the open air or swimming pools for an unknown length of time, making a negative result inconclusive.[51]

In vitro experiments are effective in demonstrating how long exposure to chlorinated swimming pool water affects drug and alcohol level markers in hair. One might assume that reduced levels of earlier samples might be the result of such exposure. However, decreasing levels from earlier time windows might be relevant to provide important information on decreasing consumption.

3.3.3 Dreadlocks

The issue of testing dreadlocked hair is that the section tested could contain strands of hair from previous periods. The interpretation of the results in such cases may differ from samples of hair that were not dreadlocked. Dreadlocked hair can only provide information on whether or not the individual used drugs over an unspecific period.

Table 3.2 displays the results for nine cases where dreadlocked hair was tested for specific drug and alcohol markers in segmentation to cover different periods. Samples were collected by cutting the dreadlock next to the scalp and sent to Cansford Laboratories for analysis following Court orders. Court orders frequently request month by month segmentation of dreadlocked hair. However, the period covered by each individual segment is uncertain. Before sectioning, the dreadlocks were untangled, and the strands of hair were stretched to enable the best possible representation of the length. The sections were cut according to requirements and then washed, pulverised, and extracted according to each workflow protocol before analysis by LC–MSMS.

The results for Case 1 are consistent with the declared use of home-made THC wax and CBD oil. Cases 3 and 4 confirm use of THC, however, it is not possible to specify the period when cannabis was used. The results for Cases 5, 8, and 9 confirm use of cocaine, and Case 7 confirms use of crack cocaine and MDMA. Case 6 showed low levels of cocaine in the 2–3 cm section with a wash residue to hair ratio of 0.37, indicating a great probability of external contamination, but it is not possible to rule out use. In other words, the results indicate that cocaine use by the donor is possible and indicates an association with the drug.

The results from the analysis of dreadlocked hair samples are useful to provide information on whether the individual has used any substances over an unspecified period.

3.4 Overall Conclusions

In this chapter we reviewed the two main factors that affect the results of drugs detected in the hair – external contamination (including the implications of hair washing procedures) and the effect of cosmetics and dreadlocks.

Table 3.2 Dreadlock sample testing results from nine different individuals.[a]

	Segment	THCC	CBD	CBN	THC	AEME	BZE	CE	CO	NCO	6MAM	COD	HER	DHC	MOR	MDA	MDMA	EtG	FAEE
	cm	$ng\,mg^{-1}$																$pg\,mg^{-1}$	
Case 1	0–3	ND	ND	0.04	0.06	—	—	—	—	—	—	—	—	—	—	—	—	—	—
Case 2	0–6	—	—	—	—	—	—	—	—	—	—	—	—	—	—	—	—	19	3901
Case 3	0–3	0.004	ND	0.19	1.88	—	—	—	—	—	—	—	—	—	—	—	—	—	—
	3–6	ND	ND	0.53	3.65	—	—	—	—	—	—	—	—	—	—	—	—	—	—
	0–6	—	—	—	—	—	—	—	—	—	—	—	—	—	—	—	—	ND	—
Case 4	0–1	0.008	—	0.09	1.08	—	—	—	—	—	—	—	—	—	—	—	—	—	—
	1–2	0.006	—	0.07	0.71	—	—	—	—	—	—	—	—	—	—	—	—	—	—
	2–3	0.009	—	0.07	0.81	—	—	—	—	—	—	—	—	—	—	—	—	—	—
Case 5	0–1	—	—	—	—	ND	0.44	0.05	9.4	0.1	ND	ND	ND	ND	ND	—	—	—	—
	1–2	—	—	—	—	ND	0.24	ND	9.4	0.1	ND	ND	ND	ND	ND	—	—	—	—
	2–3	—	—	—	—	ND	0.24	ND	7.5	0.1	ND	ND	ND	ND	ND	—	—	—	—
	0–3	—	—	—	—	—	—	—	—	—	—	—	—	—	—	—	—	79	ND
Case 6	0–1	ND	ND	ND	ND	ND	ND	ND	ND	ND	—	—	—	—	—	—	—	—	—
	1–2	ND	ND	ND	ND	ND	ND	ND	ND	ND	—	—	—	—	—	—	—	—	—
	2–3	ND	ND	ND	ND	ND	ND	ND	0.8	ND	—	—	—	—	—	—	—	—	—
	3–4	ND	ND	ND	ND	ND	ND	ND	ND	ND	—	—	—	—	—	—	—	—	—
	4–5	ND	ND	ND	ND	ND	ND	ND	ND	ND	—	—	—	—	—	—	—	—	—
	5–6	ND	ND	ND	ND	ND	ND	ND	ND	ND	—	—	—	—	—	—	—	—	—
	0–6	ND	ND	ND	ND	ND	ND	ND	ND	ND	—	—	—	—	—	—	—	ND	—
	W (2–3)	—	—	—	—	—	—	—	0.3	—	—	—	—	—	—	—	—	—	—
	W/H (2–3)	—	—	—	—	—	—	—	0.4	—	—	—	—	—	—	—	—	—	—

Table 3.2 (*Continued*)

	Segment	THCC	CBD	CBN	THC	AEME	BZE	CE	CO	NCO	6MAM	COD	HER	DHC	MOR	MDA	MDMA	EtG	FAEE
	cm	ng mg^{-1}																pg mg^{-1}	
Case 7	0–1	ND	ND	0.26	2.45	0.5	10.4	0.41	34.9	0.22	—	—	—	—	—	0.2	8.9	—	—
	1–2	ND	ND	0.23	2.23	ND	8.2	0.44	27.1	0.21	—	—	—	—	—	0.2	8.1	—	—
	2–3	ND	ND	0.20	2.46	0.5	8.6	0.38	30.6	0.21	—	—	—	—	—	0.2	8.7	—	—
	0–3	—	—	—	—	—	—	—	—	—	—	—	—	—	—	—	—	ND	—
Case 8	0–1	—	—	—	—	—	2.40	—	5.3	0.12	0.2	0.2	—	—	0.6	—	—	—	—
	1–2	—	—	—	—	—	2.35	—	4.8	0.08	0.2	—	—	—	0.5	—	—	—	—
	2–3	—	—	—	—	—	2.43	0.05	5.3	0.10	0.2	—	—	—	0.4	—	—	—	—
	3–4	—	—	—	—	—	2.49	0.06	5.0	0.07	0.2	—	—	—	0.3	—	—	—	—
	4–5	—	—	—	—	—	3.00	0.06	3.9	0.07	0.2	—	—	—	0.3	—	—	—	—
	5–6	—	—	—	—	—	3.45	0.08	4.6	0.07	0.3	—	—	—	0.3	—	—	—	—
Case 9	0–1	—	—	—	—	—	1.03	—	2.1	—	—	—	—	—	—	—	—	—	—
	1–2	—	—	—	—	—	1.21	—	2.7	—	—	—	—	—	—	—	—	—	—
	2–3	—	—	—	—	—	1.30	—	3.1	—	—	—	—	—	—	—	—	—	—

[a] THCC: THC-COOH; CBD: cannabidiol; CBN: cannabinol; THC: △9-tetrahydrocannabinolic; AEME: anhydroecgonine methyl ester; BZE: benzoylecgonine; CE: cocaethylene; CO: cocaine; NCO: norcocaine; 6MAM: 6-acetylmorphine; COD: codeine; HER: heroin; DHC: dihydrocodeine; MDA: 3,4-methylenedioxyamphetamine; MDMA: methylenedioxymethamphetamine; EtG: ethylglucuronide; FAEE: fatty acid ethyl esters; "ND": not detected; "—": Not tested; "W": wash residue; "H": hair.

Our data shows that drug levels are still detected in the hair of regular users even after strong chemical cosmetic hair treatments which damage hair, such as hair dyes, bleaches, and perms. However, overall levels of drug detected are reduced in such hair. The extent of drug decline following strong cosmetic treatment is dependent on its initial concentration and the properties of the hair matrix. This means that single or low drug usage may not be detected after strong chemical hair treatment with dyes, bleaches, or perms, although it is always wise to be alert to the possibility of a false negative. Long-term effects of weather (sunshine, rain, wind, chlorine) may also cause damage to the hair shaft with a subsequent impact on drug concentration. As dreadlocked hair could contain strands of hair from previous periods, hair testing can only provide information on whether the individual used drugs over an unspecific period.

However, the issue of external contamination can be overstated. Most people being tested are not contaminated and normal shampooing removes drugs on the exterior of hair very efficiently. The issue of external contamination is manageable in most cases and testing results will always need to be interpreted in the light of corroborating evidence from "clinical data" or "social context". However, results are not always conclusive and demand an honest interpretation. External contamination of hair by drugs does not need to be a limitation in the interpretation of hair analysis. It is rare that results cannot discriminate between use and external contamination. The data available up to now confirms that it is very important to be aware of alternative explanations for test results regarding the possibility and probability of external contamination when interpreting hair tests results, when the relevant metabolites are not detected.

The possibility of external contamination is relevant to hair testing as it is in all forms of chemical testing. The context matters. Testing for drug use in a workplace setting is completely different from testing in a forensic situation. In the workplace there may be no indication of poor performance or excessive absences for an individual and yet they test positive. That test result could be the result of a partner's use contaminating the person tested. It is always necessary to wash the sample irrespective of the reason for the test. The use of cut-offs in workplace testing also protects the individual from losing their job though contamination or very low-level recreational use of no consequence to the employer's business. But circumstances differ from industry to industry.

In situations of children at risk it is relevant to report any trace of drug found even where contamination may be the cause. In this situation there is a positive association with drugs, and possibly alcohol, that could be indicative of abuse or neglect.

External contamination must always be considered a possibility. It is incumbent on the analyst to consider the context of the testing and consider all possible causes of the results they find.

References

1. J. A. Siegel and K. M. Hair, in *Forensic Science*, CRC Press, Boca Raton, 4th edn, 2021.
2. D. L. Blank and D. A. Kidwell, External contamination of hair by cocaine: an issue in forensic interpretation, *Forensic Sci. Int.*, 1993, **63**, 145.
3. G. Romano, N. Barbera and I. Lombardo, Hair testing for drugs of abuse: evaluation of external cocaine contamination and risk of false positives, *Forensic Sci. Int.*, 2001, **123**, 119.
4. P. R. Stout, J. D. Ropero-Miller, M. R. Baylor and J. M. Mitchell, External contamination of hair with cocaine: evaluation of external cocaine contamination and development of performance-testing materials, *J. Anal. Toxicol.*, 2006, **30**, 490.
5. C. Chatterton, External contamination: still a debate? in *Hair Analysis in Clinical and Forensic Toxicology*, Academic Press, New York, 1st edn, 2015.
6. R. Kronstrand, Hair Drug Testing, in *Principles of Forensic Toxicology*, Springer, 5th edn, 2020.
7. A. Patzelt and J. Lademann, Drug delivery to hair follicles, *Expert Opin. Drug Delivery*, 2013, **10**, 787.
8. L. Tsanaclis, M. Andraus and J. Wicks, Hair analysis when external contamination is in question: A review of practical approach for the interpretation of results, *Forensic Sci. Int.*, 2018, **285**, 105.
9. T. Cairns, V. Hill, M. Schaffer and W. Thistle, Removing and identifying drug contamination in the analysis of human hair, *Forensic Sci. Int.*, 2004, **145**, 97.
10. V. A. Hill, M. I. Schaffer, R. B. Paulsen and G. N. Stowe, Identifying contamination versus ingestion in hair testing, with cocaine as example, *Drug Test. Anal.*, 2022, **14**, 1557.
11. C. L. Morris-Kukoski, M. A. Montgomery and R. L. Hammer, Analysis of extensively washed hair from cocaine users and drug chemists to establish new reporting criteria, *J. Anal. Toxicol.*, 2014, **38**, 628.
12. R. Wennig, Potential problems with the interpretation of hair analysis results, *Forensic Sci. Int.*, 2000, **107**, 5.
13. F. Pragst and M. A. Balikova, State of the art in hair analysis for detection of drug and alcohol abuse, *Clin. Chim. Acta*, 2006, **370**, 17.
14. K. Rygaard, K. Linnet and S. S. Johansen, A Systematic Review of Metabolite-to-Drug Ratios of Pharmaceuticals in Hair for Forensic Investigations, *Metabolites*, 2021, **11**, 686.
15. L. Tsanaclis and J. F. Wicks, Patterns in drug use in the United Kingdom as revealed through analysis of hair in a large population sample, *Forensic Sci. Int.*, 2007, **170**, 121.
16. P. Kintz, Segmental hair analysis can demonstrate external contamination in postmortem cases, *Forensic Sci. Int.*, 2012, **215**, 73.
17. P. Kintz, A. Ameline, L. Gheddar, P. Escudero, L. Ferrari and J. S. Raul, Cocaine external contamination can be documented by a hair test, *J. Anal. Toxicol.*, 2020, **44**, e4.

18. K. Kuwayama, H. Miyaguchi, T. Kanamori, K. Tsujikawa, T. Yamamuro, H. Segawa, Y. Okada and Y. T. Iwata, Distribution profiles of diphenhydramine and lidocaine in scalp, axillary, and pubic hairs measured by micro-segmental hair analysis: good indicator for discrimination between administration and external contamination of the drugs, *Forensic Toxicol.*, 2022, **40**, 64.
19. B. Moosmann, N. Roth and V. Auwärter, Hair analysis for Δ9-tetrahydrocannabinolic acid A (THCA-A) and Δ9-tetrahydrocannabinol (THC) after handling cannabis plant material, *Drug Test. Anal.*, 2016, **8**, 128.
20. C. Scholz, M. M. Madry, T. Kraemer and M. R. Baumgartner, LC–MS-MS Analysis of Δ9-THC, CBN and CBD in Hair: Investigation of Artifacts, *J. Anal. Toxicol.*, 2022, **46**, 504.
21. T. Franz, F. Scheufler, K. Stein, M. Uhl, T. Dame, G. Schwarz, H. Sachs, G. Skopp and F. Musshoff, Determination of hydroxy metabolites of cocaine from hair samples and comparison with street cocaine samples, *Forensic Sci. Int.*, 2018, **288**, 223.
22. E. D. Hart, S. Vikingsson, R. E. Winecker, A. L. Evans, E. J. Cone, J. M. Mitchell, E. D. Hayes and R. R. Flegel, Performance of Hair Testing for Cocaine Use-Comparison of Five Laboratories Using Blind Reference Specimens, *J. Anal. Toxicol.*, 2022, 1.
23. M. M. Madry, K. Y. Rust, R. Guglielmello, M. R. Baumgartner and T. Kraemer, Metabolite to parent drug concentration ratios in hair for the differentiation of tramadol intake from external contamination and passive exposure, *Forensic Sci. Int.*, 2012, **223**, 330.
24. C. M. Marsh, L. R. Crawley, S. K. Himes, R. Aranda and M. L. Miller, Discovery of Syn-/Anti-Cocaine-N-Oxide Diastereomers in Unwashed Postmortem Hair via LC–MS-MS, *J. Anal. Toxicol.*, 2014, **38**, 360.
25. M. Minoli, S. Casati, I. Angeli, A. Ravelli, P. Rota, P. Allevi and M. Orioli, Analysis of hydroxy-cocaine metabolites as evidence of cocaine consumption: Identification by parent ion search and quantitation by UHPLC-MS/MS in hair, *J. Pharm. Biomed. Anal.*, 2019, **172**, 167.
26. C. Scholz, B. B. Quednow, M. Herdener, T. Kraemer and M. R. Baumgartner, Cocaine hydroxy metabolites in hair: indicators for cocaine use versus external contamination, *J. Anal. Toxicol.*, 2019, **43**, 543.
27. M. M. Madry, M. M. Bosshard, T. Kraemer and M. R. Baumgartner, Hair analysis for opiates: hydromorphone and hydrocodone as indicators of heroin use, *Bioanalysis*, 2016, **8**, 953.
28. Society of Hair Testing (SoHT): https://www.soht.org/images/pdf/Consensus_DoA_2021.pdf
29. A. Müller and S. Iwersen-Bergmann, Determination of ethyl glucuronide in human hair samples: Decontamination *vs.* extraction, *Drug Test. Anal.*, 2020, **12**, 948.
30. J. C. Aijala, W. Wu and A. P. DeCaprio, Assessing Hair Decontamination Protocols for Diazepam, Heroin, Cocaine and Δ9-Tetrahydrocannabinol by Statistical Design of Experiments, *J. Anal. Toxicol.*, 2021, **45**, 498.

31. T. Baciu, F. Borrull, C. Aguilar and M. Calull, Recent trends in analytical methods and separation techniques for drugs of abuse in hair, *Anal. Chim. Acta*, 2015, **856**, 1.
32. E. Cuypers and R. J. Flanagan, The interpretation of hair analysis for drugs and drug metabolites, *Clin. Toxicol.*, 2018, **56**, 90.
33. D. A. Kidwell, F. P. Smith and A. R. Shepherd, Ethnic hair care products may increase false positives in hair drug testing, *Forensic Sci. Int.*, 2015, **257**, 160.
34. R. Erne, M. R. Baumgartner and T. Kraemer, Insights into the Decontamination of Cocaine-Positive Hair Samples, *J. Anal. Toxicol.*, 2021, **45**, 713.
35. M. D. Ramírez Fernández, S. M. Wille, M. Yegles and N. Samyn, Evaluation of decontamination procedures for drug testing in undamaged versus damaged hair, *Drug Test. Anal.*, 2022, **14**, 1155.
36. R. Erne, L. Bernard, A. E. Steuer, M. R. Baumgartner and T. Kraemer, Hair analysis: Contamination versus incorporation from the circulatory system-investigations on single hair samples using time-of-flight secondary ion mass spectrometry and matrix-assisted laser desorption/ionization mass spectrometry, *Anal. Chem.*, 2019, **91**, 4132.
37. D. Mantinieks, P. Wright, M. D. Rago and D. Gerostamoulos, A systematic investigation of forensic hair decontamination procedures and their limitations, *Drug Test. Anal.*, 2019, **11**, 1542.
38. M. Poetzsch, M. R. Baumgartner, A. E. Steuer and T. Kraemer, Segmental hair analysis for differentiation of tilidine intake from external contamination using LC-ESI-MS/MS and MALDI-MS/MS imaging, *Drug Test. Anal.*, 2015, **7**, 143.
39. J. Kaudewitz, H. Schwelm and V. Auwärter, Accumulation of amphetamine in hair after contact with street amphetamine, *Toxicol. Anal. Clin.*, 2022, **34**, S126.
40. W. F. Duvivier, R. J. Peeters, T. A. van Beek and M. W. Nielen, Evidence based decontamination protocols for the removal of external Δ9-tetrahydrocannabinol (THC) from contaminated hair, *Forensic Sci. Int.*, 2016, **259**, 110.
41. D. Mantinieks, D. Gerostamoulos, P. Wright and O. Drummer, The effectiveness of decontamination procedures used in forensic hair analysis, *Forensic Sci., Med., Pathol.*, 2018, **14**, 349.
42. L. Tsanaclis, J. Nutt, K. Bagley, S. Bevan and J. Wicks, Differentiation between consumption and external contamination when testing for cocaine and cannabis in hair samples, *Drug Test. Anal.*, 2014, **6**, 37.
43. T. Bassindale, Quantitative analysis of methamphetamine in hair of children removed from clandestine laboratories–Evidence of passive exposure?, *Forensic Sci. Int.*, 2012, **219**, 179.
44. L. Tsanaclis, M. Davies, S. Bevan, J. Nutt, K. Bagley and J. Wicks, Testing venous carbohydrate-deficient transferrin or capillary phosphatidylethanol with concurrent ethyl glucuronide and ethyl palmitate hair tests to assess historical and recent alcohol use, *Drug Test. Anal.*, 2021, **13**, 203.

45. R. Boscolo-Berto, D. Favretto, G. Cecchetto, M. Vincenti, R. Kronstrand, S. D. Ferrara and G. Viel, Sensitivity and specificity of EtG in hair as a marker of chronic excessive drinking: pooled analysis of raw data and meta-analysis of diagnostic accuracy studies, *Ther. Drug Monit.*, 2014, **36**, 560.
46. A. Müller and S. Iwersen-Bergmann, Determination of ethyl glucuronide in human hair samples: Decontamination vs extraction, *Drug Test. Anal.*, 2020, **12**, 948.
47. R. F. Cook, A. D. Bernstein and C. M. Andrews, Assessing Drug Use in the Workplace: A Comparison. The validity of self-reported drug use: Improving the accuracy of survey estimates, *NIDA Res. Monogr.*, 1997, **167**, 247.
48. C. Jurado, P. Kintz, M. Menendez and M. Repetto, Influence of the cosmetic treatment of hair on drug testing, *Int. J. Leg. Med.*, 1997, **110**, 159.
49. M. Luginbühl, B. Bekaert, S. Suesse and W. Weinmann, Detox shampoos for EtG and FAEE in hair–Results from *in vitro* experiments, *Drug Test. Anal.*, 2019, **11**, 870.
50. G. Miolo, G. Stocchero, S. Vogliardi, L. Menilli, S. Scrivano, M. Montisci and D. Favretto, A study on photostability of ethyl glucuronide in hair irradiated under artificial sunlight, *J. Anal. Toxicol.*, 2020, **44**, 58.
51. P. Kintz, Negative hair test result after long-term drug use. About a case involving morphine and literature review, *Clin. Chem. Lab. Med.*, 2021, **59**, 267.
52. S. Marrinan, A. Roman-Urrestarazu, D. Naughton, E. Levari, J. Collins, R. Chilcott, G. Bersani and O. Corazza, Hair analysis for the detection of drug use-is there potential for evasion?, *Hum. Psychopharmacol.: Clin. Exp.*, 2017, **32**, e2587.
53. S. Süße, C. M. Selavka, T. Mieczkowski and F. Pragst, Fatty acid ethyl ester concentrations in hair and self-reported alcohol consumption in 644 cases from different origin, *Forensic Sci. Int.*, 2010, **196**, 111.
54. F. Pragst, S. Suesse, A. Salomone, M. Vincenti, V. Cirimele, J. Hazon, L. Tsanaclis, R. Kingston, F. Sporkert and M. R. Baumgartner, Commentary on current changes of the SoHT 2016 consensus on alcohol markers in hair and further background information, *Forensic Sci. Int.*, 2017, **278**, 326.
55. E. Alladio, L. Giacomelli, G. Biosa, D. Di Corcia, E. Gerace, A. Salomone and M. Vincenti, Development and validation of a Partial Least Squares-Discriminant Analysis (PLS-DA) model based on the determination of ethyl glucuronide (EtG) and fatty acid ethyl esters (FAEEs) in hair for the diagnosis of chronic alcohol abuse, *Forensic Sci. Int.*, 2018, **282**, 221.
56. C. L. Crunelle, M. Yegles, A. L. van Nuijs, A. Covaci A, M. De Doncker, K. E. Mauden, B. Sabbe, G. Dom, W. E. Lambert, P. Michielsen and H. Neels, Hair ethyl glucuronide levels as a marker for alcohol use and abuse: a review of the current state of the art, *Drug Alcohol Depend.*, 2014, **134**, 1.

57. R. Paul, L. Tsanaclis, C. Murray, R. Boroujerdi, L. Facer and A. Corbin, Ethyl glucuronide as a long-term alcohol biomarker in fingernail and hair. Matrix comparison and evaluation of gender bias, *Alcohol Alcohol.*, 2019, **54**, 402.
58. R. Agius, Utility of coloured hair for the detection of drugs and alcohol, *Drug Test. Anal.*, 2014, **6**, 110.
59. M. Luginbühl, S. Nussbaumer and W. Weinmann, Decrease of ethyl glucuronide concentrations in hair after exposure to chlorinated swimming pool water, *Drug Test. Anal.*, 2018, **10**, 689.

CHAPTER 4

Insight Into Drug Extraction from the Keratin Matrix

CARMEN JURADO

Socorro 12, 41003 Sevilla, Spain
Email: cjurado1957@gmail.com

4.1 Introduction

One feature of hair analysis when compared with the analysis of other biological fluids, such as urine or blood, is the complexity of the analytical procedures.

According to the 2021 consensus of the Society of Hair Testing (SoHT), the preparation of hair samples involves the following steps:[1] (a) pre-analytical steps, including decontamination and homogenization by pulverization or cutting into small pieces; (b) extraction of the drugs from the hair matrix; (c) instrumental analysis, including screening and confirmation techniques.

One of the main pitfalls in hair analysis is external contamination and the possibility of reporting a false positive result. In addition to the bloodstream, drugs and xenobiotics can be incorporated from the external environment *via* other pathways, such as powder, smoke, *etc.*, which can lead to external contamination. For this reason, it is very important to distinguish between passive contamination and active consumption. Several criteria have been proposed to avoid this risk, the application of different washing procedures being one of them. Several papers have compared the efficacy of different decontamination procedures.[2–4] To date there is no consensus on a specific procedure to conduct the decontamination process. SOHT recommends including washing steps with both organic solvent and aqueous solutions.[5]

Issues in Toxicology No. 47
Perspectives and Challenges of Hair Analysis
Edited by Pascal Kintz, Alberto Salomone and Marco Vincenti

Published by the Royal Society of Chemistry, www.rsc.org

Extraction of drugs from the hair matrix is the most critical step in hair analysis. Drugs are bound to keratin, melanin, and other components of hair. During the extraction process, they must be released from these unions. The extraction conditions must be carefully selected, they must be able to extract all the xenobiotics present in the hair and, at the same time, they must not alter the chemical structure or degrade the molecules, especially the most labile ones, such as cocaine or heroin. A variety of extraction methods can be used, and the difference lies in the ability to act on the hair matrix. They range from a simple treatment with apolar solvents, such as methanol (see Section 4.2.1), or buffers with different pH (see Section 4.2.2) to the dissolution of the protein matrix by enzymatic (see Section 4.2.4) or alkaline digestion (see Section 4.2.5).

Once the compounds of interest have been extracted, some laboratories perform a preliminary or screening analysis, usually by immunoassay techniques. The pioneer authors on hair analysis used radioimmunoassay (RIA),[6] other laboratories perform the screening by enzyme-linked immunosorbent assay (ELISA)[7] or fluorescence polarization immunoassay (FPIA).[8] More recently, after the appearance of very sensitive instrumentation, some authors apply mass spectrometry techniques, such as ultrahigh-performance liquid chromatography-time-of-flight mass spectrometry (UPLC-TOF-MS),[9] for the screening. Although preliminary screening is accepted, very few laboratories apply it in their routine. This is demonstrated by the fact that only 9 out of 68 participants in the proficiency test organized by SoHT reported a preliminary screening analysis. For the confirmation techniques, currently most of the laboratories apply liquid chromatography–tandem mass spectrometry (LC–MS/MS), but some laboratories continue applying gas chromatography/mass spectrometry (GC–MS) and others use new sensitive techniques, such as UPLC-TOF-MS or liquid chromatography Orbitrap mass spectrometry (Orbitrap).

This chapter will focus on the extraction of drugs from the hair matrix; decontamination procedures and instrumental analyses will not be discussed. The different extraction solvents will be reviewed and their efficiency in extracting different groups of compounds will be compared based on the literature and the data from proficiency test programs.

4.2 Extraction of Drugs from the Hair Matrix

Some circumstances must be considered for the successful development of an extraction procedure:

- Type of method: multianalyte/screening or target analysis: although the ideal situation is to develop a method focused on one or more targeted compounds; the request for multi-analyte or generally unknown methods is very common in laboratories performing hair analysis. A universal extraction solution is recommended in these cases although sensitivity must be sacrificed.

- Type of drug to be analyzed: drug stability is of paramount importance when selecting an extraction solvent, especially in target analysis. Strong conditions, such as alkaline digestion, should be avoided for the analysis of labile compounds.
- Drug pharmacokinetics: drug metabolites should be included on the analytical protocol since some of them are markers of drug consumption. For example, the presence of 11-nor-9-carboxy-Δ^9-tetrahydrocannabinol (THC-COOH), metabolite of Δ^9-tetrahydrocannabinol (THC), strongly supports cannabis intake.[1]
- Method application and cut-off values to be achieved: depending on the application, hair can be used both to establish single or very occasional drug use, and to establish chronic drug consumption. In the case of single-dose analysis, method sensitivity must be optimized, in terms of extraction and instrumentation, to be able to detect very low concentrations of drugs. When chronic consumption is to be determined, although the extraction method can be the same, high resolution mass spectrometry is not required. Cut-off values are different in both circumstances. Specific cut-offs for chronic drug consumption have been established by SoHT[1] and the European Workplace Drug Testing Society (EWDTS)[10] among others. Cut-off values have not been established in cases of single consumption, but much lower limits of quantification (LOQ) are required. All laboratories performing hair analysis must comply with the cut-off values. If the limit of detection (LOD) of the method is higher than the cut-off, there is a high risk of reporting false negative results.

Different extraction procedures are described below.

4.2.1 Extraction with Methanol

This extraction procedure consists of incubating hair samples with methanol. During the incubation, methanol can penetrate the hair, producing swelling and consequent release of most substances. The incubation time varies considerably depending on whether it is performed under sonication or not. Methods applying sonication can perform the extraction in 1 to 4 hours,[11–15] or in as little as 10 min;[16] while those which do not apply sonication require overnight incubation to perform the extraction.[17–20] The methanolic extracts obtained after the incubation can be directly analyzed, without any further clean-up step,[13,15] or can be submitted to a further purification step by liquid–liquid extraction,[7,21] solid phase extraction (SPE),[9,10,12] filtration,[18] *etc.*

The main advantage of this method is that it can be applied for the extraction of almost all the compounds of toxicological interest. It is commonly applied not only to multianalyte and/or general unknown analyses, but also to target analysis, as demonstrated by the fact that some laboratories have reported the same methanolic extraction procedure for the analysis of chemically different compounds. This is the case for a laboratory, which employed a method based on the incubation with methanol at 45 °C under sonication for the analysis of different compounds. The only differences were found in the incubation time and the purification procedure. For the analysis of 33 basic drugs, including amphetamines, cocaine, opiates, opioids, and metabolites, the extraction was performed on powdered hair for 4 hours,[15] followed by SPE and analysis by LC–MS/MS. To demonstrate the efficacy of this method, they analyzed proficiency test samples, and the *z*-scores were lower than one for most of the compounds. For the analysis of 29 benzodiazepines and *z*-drugs,[12] powdered hair was sonicated for 2 hours, and the purification consisted of a liquid–liquid extraction with 1-chlorobutane. The same extraction protocol was also reported for the target analysis of dihydrocodeine.[22] Powdered hair was incubated with methanol, under sonication at 45 °C, for 4 hours, followed by SPE and analysis by LC–MS/MS.

Many of the recently published papers based on methanol extraction are related to multianalyte analyses; even in some cases compounds present at low concentrations and requiring sensitive methods are included, as well. Shin *et al.*[14] developed a method for the simultaneous determination of 75 drugs of abuse, including lysergic acid diethylamide (LSD), one of the most difficult compounds to be analyzed in hair. Hair samples were pulverized and extracted with methanol in an ultrasonic bath for 60 min at 50 °C. The extracts were purified with dispersive solid-phase extraction, an alternative approach to conventional SPE in which the sorbent is dispersed into the sample. Analysis was performed by LC–MS/MS and lower LOQ ranged from 2 to 200 pg mg^{-1} depending on the drug. Cardoso *et al.*[23] investigated 14 of the most frequently consumed drugs of abuse, including THC, which is usually analyzed in basic conditions. Previously powdered hair was incubated in methanol in an ultrasonic bath at 45 °C for 2 h. The mixture was centrifuged and injected into the LC–MS/MS. LOQ ranged from 0.025 ng mg^{-1} for THC and cocaethylene (CE) to 0.25 ng mg^{-1} for cocaine and mazindol. The method was later applied to the analysis of 104 hair samples derived from suicide attempt investigations. Di Corcia *et al.*[18] reported a method, with minimal sample workup, for the simultaneous determination of multi-class drugs of abuse, which also included THC. The analytes were incubated overnight with methanol at 55 °C. An aliquot of the extract was directly injected into the LC–MS/MS system. The authors recommend this screening method for workplace drug-screening controls and forensic investigations. The same method was applied for the analysis of 13 synthetic opioids, including fentanyl analogs.[20] The method was applied to 34 real hair samples from subjects who had reported past-year non-medical opioid and/or heroin use. The study confirmed the increasing diffusion of new synthetic opioids and fentanyl analogs among this population.

Methanol extraction has also been applied in the analysis of a single drug-class. Wen *et al.*[16] analyzed seven barbiturates in the context of drug facilitated crimes. Powdered hair was extracted with methanol under sonication for 10 min. Analysis was directly performed by ultrahigh-performance liquid chromatography-high resolution mass spectrometry (UHPLC-HRMS). LOQ was 0.02 ng mg^{-1} for all the compounds. The method was successfully applied to determine barbiturates in human hair after a single-dose exposure. Lee *et al.*[17] developed a method for the analysis of sedative-hypnotics (lorazepam, midazolam, zolpidem and diazepam, and metabolites). Hair samples were cut into small pieces and incubated in methanol at 38 °C for 16 h. After cleaning-up by SPE, the extracts were derivatized and analyzed by GC–M.

4.2.2 Extraction with Buffer Solutions

The mild conditions of some buffers allow the application of this extraction procedure to multi-analyte analysis. In this sense, Vogliardi *et al.*[24] developed a method for the determination of 28 benzodiazepines, including six metabolites. Hair samples were incubated overnight in a phosphate buffer pH 8.4 and the extracts were subsequently submitted to a liquid–liquid extraction with methylene chloride/diethyl ether (90 : 10). Analysis was performed in an Orbitrap mass spectrometer. The lowest LOQ were in the range 1–10 pg mg^{-1}. The method was applied in clinical and forensic toxicology cases. A slightly lower pH was used by Maublanc *et al.*[25] for the simultaneous determination of 35 psychotropic drugs. Hair was incubated overnight at room temperature in a phosphate buffer pH 5 solution followed by a clean-up by liquid–liquid extraction with a dichloromethane/ether mixture (70 : 30). Analysis was performed by LC–MS/MS and the LOD and LOQ were in the range 0.5–10 pg mg^{-1}. The same authors applied these extraction conditions to document LSD[26] consumption in a 24-year-old man found dead after a party. The LOQ was 0.5 pg mg^{-1} and they found an LSD concentration of 0.66 pg mg^{-1} in pubic hair. They also found LSD in the post-mortem blood and urine of the victim.

Martin *et al.*[27] used Soerensen buffer pH 7.0 for the analysis of mephedrone. Hair samples were incubated overnight at 40 °C in this buffer. The extracts were purified by liquid–liquid extraction with ethyl acetate after alkalinization with 1 N NaOH. Analysis was performed by GC–MS-SIM after derivatization with a mixture of heptafluorobutyric anhydride/ethyl acetate (100 : 50). The method was applied to 67 hair samples, 13 of them were positive for mephedrone with concentrations ranging from 0.2 to 313.2 ng mg^{-1} (mean 26.8 ng mg^{-1}).

Another short extraction procedure was proposed by Barreto *et al.*[28] for the analysis of ketamine and norketamine. Hair samples were extensively powdered, and the extraction was achieved in a phosphate buffer (pH 6) solution after 10 min vortex agitation at room temperature followed by a SPE clean-up procedure. The method was sensitive enough to detect the drugs after single (1 mg kg^{-1}) intravenous doses received by patients submitted to general anesthesia before surgical procedures. A similar buffer pH was used by Zhu *et al.*[29] for the analysis of two anticoagulant drugs, bromadiolone and brodifacoum. Phosphate buffer solution (pH 6.8) was added to pulverized hair followed by ultrasonication and liquid–liquid extraction. Analysis was performed by LC–MS/MS and LOD for bromadiolone and brodifacoum were 0.010 and 0.025 ng mg^{-1}, respectively.

Some laboratories reported the same extraction method for the analysis of different analytes. Incubation of powdered hair with phosphate buffer (pH 5) for 10 min was applied for the analysis of atropine and scopolamine,[30] baclofen[31] to assess compliance in alcohol dependent patients or cathinones (4-methylethcathinone and 3,4-methylenedioxypyrovalerone);[32] even 20 anabolic steroids[33] were analyzed with the same extraction procedure. The only difference lay in the liquid–liquid extraction used in the clean-up step, heptane for anabolic steroids and hexane/ethyl acetate (1/1) for atropine, scopolamine, and others.

Some authors reported the extraction with mixtures of buffer solutions and solvents. Ramírez *et al.*[34] reported a method for the analysis of 24 antidepressants based on the extraction with methanol/acetonitrile/ammonium formate buffer 1 mmol L^{-1} (25 : 25 : 50) followed by SPE clean-up. The method was then successfully applied to the analysis of hair samples from workplace drug testing. The same method was later applied to studying the influence of bleaching and thermal straightening on endogenous gamma hydroxybutyric acid (GHB) concentrations in hair. Methanol/acetonitrile/ammonium formate buffer 1 mM (25 : 25 : 50, v/v/v) was used to extract the drug from the hair matrix in a water bath for 1.5 h at 37 °C. The extracts were filtered and analyzed by LC–MS/MS.

A similar extraction mixture was used by Wikström *et al.*[35] for the analysis of methedrone. The samples were incubated in a water bath (with orbital shaking) with acetonitrile/methanol/20 mM ammonium formate buffer (pH 3) (10 : 10 : 80) at 37 °C for 18 h. The extracts were directly injected into the LC–MS/MS system. The same method was applied to the screening of 28 drugs:[36] in this case, analysis was performed by UPHLC-TOF-MS. LOQs ranged from 0.02 to 0.1 ng mg^{-1}.

Mixtures with 2 mM ammonium formate have been reported for multianalyte analyses. Nielsen *et al.*[37] incubated hair with methanol:acetonitrile:ammonium formate (2 mM, 8% acetonitrile, pH 5.3) overnight at 37 °C, for the analysis of 52 common pharmaceuticals and drugs. Analysis was performed by UPLC-TOF-MS and LOQ was 0.05 ng mg^{-1} for 87% of the analytes. The same extraction mixture and conditions were applied by Wang *et al.*[38] to the analysis of 116 drugs by UPLC–MS/MS and later to the analysis of 23 antidepressants and antipsychotics[39] in the hair of schizophrenic patients.

4.2.3 Extraction with Acidic Solutions

This extraction procedure consists of the incubation of the hair with diluted acidic solutions overnight. Some authors reporting this extraction procedure perform the extraction with diluted hydrochloric acid (HCl) mainly 0.1 M HCl followed by a clean-up step by liquid–liquid extraction or SPE; while others perform the extraction with methanol acidified with HCl or trifluoroacetic acid (TFA) with or without purification of the extracts.

Concentrated acidic solutions should be avoided because they can lead to the hydrolysis of very labile compounds. For example, 6-monoacetylmorphine (6-AM) can be hydrolyzed to morphine or cocaine can be hydrolyzed to benzoylecgonine (BE). Skopp *et al.*[40] during the development of a method for the analysis of buprenorphine (BUP) and norbuprenorphine (NBUP), studied the *in vitro* degradation of these compounds in diluted acid. With this interest, BUP and NBUP were incubated in diluted HCl at 60 °C for up to 24 h. Degradation of BUP and NBUP was dependent on time and the metabolite NBUP was hydrolyzed faster than BUP. The authors conclude that the concentration of these compounds could be underestimated if the extraction is performed with acidic extraction.

4.2.3.1 Extraction with Diluted Hydrochloric Acid

Diluted HCl has been used for the extraction of cocaine and its metabolites BE and CE.[41] After overnight incubation, the extract was diluted with distilled water (1 : 10) and analyzed by LC–MS/MS. Quintela *et al.*[42] included the analysis of another metabolite ecgonine methyl ester (EME). Cocaine compounds were extracted by incubation with 0.1 M HCl at 50 °C for 12 h and purified by SPE. Analysis was performed by LC–MS/MS. Lower LOQs ranged from 0.020 ng mg^{-1} for COC, 0.010 ng mg^{-1} BE and CE, to 0.005 ng mg^{-1} for EME.

Freni *et al.*[43] developed a method for the simultaneous analysis of 16 synthetic cathinones. Hair samples were ultrasonicated with 0.1 M HCl followed by clean-up by SPE. Analysis was performed by LC–MS/MS. LOQs ranged from 1 pg mg^{-1} (4-methylethcathinone, methedrone, alpha-pyrrolidinopentiophenone, alpha-pyrrolidinohexiophenone, methylenedioxypyrovalerone, butylone, ethylone, 3,4-dimethylmethcathinone, pentedrone and pentylone) to 5 pg mg^{-1} (4-fluoromethcathinone, buphedrone, ethcathinone, methcathinone, mephedrone and naphyrone).

Some authors reported acidic extraction methods for target analysis. Verri *et al.*[44] reported a method for the analysis of tramadol and metabolites (*O*-desmethyltramadol, *N*-desmethyltramadol and *N,O*-didesmethyltramadol) with the aim of differentiating chronic therapeutic use and a binge pattern of tramadol. Hair samples were incubated overnight in diluted HCl; the extracts were purified by SPE and analyzed by LC–MS/MS. They found different metabolites to parent drug concentration ratios in chronic therapeutic consumption and in the binge pattern. Hongkun *et al.*[45] incubated the hair with diluted HCl for the extraction of etaqualone. The extracts were purified by liquid–liquid extraction with ethyl ether. Analysis was performed by gas chromatography tandem mass spectrometry (GC–MS/MS) without derivatization. The lower LOQ was 1 pg mg^{-1}.

Diluted HCl was used by Shu *et al.*[46] for the determination of the methamphetamine enantiomeric composition. Previously powdered hair was incubated overnight at 53 °C in 0.1 M HCl, and then subjected to a SPE. Analysis was performed by LC–MS/MS after derivatization with Marfey's reagent. They compared the efficacy of this method with another involving derivatization with (*S*)-(−)-*N*-trifluoroacetylprolyl chloride and analysis by GC–MS. They found satisfactory correlation and the signal to noise ratio of the chromatograms was improved after derivatization with Marfey's reagent. Porpiglia *et al.*[47] applied a chiral separation of ketamine and norketamine by capillary electrophoresis to discriminate between the single enantiomer and the racemic drug. Ketamine is usually available as a racemic mixture, but it is also available in the form of the *S*-enantiomer with higher anesthetic potency. Hair was sonicated for 1 h with 0.1 M HCl and incubated overnight in a water bath at 37 °C. The aqueous phase was extracted with hexane/ethyl acetate (50 : 50) and analyzed by capillary electrophoresis. A complete separation of both racemic ketamine and norketamine in the respective enantiomers was obtained in less than 10 min. LOQs were 0.25 ng mg^{-1}.

4.2.3.2 Extraction with Acidic Methanolic Solutions

Other authors reported the extraction with acidic methanolic solutions. Ketamine and norketamine and six amphetamine related compounds: amphetamine (AP), methamphetamine (MA), methylenedioxyamphetamine (MDA), methylenedioxymethamphetamine (MDMA), *para*-methoxyamphetamine (PMA) and *para*-methoxy-*N*-methylamphetamine (PMMA) were analyzed by Rhee *et al.*[48] Hair samples were incubated with 1% HCl in methanol for 16 h, with a magnetic stirrer. The extracts were evaporated and analyzed by LC–MS/MS. LOQ for eight analytes in the range 0.003–0.01 ng in total. Lee *et al.*[49] developed an analytical method for the simultaneous detection of phentermine, phendimetrazine, amfepramone, fenfluramine, mazindol, MA, and MDMA, as well as their metabolites. The different compounds were simultaneously extracted using 1% HCl in methanol. The extracts were then filtered and analyzed by LC–MS/MS with electrospray ionization in positive mode. LOQs were 0.025 ng mg^{-1} hair for all analytes except mazindol and phentermine (1 ng mg^{-1}). Han *et al.*[50] analyzed MA in hair to establish the pattern of drug consumption. Hair samples were incubated in 1 mL methanol containing 1% HCl for 20 h. The extracts were evaporated, derivatized with trifluoroacetic anhydride in ethylacetate at 65 °C for 30 min and analyzed by GC–MS. MA concentrations in the whole hair were in the range 3.00–105.10 ng mg^{-1} (mean, 34.53 ng mg^{-1}) and the concentration of AP, the main metabolite, was in the range 0.05–4.76 ng mg^{-1} (mean, 2.42 ng mg^{-1}).

Martin *et al.*[51] analyzed hallucinogenic substances, such as psilocin, bufotenine, a position isomer of psilocin, and lysergic acid diethylamide (LSD) and related compounds iso-LSD, nor-LSD, nor-iso-LSD and O–H-LSD. Hair samples were extracted with a mixture of HCl (0.01 M) and methanol (1 : 1) in an ultrasonic bath for 6 h at a temperature below 50 °C, due to the instability of psilocin. The extracts were purified by SPE and analyzed by LC–MS/MS. The method was validated successfully for all analytes except for O–H-LSD, for which the LOQ was higher than the lowest calibrator.

Some authors have applied the extraction with acidic methanol solutions for the analysis of new psychoactive substances (NPS). Nieddu *et al.*[52] analyzed 11 phenethylamines: PMA; PMMA; 3,4,5-trimethoxyamphetamine; 2,5-dimethoxyamphetamine; 2,5-dimethoxy-4-methylamphetamine; 2,5-dimethoxy-4-ethylamphetamine; 2,5-dimethoxy-4-bromoamphetamine; 2,5-dimethoxy-4-bromophenethylamine; 2,5-dimethoxy-4-iodophenethylamine; 2,5-dimethoxy-4-ethylthiophenethylamine, and 2,5-dimethoxy-4-*n*-propylthiophenethylamine. Hair samples were incubated with 1% HCl in methanol at 45 °C for 24 h with continuous stirring. The extracts were filtered, evaporated, and directly analyzed by LC–MS–MS. LOQs ranged from 0.09 (DOM) to 0.20 ng mg^{-1} (2C-B).

Lendoiro *et al.*[53] analyzed amphetamine type stimulants, including AP, MA, MDA, MDMA, synthetic cathinones (methylone, methedrone, mephedrone, 3,4-methylenedioxypyrovalerone, (±)-4-fluoromethamphetamine and 4-fluoromethcathinone), synthetic piperazines (1-(3-chlorophenyl)piperazine (mCPP) and 3-trifluoromethylphenylpiperazine), and trazodone and phenazone that produce mCPP as a metabolite. Hair samples were incubated in methanol with 0.1% HCl. The extracts were cleaned-up by SPE and analyzed by LC–MS/MS. LOQs were in the range 2–20 pg mg^{-1}. The detection of trazodone was important to determine the licit origin of mCPP. A large number of NPS, up to 132, were included in the screening method proposed by Boumba *et al.*[54] The method involved direct extraction by pulverization in acidified methanol and analysis by LC–MS/MS. The method has proven to be efficient for the extraction of chemically different NPS from hair; from unstable and volatile compounds, such as most of the cathinones, to hydrophobic compounds such as synthetic cannabinoids.

In some cases, methanol was acidified with TFA. Huang *et al.*[55] developed a method for the analysis of the antiretroviral nevirapine. Extraction was performed with methanol/TFA (v/v, 9 : 1) under agitation at 37 °C in a water bath overnight, followed by liquid–liquid extraction under alkaline conditions. Analysis was performed by LC-MS/MS and the LOQ was 0.25 ng mg^{-1} using 2 mg of hair. The method was applied to analyze more than 1000 hair samples of HIV-infected patients. Frison *et al.*[56] reported a method for the analysis 3-methylmethcathinone (3-MMC) and its metabolites. They performed the extraction, in previously powdered hair, with methanol : TFA (9 : 1) at 45 °C overnight. The extracts were analyzed by Orbitrap. They found 3-MMC in pubic hair (25.8 ng mg^{-1}) together with the metabolites 3-methylephedrines and 3-methylnorephedrines, thus proving 3-MMC consumption. Chang *et al.*[57] used the same acidified methanol for the analysis of amphetamines, ketamine, opiates, and their metabolites. They considerably shortened the analysis by using microwave-assisted extraction. Hair samples were incubated for 3 min with methanol:TFA during microwave-assisted extraction (MAE) at 700 W. When the 3 min MAE method was compared with overnight incubation, the extraction yields were similar or even greater.

4.2.4 Digestion with Enzymes

The application of enzymatic treatments allows the release of all the compounds present in the hair, due to the hydrolytic activity of the enzymes. The main advantages are the wide range of compounds which can be extracted, and the high recovery achieved with this procedure. Other advantages of this extraction are that the extracts obtained can be directly analyzed by immunochemical methods, after inactivation of the enzyme. Cheong *et al.*[8] incubated the hair with a mixture of 1 M tris[hydroxymethyl]aminomethane hydrochloride:1 M dithiothreitol:proteinase K 4 h at 56 °C for 4 h, for the analysis of MA. Once the hair sample was digested, the temperature in the incubator was increased to 100 °C and held for 1 min to denature the enzyme. The solution was cooled, centrifuged and analyzed by FPIA.

Enzymatic digestion was mainly used in the 1980s and 1990s; currently it is not so common. In fact, very few recent papers apply this type of extraction. Baptista *et al.*[58] used β-glucuronidase/arylsulfatase for the analysis of cannabis compounds (THC, cannabinol, cannabidiol, and THC-COOH). Hair samples were incubated with the enzyme at 40 °C for 2 h. The extracts were purified by liquid–liquid extraction, derivatized and analyzed by GC–MS–NCI.

Proteinase K was used by Breidi *et al.*[59] for the analysis of THC and cocaine followed by extraction with pentane and analysis by GC–MS and pronase E was used by Míguez-Framil *et al.*[60,61] for the analysis of opiates and cocaine.

4.2.5 Digestion with Alkaline Solutions

With this extraction procedure the hair is completely dissolved by the alkaline solution. Almost all authors who use this extraction method, digest the hair with sodium hydroxide (NaOH) solutions at different concentrations. In all cases a subsequent clean-up step is required before instrumental analysis.

This method is advantageous for non-labile compounds, such as amphetamines and cannabis compounds, or GHB; while it should be avoided for the analysis of labile compounds, like cocaine, heroin, or benzodiazepines. Al Jaber *et al.*[62] compared the stability of 20 basic or weak basic drugs in hair during alkaline and methanolic extraction conditions. Analytes included mainly benzodiazepines, but also antidepressants and antipsychotics. For methanolic extraction, hair was added with a mixture methanol : acetonitrile : 1% formic acid (60 : 10 : 30) under sonication for 16 h at 40 °C. For alkaline extraction, hair was incubated with 1 M NaOH for 3 h at 50 °C. Results demonstrated that all the analytes were stable with the methanolic extraction method, with a mean recovery of 89.6% ± 4.2%. The least stable was oxazepam with 56.1% recovery. While the stability of analytes under alkaline conditions was remarkably lower with a mean recovery of 51.7% ± 19.0%. The least stable compounds were 7-aminoclonazepam (17.9%), amisulpiride (25.0%) or citalopram (28.3%).

Alkaline extraction has been successfully applied for the analysis of THC-COOH, achieving the LOQ proposed by SoHT[1] (0.2 pg mg^{-1}) or even lower. Han *et al.*[63] analyzed THC-COOH in hair to evaluate patterns of cannabis use. Hair samples were digested with 1 mL of 1 M NaOH at 85 °C for 30 min. The extracts were submitted to a liquid–liquid extraction with 2 mL of *n*-hexane–ethyl acetate (9 : 1) twice after adding 1 mL of 0.1 N sodium acetate buffer (pH = 4.5) and 200 μL of acetic acid. Analysis was performed by GC–MS/MS after derivatization and LOQ was 0.05 pg mg^{-1}. A similar procedure was applied by Cho *et al.*,[64] but they performed the analysis by LC–MS/MS without derivatization and the LOQ was slightly higher (0.1 pg mg^{-1}), but within the limit proposed by SoHT. A very sensitive method was proposed by Minoli *et al.*[65] for the simultaneous analysis of THC and THC-COOH. Hair was digested in 1 mL of 1 M NaOH at 90 °C for 15 min. Then, THC and THC-COOH were sequentially extracted. First, THC was extracted with *n*-hexane : ethyl acetate (9 : 1); next the aqueous solution was acidified (pH 4) adding concentrated acetic acid and THCCOOH was extracted with the same mixture. Analysis was performed by GC–MS/MS and LOQ was 0.04 pg mg^{-1} for THC-COOH. Higher LOQs (3 pg mg^{-1} for both THC and THC-COOH) were achieved when the extraction conditions changed. Mercolini *et al.*[66] dissolved the hair in 2.5 M NaOH at 60 °C, THC and THC-COOH were simultaneously extracted with ethyl aetate after neutralization with formic acid and analyzed by LC–MS/MS.

The sensitivity decreased when more cannabis compounds were included in the analytical protocol. Cobo-Golpe *et al.*[67] analyzed six cannabinoids in hair samples: cannabidiol (CBD), cannabinol (CBN) and THC, and the metabolites THC-COOH, 11-hydroxy-THC (OHTHC) and 8-β-11-dihydroxyTHC (diOHTHC). They digested previously powdered hair with 1 mL NaOH 1 N for 15 min at 95 °C. The extracts were submitted to SPE with OASIS MAC cartridges. Analysis was performed by LC–MS/MS and LOQs ranged from 40 pg mg^{-1} to 100 pg mg^{-1}.

Some authors have used alkaline digestion for the analysis of gamma hydroxybutyric acid (GHB). Shi *et al.*[68] analyzed endogenous GHB in the hair of the Chinese population. Hair was digested with 1 mol L^{-1} NaOH at 90 °C for 10 min. Before derivatization with *N,O*-bis(trimethylsilyl)trifluoroacetamide (BSTFA), a liquid–liquid extraction with ethyl acetate was performed under acidic conditions. Analysis was performed by GC/MS/MS and the LOD was 0.02 g mg^{-1}. A similar extraction procedure was applied by Bertol *et al.*[69] They compared the efficacy of GC–MS *versus* LC–MS/MS for the analysis of GHB. The GC–MS-SIM method involved digestion of hair with 1 N NaOH at 56 °C overnight, extraction with ethyl acetate and derivatization with BSTFA. The extraction time of LC–MS/MS method was shorter: hair was digested with NaOH at 75 °C for 40 min. Analysis was performed by LC–MS/MS after extraction with ethyl acetate. LOQs were 0.6 ng mg^{-1} with both methods and performances were also similar.

Amphetamines are very stable compounds and can be extracted from the hair with alkaline solutions. Jakobsson *et al.*[9] developed a method for the analysis of AP, MA, MDA, and MDMA in hair samples. Hair was incubated with 2 M NaOH at 80 °C for 10 min, followed by liquid–liquid extraction with isooctane and analyzed by LC–MS/MS. LOQ was 0.02 ng mg^{-1}. A similar method, including two more amphetamines (3,4-methylenedioxyethamphetamine, MDEA, and *N*-methyl-1-(3,4-methylenedioxyphenyl)-2-butanamine, MBDB) was reported by Chèze *et al.*[70] It involved hair that was digested with 1 M NaOH at 80 °C for 15 min followed by liquid–liquid extraction with hexane/ethyl acetate (2 : 1). Analysis was performed by LC–MS/MS with LOQs of <5 pg mg^{-1} for MA, MDMA, MDEA and MBDB, 14.7 pg mg^{-1} for AP and 15.7 pg mg^{-1} for MDA.

Doping agents, such as nandrolone and stanozolol, have also been analyzed with this extraction procedure. Deshmukha *et al.*[71] analyzed hair from athletes previously screened by ELISA and with positive results for these compounds. Hair was incubated with 1 M NaOH at 95 °C for 15 min. The extracts were submitted to a liquid–liquid extraction with pentane and analyzed by LC–MS/MS. LOQs were 0.5 pg mg^{-1} for stanozolol and 3.0 pg mg^{-1} for nandrolone.

Alkaline digestion is being applied to the analysis of new psychoactive substances (NPS). Barroso *et al.*[72] reported the analysis of three piperazines 1-(3-trifuoromethylphenyl) piperazine (TFMPP), 1-(3-chlorophenyl)piperazine (mCPP) and 1-(4-methoxyphenyl)piperazine (MeOPP). Extraction was performed by incubation with 1 M NaOH at 50 °C for 40 min, followed by mixed-mode solid-phase extraction. The analytes were derivatized with *N*-methyl-*N*-(trimethylsilyl) trifluoroacetamide with 5% trimethylchlorosilane and analyzed by GC–MS-SIM. Lower LOQ was 0.05 ng mg^{-1}.

4.2.6 Assisted Extraction Methods

There are some assisted extraction methods that considerably reduce the time of analysis and at the same time can improve the extraction yield. They include ultrasonic-assisted extraction (UAE), microwave assisted extraction (MAE), pressurized fluid extraction (PFE), matrix solid-phase dispersion (MSPD) and micropulverized extraction (MPE). Supercritical fluid extraction was applied in the past.

UAE is the most widely used assisted extraction method. Ultrasonic treatment alters the hair structure, enhancing the release/extraction of the compounds. As previously explained, the incubation time can be reduced from 18 h (one night) to 1–4 h. UAE is mainly applied in the methanolic extraction (see Section 4.2.1), but there are some papers reporting UAE with buffer solutions,[33] and with acidic solutions.[47,51,55]

In the MAE method, the solvent will penetrate the hair matrix and cause the hair to swell. At the same time the microwaves act selectively on the hair cells, causing them to break and release the compounds present, which are dissolved in the extraction solvent. One of the issues of this extraction assisted method is the stability of the analytes under the microwave conditions. Chang *et al.*[57] studied the stability of 6-AM in methanol : TFA

(8.5 : 1.5) under incubation for 1–6 min in a microwave at 700 W. The extent of degradation increased with time. More than 14% was observed after 4 min, but less than 5% was found after 3 min. In the same study, they also compared the recovery after overnight incubation and MAE for 3 min for MA, MDMA, ketamine and metabolites and opiates (6-AM, morphine, and codeine). 0.5 mL of methanol : TFA (8.5 : 1.5) were added to the hair samples. One group was incubated at 45 °C overnight and the other was submitted to MAE at 700 W for 3 min. In both cases the extraction solution was evaporated to dryness, reconstituted in water : methanol (99 : 1) and injected into a LC–MS/MS. The efficiency of extraction after MAE was greater than 100% for all the analytes, even for MDMA, morphine, dihydro-norketamine, and codeine was significantly greater than 150%. MAE proved to be able to not only reduce the time, but also increase the extraction yield. Wietecha-Posłuszny *et al.*[73] developed a method based on simultaneous alkaline hair microwave-assisted extraction and clean-up for the analysis of tricyclic antidepressants. Hair samples were added with 4 mL 1 M NaOH, as extraction solution, followed by 5 mL of *n*-hexane–isoamyl alcohol (99 : 1), as purification solvent. Incubation was performed in a microwave at 60 °C for 40 min. The organic layer was separated by centrifugation and evaporated to dryness. LOQs ranged from 0.9 to 4.0 ng mg^{-1} for the different drugs. The same authors later applied the same method to the analysis of six benzodiazepines,[74] but in this case the extraction was performed with a borate buffer and no simultaneous clean-up was required.

PFE uses elevated temperature and pressure to reduce the time of analysis and increase the efficiency of the extraction procedure. With PFE, samples are extracted with solvent in a pressurized cell. Samples are placed in the cell body along with an inert granular matrix. Solvent is pumped into the cell until it is filled, and then allowed to equilibrate under static conditions. Then, the extract is pumped out of the cell and is collected for analysis.[75] Vincenti *et al.*[76] reported a method for the analysis of ethyl-glucuronide (EtG) based on decontamination and extraction by PFE. The cell was filled with diatomaceous earth and decontamination was firstly performed with a sequence of phosphate buffer 0.25 mM at pH 6.5, 2-propanol and acetone. After decontamination, the extraction was performed within a single cycle using a mixture of water/methanol (80/20). They compared the efficacy of this PFE with an overnight incubation at room temperature with the same extraction solvent. An average positive difference of +32% was observed when the extraction was performed by PLE. They also studied the effect of pulverization, and no differences were found between cut and powdered hair when the extraction is by PFE, thus suggesting that high pressure and temperature used in PLE facilitate the extraction of the analytes from the hair, independently if it has been cut or powdered. The same authors applied PFE to the analysis of new psychoactive substances including cathinones and synthetic cannabinoids.[77] The use of PFE significantly reduced the time of analysis: decontamination, incubation, clean-up, and liquid chromatography–mass spectrometry (LC–MS) analyses were carried out in approximately 45 min. They also applied this extraction assisted method to the analysis of cannabis compounds (THC, CBD, CBN, and THC-COOH). The eluent phase, for these analytes, was water modified with the surfactant sodium dodecyl sulphate. LOQs were 0.1 pg mg^{-1} for THC-COOH and 2 pg mg^{-1} for the other compounds.

MSPD consists of blending or dispersing the hair sample with an appropriate sorbent or solid support to obtain a homogeneous mixture. The mixture is then transferred and packed inside a cartridge, or a disposable syringe. The compounds of interest are subsequently eluted with the extraction solution. Different silica-, magnesium-, or aluminum-based materials can be used as dispersing agent. Míguez-Framil *et al.*[61] reported a method for the analysis of cocaine and opiate compounds based on enzymatic hydrolysis assisted by MSPD. Hair samples (50 mg) were dispersed/blended with alumina (2.25 g). The mixture was packaged inside a disposable MSPD syringe, and the extraction was achieved by passing an enzymatic solution (6 mL of 1 mg mL^{-1} pronase E in 1.4 M/1.4 M Tris/HCl, pH 7.3). They simultaneously performed an on-column clean-up step, by attaching a previously conditioned SPE cartridge to the end of the MSPD syringe. The whole extraction procedure, including sample blending/dispersion, MSPD syringe package, elution with enzymatic solution, and extract clean-up took only 30 min. A different extraction solvent was also applied for the analysis of cocainics and opiates.[78] In this case, the MSPD approach consisted of using alumina (1.80 g) as a dispersing agent and 0.6 M HCl (4 mL) as extracting solvent.

In MPE powdering and extraction of drugs from the hair are simultaneously performed, which leads to considerable time saving, with the same extraction yield or even better in most of the cases. For example, Miyaguchi *et al.*[79] during the development of a method for the determination of zolpidem and other psychoactive drugs, compared the efficacy of MPE for 10 min with 45% aqueous ammonium phosphate (pH 8.4) *versus* overnight incubation with methanol. The highest extraction efficiencies for all the analytes were achieved using MPE, even though the extraction time was shorter. The same method[80] was later applied to the enantioselective determination of (*R*)-zopiclone and (*S*)-zopiclone. In this case the clean-up procedure involved a salting-out assisted liquid–liquid extraction with acetonitrile to minimize the degradation of zopiclone. They compared the efficacy of the MPE with the overnight incubation with the same buffer and the recovery of the longer incubation time was only $0.58 \pm 0.12\%$ of the maximum recovery achieved by the MPE method (10 min incubation) due to the decomposition in the phosphate buffer. Another advantage of MPE is the small amount of hair sample required for the analysis, when applying very sensitive instrumentation, such as LC–MS/MS, Orbitrap, *etc.*

4.2.7 Non-extraction Methods

Currently there are very sensitive and advanced instruments, which allow the direct detection of drugs in the integer hair shaft, thus avoiding the extraction of the drugs from the hair matrix. They include the matrix-assisted laser desorption/ionization (MALDI) or the direct analysis in real time (DART) coupled to different mass spectrometers in both cases.

MALDI couple to ionization-mass spectrometry imaging (MALDI-MSI) has been applied to the analysis of cocaine,[81] tilidine,[82] MA,[83] olanzapine,[84] or even synthetic cannabinoids,[85] MALDI coupled to Fourier transform mass spectrometry (MALDI-FTMS) has been applied to the analysis of cocaine and metabolites and cannabinoids[86] or to the analysis of ketamine;[87] while some papers report the analysis of methoxyphenamine[88] or MA[89] by MALDI coupled to time-of-flight mass spectrometry (MALDI-TOF-MS).

More recently, the direct analysis in real time mass spectrometry (DART-MS) technique has also been applied in hair analysis.[90,91]

The drawback of these techniques is that they require very expensive instruments, which are not available in most of the laboratories. In addition, they are not yet sufficiently well established, as evidenced by the fact that laboratories reporting analyses with these techniques are applying the classical sample preparation methods in their routine.

4.3 Comparison of Extraction Methods

During method development several parameters must be evaluated and optimized, one of them being the extraction procedure. The efficacy of different extraction solvents and conditions should be compared before selecting the best extraction protocol.

Kim *et al.*[92] compared the efficacy of three extraction procedures during the development of an analytical method for antiepileptic drugs, including barbiturates, phenytoin and topiramate. Hair was finely cut, and the extraction was performed with: (a) methanol, (b) methanol containing 1% HCl, and (c) an acidic solution. There was no significant difference in the matrix effects among these three methods. Nevertheless, they chose methanol extraction because recoveries were clearly higher with this method. Analysis was performed by LC–MS/MS and the LOQs were 0.02 $ng\,mg^{-1}$ for topiramate and 0.5–1 $ng\,mg^{-1}$ for the others.

Tassoni *et al.*[7] compared the concentrations of THC, CBD, and CBN in two aliquots of the same samples after performing two extraction procedures. (A) Hair samples were subjected to basic hydrolysis with NaOH 1 M, at 95 °C for 15 min; (B) the hair was previously incubated overnight in 0.1 M HCl at 50 °C, followed by the same basic hydrolysis. The results showed that there were no statistical differences between the direct basic digestion and the hair previously submitted to acidic extraction. THC mean concentrations were 0.73 ± 0.52 $ng\,mg^{-1}$ and 0.77 ± 0.54 $ng\,mg^{-1}$ for groups A and B, respectively. The same profile was found for CBN and CBD. The absence of statistical differences in the concentrations of the two groups indicated that cannabis compounds were not extracted in acidic conditions. The authors concluded that diluted HCl should be avoided for the analysis of cannabis compounds.

Pujadas *et al.*[93] compared four extraction solvents during the development of a method for the analysis of AP MA and methylenedioxy derivatives. Two alkaline digestions with 1 M sodium sulfide (Na_2S) and 1 M NaOH and two extractions in acidic conditions with methanol : 5 M HCl (20 : 1) or methanol : TFA (9 : 1). Digestion with 1 M sodium sulfide was selected because it gave the best results on terms of recovery of analytes from hair matrix, clean-up of extracts and absence of chromatographic interferences.

Proficiency test programs are very useful to compare the efficiency of one extraction method *versus* others.

4.3.1 Comparison of Extraction Methods for Cocaine Compounds, Opiates, Amphetamines, Cannabis, Benzodiazepines, Methadone, and Tramadol Based on Data Obtained from Proficiency Tests

This comparison will be based on the data of the proficiency tests organized by SoHT in 2021.[94] That year 68 laboratories participated in the two rounds organized. They received three real hair samples, obtained from drug consumers, per round (A1, B1, C1, A2, B2 and C2) along with a result form, where they were asked to provide information related to: experience in hair testing, extraction methods, clean-up procedures and instrumentation used for screening, and confirmation analysis.

Participants applied different extraction procedures to extract the compounds of interest from the hair matrix. Table 4.1 shows the number of laboratories applying each extraction method for each compound and the number which reported results, because not all the laboratories analyzed all the compounds.

The methods mostly applied were the extractions with methanol, buffers, and acidic solutions. Only two laboratories used an enzymatic digestion, while alkaline digestion was mainly used for the analysis of cannabis compounds. Six participants did not provide information on their methods. Most of the laboratories extracted simultaneously cocaine compounds, opiates, methadone, tramadol, benzodiazepines, and amphetamines.

4.3.1.1 Qualitative Results

Table 4.2 shows the efficacy of each extraction method to identify the presence of the different compounds. it shows false results, both positive and negative, obtained with each method. No false positive results were reported. All participants successfully identified negative samples.

The same did not happen with false negatives (FN). Some participants did not identify some compounds that were present, especially when concentrations were low. Nevertheless, one of the questions to be addressed when interpreting qualitative results is the cut-off values. Some participants reported higher cut-offs than the median concentration and, in those cases,

Table 4.1 Number of laboratories applying each extraction procedure.

	Extraction procedure							
	Compounds	N^a	Methanol	Buffer solutions	Acidic solutions	Enzyme digestion	Alkaline digestion	Non-reported
Cocaine compounds	Cocaine	68	26	18	16	2	0	6
	Benzoylecgonine	67	25	18	16	2	0	6
	Cocaethylene	58	23	16	12	2	0	5
	Norcocaine	48	18	13	11	1	0	5
Opiate compounds	6-Acetylmorphine	66	25	18	15	2	0	6
	Morphine	67	26	18	15	2	0	6
	Codeine	67	26	18	15	2	0	6
Amphetamine compounds	Amphetamine	66	24	17	14	2	3	6
	Methamphetamine	66	24	17	14	2	3	6
	MDMA	66	24	17	14	2	3	6
Methadone and metabolite	Methadone	55	22	16	10	2	0	5
	EDDP	38	23	14	7	2	0	5
Benzodiazepines	Diazepam	48	20	15	8	2	0	3
	Nordiazepam	47	20	14	8	2	0	3
Tramadol	Tramadol	40	17	13	5	1	0	4
Cannabis compounds	THC	53	15	7	4	0	20	7
	THC-COOH	29	8	1	0	0	15	5

[a] Number of laboratories reporting results for this compound.

Table 4.2 Qualitative results for all the compounds. Incidence of false results.

	Compounds	Sample A1 median[a] FR (method)[b]	Sample B1 median FR (method)	Sample C1 median FR (method)	Sample A2 median FR (method)	Sample B2 median FR (method)	Sample C2 median FR (method)
Cocaine compounds	Cocaine	3.66 ng mg^{-1}	2.51 ng mg^{-1}	3.50 ng mg^{-1}	4.46 ng mg^{-1}	<0.50 ng mg^{-1}	2.46 ng mg^{-1}
	Benzoylecgonine	6.11 ng mg^{-1}	4.92 ng mg^{-1}	5.01 ng mg^{-1}	6.23 ng mg^{-1}	0.17 ng mg^{-1} 3(B) 3(C) 1, **1**(E)	9.23 ng mg^{-1} 1(A)
	Cocaethylene	0.06 ng mg^{-1} 2, **1**(A) 2(B) 2(C) 1, **1**(E) 1, **1**(F)	<0.05 ng mg^{-1}	<0.05 ng mg^{-1}	<0.05 ng mg^{-1}	<0.05 ng mg^{-1}	<0.05 ng mg^{-1}
	Norcocaine	0.05 ng mg^{-1} 2(A) 3, **1**(B) 1(C) 2, **1**(F)	<0.05 ng mg^{-1}	0.05 ng mg^{-1} 2(A) 3, **1**(B) 3(C) 2, **1**(F)	0.09 ng mg^{-1} 3(A) 1(B) 1, **1**(F)	<0.05 ng mg^{-1}	0.07 ng mg^{-1} 4(A) 2(B) 1(C) 1, **1**(F)
Opiate compounds	6-Acetylmorphine	10.65 ng mg^{-1}	4.84 ng mg^{-1} 1(A)	13.26 ng mg^{-1}	11.40 ng mg^{-1}	1.84 ng mg^{-1}	4.23 ng mg^{-1}
	Morphine	3.20 ng mg^{-1}	1.53 ng mg^{-1}	2.24 ng mg^{-1}	2.74 ng mg^{-1}	0.49 ng mg^{-1} 1(B)	1.92 ng mg^{-1}
	Codeine	0.97 ng mg^{-1}	0.36 ng mg^{-1} 1, **1**(A) 1(B)	0.46 ng mg^{-1} 1(B)	0.76 ng mg^{-1} 1(B)	0.18 ng mg^{-1} 4, **4**(A) 7, **6**(B) 1, **1**(C) 1, **1**(F)	0.42 ng mg^{-1} 2(B)

Amphetamine compounds	Amphetamine	<0.2 ng mg^{-1}	<0.2 ng mg^{-1}	0.70 ng mg^{-1} 1(B)	0.70 ng mg^{-1} 1(A) 1(B) 1(E)	<0.2 ng mg^{-1}	<0.2 ng mg^{-1}
	Methamphetamine	<0.2 ng mg^{-1}	<0.2 ng mg^{-1}	0.45 ng mg^{-1} 1(B)	0.54 ng mg^{-1} 1(A) 1(B)	<0.2 ng mg^{-1}	<0.2 ng mg^{-1}
	MDMA	<0.2 ng mg^{-1}	<0.2 ng mg^{-1}	<0.20 ng mg^{-1}	0.27 ng mg^{-1} 3, **1**(A) 1(B)	<0.2 ng mg^{-1}	<0.2 ng mg^{-1}
Methadone & metabolite	Methadone	2.11 ng mg^{-1}	0.49 ng mg^{-1} 1, **1**(A)	0.50 ng mg^{-1} 1(A)	0.78 ng mg^{-1}	1.64 ng mg^{-1}	2.01 ng mg^{-1}
	EDDP	0.20 ng mg^{-1} 1(A) 1(B)	0.05 ng mg^{-1} 4, **2**(A) 2, **2**(B) 1, **1**(F)	0.07 ng mg^{-1} 3, **1**(A) 2, **2**(B) 1, **1**(F)	0.076 ng mg^{-1} 2, **1**(A) 2, **1**(B) 1, **1**(C)	0.15 ng mg^{-1} 1, **1**(A) 2, **1**(B) 1, **1**(C)	0.16 ng mg^{-1} 1, **1**(B) 1, **1**(C)
Benzodiazepines	Diazepam	0.07 ng mg^{-1} 1, **1**(A) 2, **2**(B) 3, **2**(C)	0.032 ng mg^{-1} 6, **6**(A) 3, **2**(B) 3, **2**(C) 2, **1**(F)	0.51 ng mg^{-1}	0.087 ng mg^{-1} 1(A) 2, **1**(B) 2, **2**(C)	0.049 ng mg^{-1} 3, **3**(A) 3, **2**(B) 3, **3**(C)	0.037 ng mg^{-1} 4, **4**(A) 3, **2**(B) 4, **3**(C) 1, **1**(F)
	Nordiazepam	0.04 ng mg^{-1} 3, **3**(A) 3, **2**(B) 3, **1**(C)	0.017 ng mg^{-1} 7, **6**(A) 5, **2**(B) 5, **2**(C) 2, **2**(F)	0.04 ng mg^{-1} 4, **3**(A) 1, **1**(B) 3, **1**(C) 1, **1**(F)	0.03 ng mg^{-1} 4, **4**(A) 3, **3**(B) 3, **2**(C) 1, **1**(F)	0.031 ng mg^{-1} 4, **4**(A) 2, **2**(B) 4, **4**(C) 1, **1**(F)	0.023 ng mg^{-1} 4, **4**(A) 5, **4**(B) 4, **2**(C) 1, **1**(F)

Table 4.2 (*Continued*)

	Compounds	Sample A1 median[a] FR (method)[b]	Sample B1 median FR (method)	Sample C1 median FR (method)	Sample A2 median FR (method)	Sample B2 median FR (method)	Sample C2 median FR (method)
Tramadol	Tramadol	0.87 ng mg^{-1}	0.43 ng mg^{-1}	0.23 ng mg^{-1} 1(A) 1(B) 1, **1**(C)	0.54 ng mg^{-1}	0.34 ng mg^{-1} 2(B) 1, **1**(C)	0.24 ng mg^{-1} 3(B) 1, **1**(C)
Cannabis compounds	THC	<0.05 ng mg^{-1}	0.10 ng mg^{-1}	0.12 ng mg^{-1} 1(B)	0.13 ng mg^{-1} 1(A)	<0.05 ng mg^{-1}	<0.05 ng mg^{-1}
	THC-COOH	<0.02 pg mg^{-1}	0.29 pg mg^{-1} 3, **2**(A) 1(D) 2, **1**(F)	0.42 pg mg^{-1} 2, **2**(A) 1(D) 2, **1**(F)	0.32 pg mg^{-1} 2, **2**(A) 2(D) 2, **1**(F)	0.30 pg mg^{-1} 2, **2**(A) 3(D) 3, **2**(F)	<0.02 pg mg^{-1}

[a] Median: median concentration (ng mg^{-1}) reported by the participants.
[b] FR (method): false result (extraction method: method A: methanol extraction; method B: buffer extraction; method C: acidic solutions; D: alkaline digestion; E: enzymatic digestion; F: non-reported method). Results from laboratories reporting cut-offs higher than the median concentrations are in bold.

the negative result cannot be considered false negative (they appear in bold in Table 4.2).

All samples, but B2, contained cocaine at a concentration higher than 0.5 ng mg^{-1}. All participants, applying different extraction methods, successfully identified cocaine in all the samples. So, no method was better than any other. BE was present in all the samples at concentrations higher than 0.05 ng mg^{-1}, even in sample B2, where cocaine concentration was lower than 0.5 ng mg^{-1}. For this reason, negative results in this sample cannot be considered FN because participants did not report a positive result for the metabolite in absence of the main compound. All the extraction methods were also useful to extract BE from the hair, since only one FN was reported in sample C2 after methanolic extraction.

Opiate compounds were present in all the samples. All extraction solvents were able to extract 6-AM and morphine. Only one FN for 6-AM was reported for sample B1 after methanolic extraction and another FN for morphine in sample B2 after buffer extraction.

Positive results for AP and MA at higher concentration than 0.2 ng mg^{-1} were only found in samples C1 and A2, while MDMA was only present in sample A2. No false results were reported after acidic extraction (14 laboratories) and alkaline digestion (3 laboratories) and the incidence of FN after applying the other extraction methods was very low.

Methadone was positive in all the samples. Only two FN were reported, and they came from the same laboratory, which applies a cut-off of 0.5 ng mg^{-1}, higher than that recommended by SoHT (0.2 ng mg^{-1}). 2-Ethylidene-1,5-dimethyl-3,3-diphenylpyrrolidine (EDDP) concentrations were equal to or higher than 0.05 ng mg^{-1}, the recommended cut-off by EWDTS.[10] None of the extraction methods demonstrated greater efficacy at low concentrations. In addition, most of the negative cases, applied high cut-offs.

SoHT does not propose cut-off values for benzodiazepines, but EWDTS recommends a concentration of 0.05 ng mg^{-1} for the confirmation tests for all the benzodiazepines. Diazepam concentrations higher than 0.05 ng mg^{-1} were found in samples A1, C1, and A2. All the laboratories, independently of the extraction method, identified diazepam at high concentration (sample C1). Almost all the cases with FN applied high cut-off values. Nevertheless, from the results it could be deduced that the lower efficiency was found with acidic solutions, 3FN out of 8 in sample A1, and the best data were obtained after methanol extraction, 1FN out of 20 in sample A1. Nordazepam concentrations were below 0.05 ng mg^{-1} in all the samples.

Tramadol median concentrations were higher than 0.2 ng mg^{-1}, the recommended cut-off by SoHT. As happened with the previous compounds, all the laboratories identified tramadol at high concentrations; the differences appeared when the concentrations were close to the cut-off. Only one laboratory out of 17, reported a FN result after methanolic extraction. Up to three FN results were reported by 13 laboratories performing buffer extraction. Only five laboratories applying acidic extraction analyzed tramadol and one of them reported FN results for samples C1, B2, and C2.

Positive results for THC at higher concentrations than 0.05 ng mg^{-1} were found only in two samples (C1 and A2). All the participants identified THC in both samples, except for one laboratory performing buffer extraction (sample C1) and other using methanolic solutions (sample C1). No FN were obtained after acidic extraction (four laboratories applied this method) and alkaline digestion (20 laboratories).

THC-COOH analysis can be extremely challenging due to the low concentrations typically found in hair. In fact, only 29 out of 68 participants reported results for this compound. Seven laboratories applied a cut-off higher than 0.2 pg mg^{-1} and all of them reported negative results for all the samples despite the concentration. These data highlight the importance of applying adequate cut-offs to avoid false negative results. Extraction was mainly performed with alkaline digestion (15 laboratories) and methanolic extraction (8 laboratories). Data obtained showed that the risk of reporting FN results for THC-COOH was lower after alkaline digestion than after methanol extraction (see Figure 4.1).

4.3.1.2 *Quantitative Results*

In the evaluation of quantitative results, enzymatic extraction (only two laboratories applied this extraction method) and non-reported methods have not been considered.

Cocaine concentrations reported after applying the three extraction procedures (methanol, buffer, and acidic solutions) were very close (see Figure 4.2A), although recoveries are slightly higher after acid extraction and lower after methanol extraction, especially in sample C2. Even closer are the results obtained for BE (see Figure 4.2B), but the best results were obtained after buffer extraction, except sample C2, with scattered results. The same profiles were found in norcocaine and cocaethylene (Figure 4.2C).

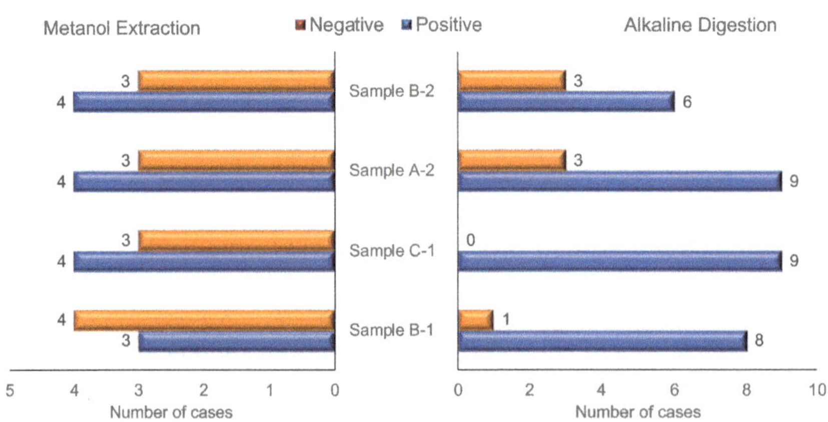

Figure 4.1 Efficiency of methanol incubation *versus* alkaline digestion for the extraction of THC-COOH from hair samples. Number of positive and negative results obtained with each method.

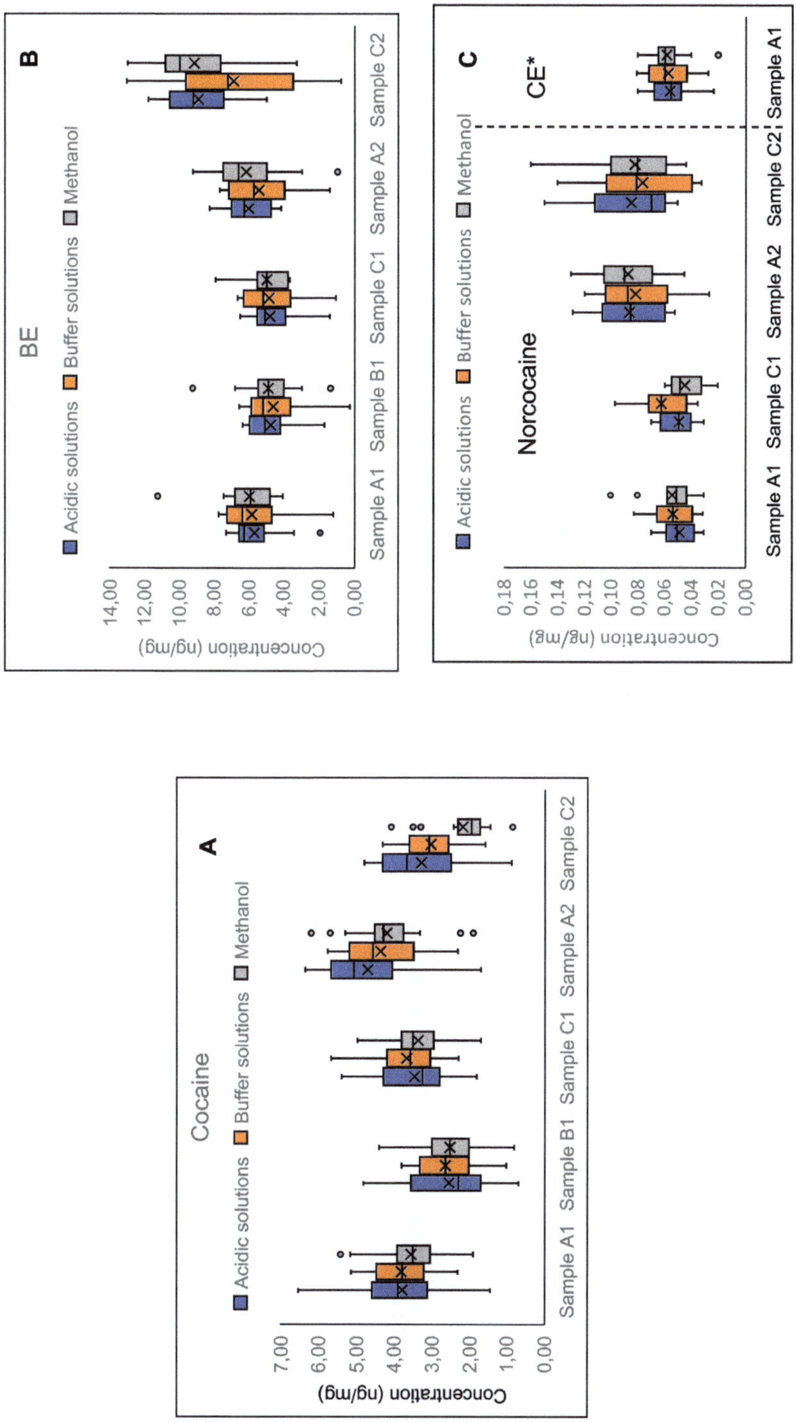

Figure 4.2 Comparison of different extraction solutions (acidic, buffer, and methanol) for the analysis of cocaine compounds.

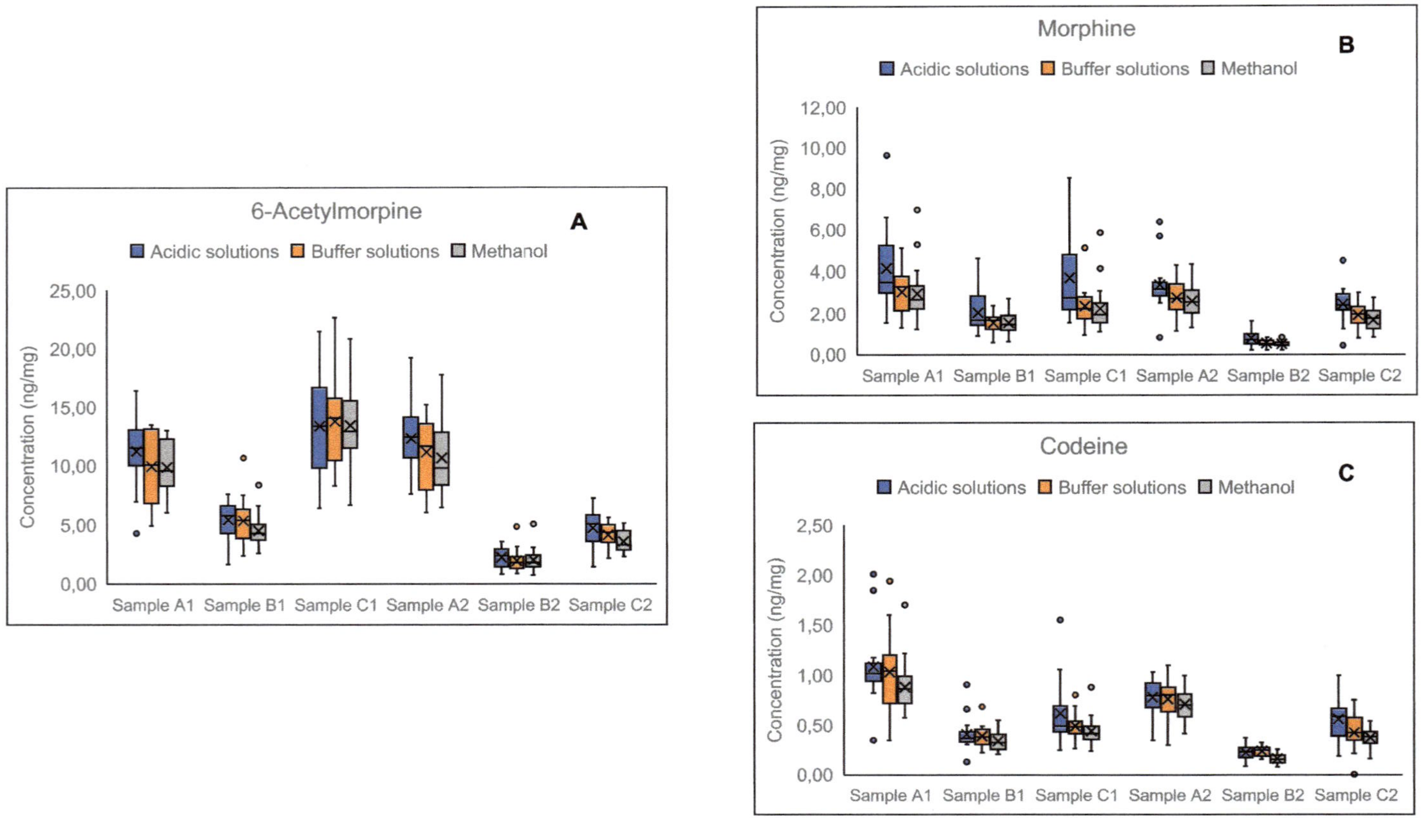

Figure 4.3 Comparison of different extraction solutions (acidic, buffer, and methanol) for the analysis of opiates.

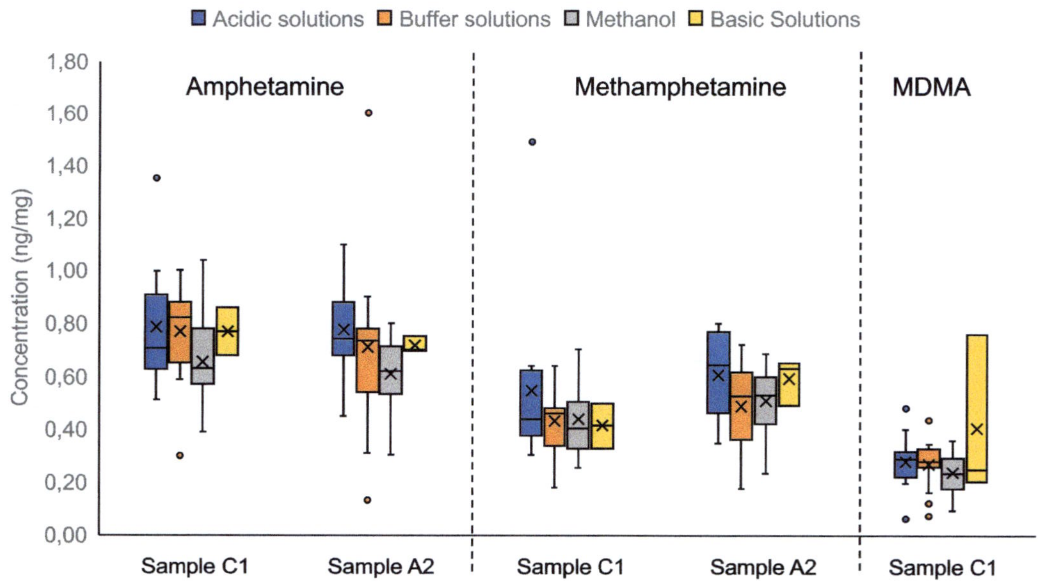

Figure 4.4 Comparison of different extraction solutions (acidic, buffer, and methanol) for the analysis of amphetamine related compounds (amphetamine, methamphetamine, and MDMA).

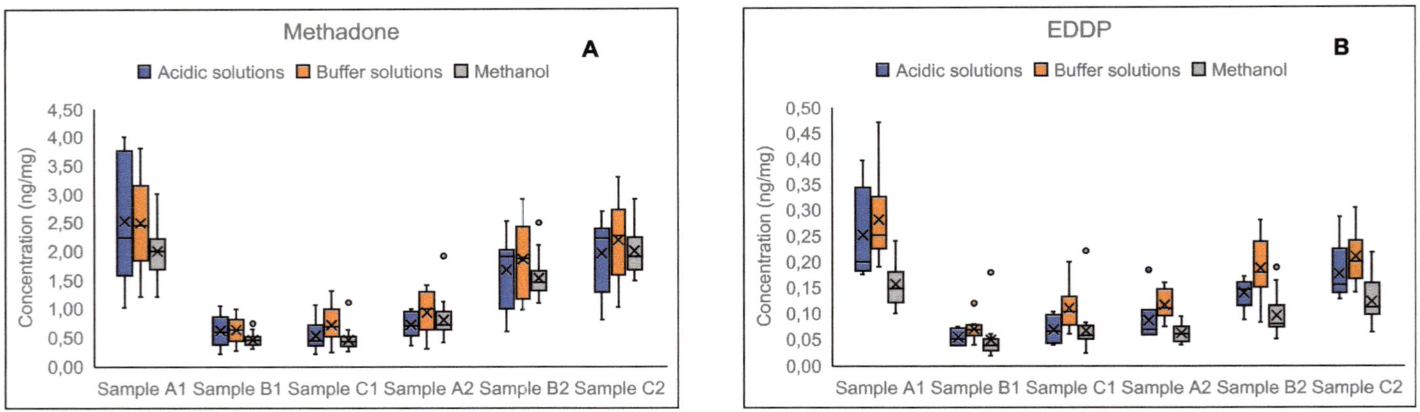

Figure 4.5 Comparison of different extraction solutions (acidic, buffer, and methanol) for the analysis of methadone and its metabolite EDDP.

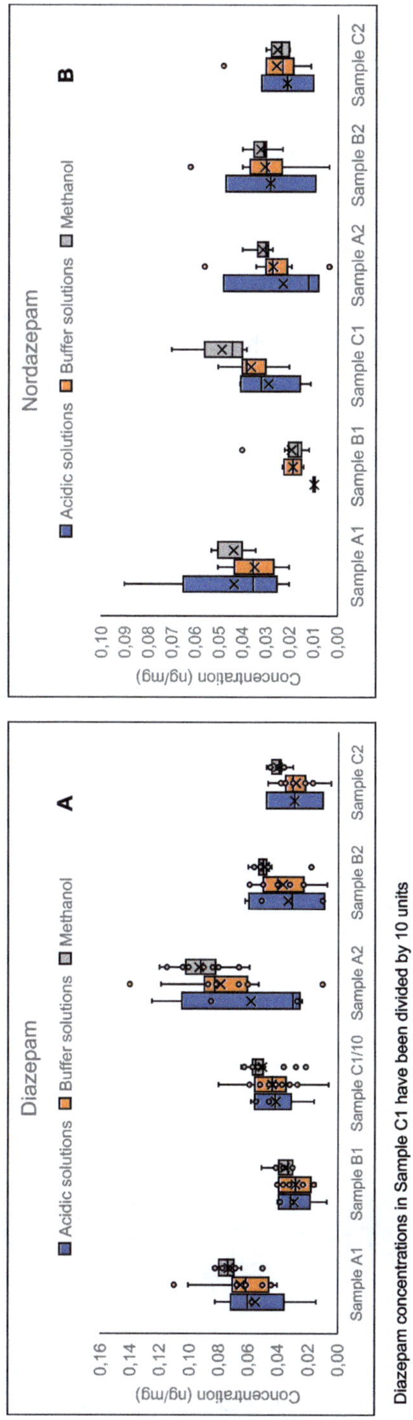

Figure 4.6 Comparison of different extraction solutions (acidic, buffer, and methanol) for the analysis of diazepam and nordazepam.

In the study of opiates, 6-AM concentrations were also very close, thus demonstrating that no method is better than another to extract this compound. The highest recoveries of morphine were obtained after acidic extraction, although they were scattered in samples A1 and C1. It could be thought that these higher concentrations derive from the hydrolysis of 6-AM in the acidic solution. This hypothesis would lead to a decrease in 6-AM, but this did not occur (see Figure 4.3A).

Alkaline digestion is expected to be more effective than other extraction methods for the analysis of amphetamine related compounds. For this reason, it has been included in the comparison, although only three laboratories applied this extraction solvent. However, the results were not better than those obtained with the other methods. In addition, they were very scattered in the sample C1 (see Figure 4.4).

Extraction with buffer solutions proved to be the most effective for the analysis of methadone and its metabolite EDDP (see Figure 4.5) and the lowest concentrations were found after methanol extraction.

Extraction with acidic solutions is not useful for the analysis of benzodiazepines. Concentrations reported were very scattered although only 3–4 laboratories applied this extraction method. Recoveries of diazepam and nordazepam were, in general, higher after methanol extraction than after buffer extraction (see Figure 4.6).

A similar profile to benzodiazepines was found with tramadol. Only five laboratories extracted this compound with acidic solutions and the results were very scattered. Concentrations were very close, but recoveries were slightly better after extracting with buffer solutions than with methanol (see Figure 4.7).

Most of the laboratories performed an alkaline digestion for the analysis of THC (22 out of 53), but others perform a simultaneous extraction with the other groups of compounds. Figure 4.8A shows that concentrations from

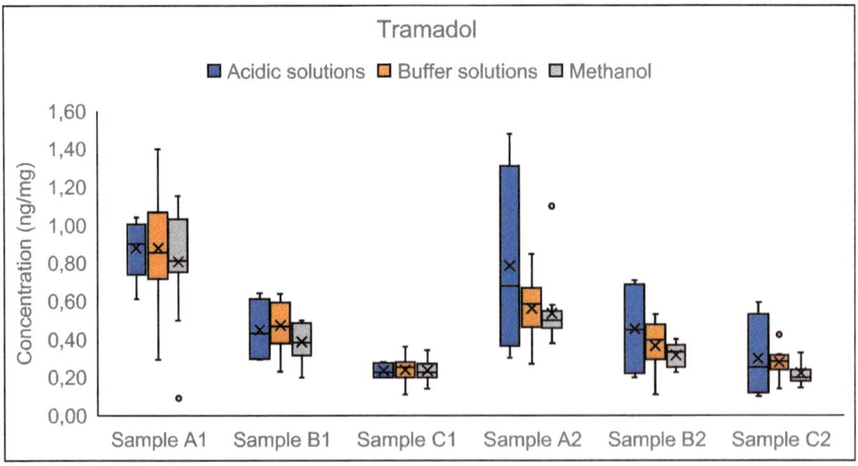

Figure 4.7 Comparison of different extraction solutions (acidic, buffer, and methanol) for the analysis of tramadol.

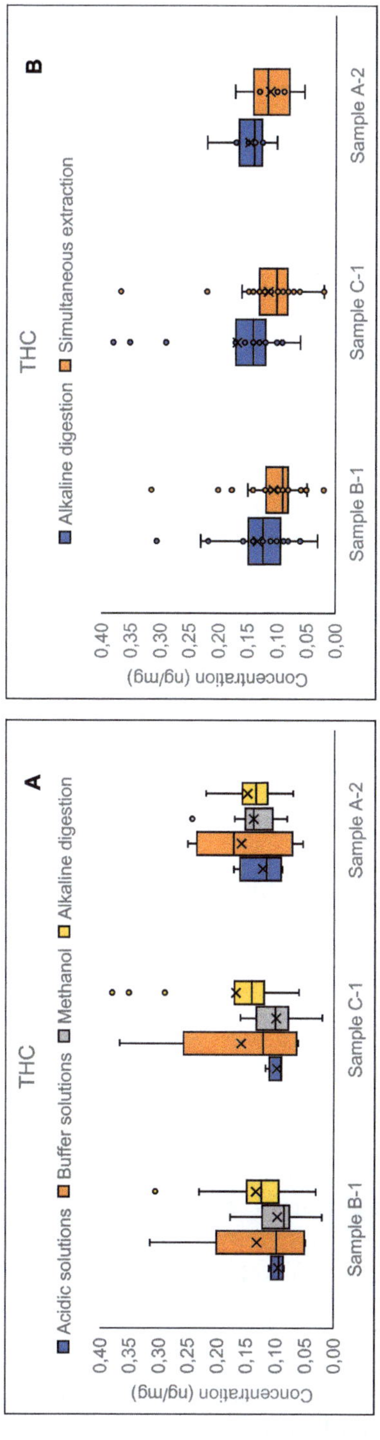

Figure 4.8 Comparison of different extraction procedures for the analysis of THC: (A) acidic solutions, buffer solutions, methanol, and alkaline digestion; (B) target analysis by alkaline digestion *versus* simultaneous analysis with other extraction solvents.

Table 4.3 Quantitative results for THC-COOH.

	Alkaline digestion				Methanol extraction			
	Concentration (pg mg^{-1})				Concentration (pg mg^{-1})			
	N^a	Median	Range	CV%	N^a	Median	Range	CV%
Sample B1	9	0.31	0.18–0.68	41.4	3	0.21	0.18–0.45	43.3
Sample C1	9	0.54	0.30–1.29	50.5	4	0.65	0.3–0.92	34.9
Sample A2	10	0.38	0.24–0.62	27.7	4	0.32	0.16–0.72	56.5
Sample B2	7	0.36	0.19–0.48	29.9	4	0.28	0.10–0.50	49.7

[a] Number of laboratories reporting positive results.

laboratories applying buffer extraction were very scattered. As expected, alkaline digestion provided higher recoveries than the extraction with methanol and acidic solutions, except for sample C1, where the concentrations were very close. When comparing the concentrations reported after the simultaneous extraction *versus* target analysis by alkaline digestion, the difference is even more evident (see Figure 4.8B).

The low number of participants reporting results for THC-COOH make the comparison of extraction methods very difficult. Nevertheless, concentrations were slightly higher after alkaline digestion than after methanol extraction (see Table 4.3), except for sample C1. Another advantage of the alkaline extraction is the lower coefficients of variations, in spite of the number of laboratories being higher.

Abbreviations

6-AM	6-Monoacetylmorphine
AP	Amphetamine
BE	Benzoylecgonine
BSTFA	*N*,*O*-Bis(trimethylsilyl)trifluoroacetamide
BUP	Buprenorphine
CBD	Cannabidiol
CBN	Cannabinol
CE	Cocaethylene
DART	Direct analysis in real time
diOHTHC	8-β-11-DihydroxyTHC
EDDP	2-Ethylidene-1,5-dimethyl-3,3-diphenylpyrrolidine
EME	Ecgonine methyl ester
EtG	Ethyl-glucuronide
ELISA	Enzyme-linked immunosorbent assay
EWDTS	European Workplace Drug Testing Society
FN	False negative
FPIA	Fluorescence polarization immunoassay
GC–MS	Gas chromatography/mass spectrometry
GC–MS/MS	Gas chromatography tandem mass spectrometry
GHB	Gamma hydroxybutyric acid

HCl	Hydrochloric acid
LC–MS/MS	Liquid chromatography–tandem mass spectrometry
LOD	Limit of detection
LOQ	Limit of quantification
LSD	Lysergic acid diethylamide
MA	Methamphetamine
MAE	Microwave assisted extraction
MALDI	Matrix-assisted laser desorption/ionization
MDA	Methylenedioxyamphetamine
mCPP	1-(3-Chlorophenyl)piperazine
MDMA	Methylenedioxymethamphetamine
MPE	Micropulverized extraction
MSPD	Matrix solid-phase dispersion
NaOH	Sodium hydroxide
NBUP	Norbuprenorphine
OH-THC	11-Hydroxy-THC
Orbitrap	Liquid chromatography Orbitrap mass spectrometry
PFE	Pressurized fluid extraction
PMA	*para*-Methoxyamphetamine
PMMA	*para*-Methoxy-*N*-methylamphetamine
RIA	Radioimmunoassay
SoHT	Society of Hair Testing
SPE	Solid phase extraction
TFA	Trifluoroacetic acid
THC-COOH	11-Nor-9-carboxy-Δ^9-tetrahydrocannabinol
THC	Δ^9-Tetrahydrocannabinol
UAE	Ultrasonic-assisted extraction
UHPLC/HRMS	Ultrahigh-performance liquid chromatography-high resolution mass spectrometry
UPLC-TOF-MS	Ultrahigh-performance liquid chromatography-time-of-flight mass spectrometry

References

1. Society of Hair Testing, 2021 SoHT Consensus on drugs of abuse (DoA) testing in hair, available from https://www.soht.org/images/pdf/Consensus_DoA_2021.pdf, accessed November 11, 2022.
2. J. C. Aijala, W. Wu and A. P. DeCaprio, Assessing Hair Decontamination Protocols for Diazepam, Heroin, Cocaine and delta9-Tetrahydrocannabinol by Statistical Design of Experiments, *J. Anal. Toxicol.*, 2021, **45**(5), 498–505.
3. D. Mantinieks, P. Wright, M. Di Rago and D. Gerostamoulos, A systematic investigation of forensic hair decontamination procedures and their limitations, *Drug Test. Anal.*, 2019, **11**(10), 1542–1555.
4. R. Erne, M. R. Baumgartner and T. Kraemer, Insights into the Decontamination of Cocaine-Positive Hair Samples, *J. Anal. Toxicol.*, 2021, **45**, 713–721.

5. G. A. A. Copper, R. Kronstrand and P. Kintz, Society of Hair Testing guidelines for drug testing in hair, *Forensic Sci. Int.*, 2012, **218**, 20–24.
6. A. M. Baumgartner, P. F. Jones, W. A. Baumgartner and C. T. Black, Radioimmunoassay of hair for determining opiate-abuse histories, *J. Nucl. Med.*, 1979, **20**(7), 748–752.
7. G. Tassoni, M. Cippitelli, G. Ottaviani, R. Froldi and M. Cingolani, Detection of Cannabinoids by ELISA and GC–MS Methods in a Hair Sample Previously Used to Detect Other Drugs of Abuse, *J. Anal. Toxicol.*, 2016, **40**(6), 408–413.
8. J. C. Cheong, S. Suh, B. J. Ko, J. Lee, J. Y. Kim, Y. J. Suh and M. K. In, Screening Method for the Detection of Methamphetamine in Hair Using Fluorescence Polarization Immunoassay, *J. Anal. Toxicol.*, 2013, **37**, 217–221.
9. G. Jakobsson and R. Kronstrand, Segmental analysis of amphetamines in hair using a sensitive UHPLC–MS/MS method, *Drug Test. Anal.*, 2014, **6**(Suppl. 1), 22–29.
10. European Workplace Drug Testing Society, European Guidelines for Workplace Drug and Alcohol Testing in Hair, available from http://www.ewdts.org/data/uploads/documents/ewdts-guideline-hair-v2.0.pdf, accessed November 11, 2022.
11. N. H. Kwon, Y. R. Lee, H. S. Kim, J. C. Cheong and J. Y. Kim, Hybrid Solid-Phase Extraction for Selective Determination of Methamphetamine and Amphetamine in Dyed Hair by Using Gas Chromatography–Mass Spectrometry, *Molecules*, 2019, **24**(13), 2501.
12. M. M. Ramírez-Fernández, S. M. Wille, V. di Fazio, N. Kummer, V. Hill and N. Samyn, Detection of Benzodiazepines and z-Drugs in Hair Using an UHPLC–MS/MS Validated Method: Application to Workplace Drug Testing, *Ther. Drug Monit.*, 2015, **37**(5), 600–608.
13. P. Kintz, L. Gheddar, A. Blanchot, A. Ameline and J. S. Raul, In a Case of Death Involving Steroids, Hair Testing is More Informative than Blood or Urine Testing, *J. Anal. Toxicol.*, 2021, **45**(8), 829–834.
14. Y. Shin, T. Y. Kong, J. C. Cheong, J. Y. Kim, J. I. Lee and H. S. Lee, Simultaneous determination of 75 abuse drugs including amphetamines, benzodiazepines, cocaine, opioids, piperazines, zolpidem and metabolites in human hair samples using liquid chromatography-tandem mass spectrometry, *Biomed. Chromatogr.*, 2019, **33**(9), e4600.
15. M. M. Ramírez-Fernández, V. Di Fazio, S. M. Wille, N. Kummer and N. Samyn, A quantitative, selective and fast ultra-high performance liquid chromatography tandem mass spectrometry method for the simultaneous analysis of 33 basic drugs in hair (amphetamines, cocaine, opiates, opioids and metabolites), *J. Chromatogr. B*, 2014, **965**, 7–18.
16. D. Wen, Y. Shi, X. Zhang, B. Xie, W. Liu, F. Yu, P. Xiang, B. Cong and C. Ma, Determination of barbiturates in hair samples by using a validated UHPLC-HRMS method: application in investigation of drug-facilitated sexual assault, *Forensic Sci. Prog.*, 2019, **7**(1), 78–87.

17. S. Lee, E. Han, S. In, H. Choi, H. Chung and K. H. Chung, Determination of Illegally Abused Sedative-Hypnotics in Hair Samples from Drug Offenders, *J. Anal. Toxicol.*, 2011, **35**(5), 312–315.
18. D. Di Corcia, F. D'Urso, E. Gerace, A. Salomone and M. Vincenti, Simultaneous determination in hair of multiclass drugs of abuse (including THC) by ultra-high performance liquid chromatography-tandem mass spectrometry, *J. Chromatogr. B*, 2012, **899**, 154–159.
19. M. Licata, C. Rustichelli, F. Palazzoli, A. Ferrari, C. Baraldi, D. Vandelli, P. Verri, F. Marchesi and E. Silingardi, Hair testing in clinical setting: Simultaneous determination of 50 psychoactive drugs and metabolites in headache patients by LC tandem MS, *J. Pharm. Biomed. Anal.*, 2016, **126**, 14–25.
20. A. Salomone, J. J. Palamar, R. Bigiarini, E. Gerace, D. Di Corcia and M. Vincenti, Detection of Fentanyl Analogs and Synthetic Opioids in Real Hair Samples, *J. Anal. Toxicol.*, 2019, **43**(4), 259–265.
21. S. A. Shah, R. Mullin, G. Jones, I. Shah, J. Barker, A. Petroczi and D. P. Naughton, Simultaneous analysis of antiretroviral drugs abacavir and tenofovir in human hair by liquid chromatography-tandem mass spectrometry, *J. Pharm. Biomed. Anal.*, 2013, **74**, 308–313.
22. M. M. Ramírez Fernández, S. M. R. Wille, V. Di Fazio and N. Samyn, Time course detection of dihydrocodeine in body hair after a single dose, *Forensic Sci. Int.*, 2019, **302**, 109864.
23. M. S. Cardoso, R. Lanaro, R. C. Dolores, D. R. Morais, A. C. F. Arantes, K. D. Oliveira and J. L. Costa, Determination of Drugs of Abuse in Hair by LC–MS–MS: Application to Suicide Attempts Investigation, *J. Anal. Toxicol.*, 2022, **46**(5), 577–581.
24. S. Vogliardi, D. Favretto, M. Tucci, G. Stocchero and S. D. Ferrara, Simultaneous LC-HRMS determination of 28 benzodiazepines and metabolites in hair, *Anal. Bioanal. Chem.*, 2011, **400**(1), 51–67.
25. J. Maublanc, S. Dulaurent, J. Morichon, G. Lachâtre and J. M. Gaulier, Identification and quantification of 35 psychotropic drugs and metabolites in hair by LC–MS/MS: application in forensic toxicology, *Int. J. Legal Med.*, 2015, **129**(2), 259–268.
26. J. M. Gaulier, J. Maublanc, F. Lamballais, S. Bargel and G. Gérard Lachâtre, LSD in pubic hair in a fatality, *Forensic Sci. Int.*, 2012, **218**(1–3), 25–27.
27. M. Martin, J. F. Muller, K. Turner, M. Duez and V. Cirimele, Evidence of mephedrone chronic abuse through hair analysis using GC/MS, *Forensic Sci. Int.*, 2012, **218**(1–3), 44–48.
28. A. S. Barreto, V. F. Brant, E. Spinelli and S. V. Rodrigues, Validation of a SPE-LC–MS/MS method for the determination of ketamine and norketamine in micropulverized hair after a single IV dose, *J. Chromatogr. B*, 2016, **1033–1034**, 200–209.
29. L. Lin Zhu, H. Yan, B. Shen, Y. Shi, M. Shen and P. Xiang, Determination of bromadiolone and brodifacoum in human hair by liquid chromatography/tandem mass spectrometry and its application to poisoning cases, *Rapid Commun. Mass Spectrom.*, 2013, **27**(4), 513–520.

30. F. Ricard, E. Abe, C. Duverneuil-Mayer, P. Charlier, G. de la Grandmaison and J. C. Alvarez, Measurement of atropine and scopolamine in hair by LC–MS/MS after Datura stramonium chronic exposure, *Forensic Sci. Int.*, 2012, **223**(1–3), 256–260.
31. A. Larabi, N. Fabresse, A. Knapp, M. Forcet, F. J. Baud, G. Lorin de la Grandmaison and J. C. Alvarez, LC–MS/MS method for quantification of baclofen in hair: A useful tool to assess compliance in alcohol dependent patients?, *Drug Test. Anal.*, 2018, **10**(4), 694–700.
32. J. C. Alvarez, I. Etting, E. Abe, A. Villa and N. Fabresse, Identification and quantification of 4-methylethcathinone (4-MEC) and 3,4-methylenedioxypyrovalerone (MDPV) in hair by LC–MS/MS after chronic administration, *Forensic Sci. Int.*, 2017, **270**, 39–45.
33. N. Fabresse, S. Grassin-Delyle, I. Etting and J. C. Alvarez, Detection and quantification of 12 anabolic steroids and analogs in human whole blood and 20 in hair using LC-HRMS/MS: application to real cases, *Int. J. Legal Med.*, 2017, **131**(4), 989–999.
34. M. M. Ramírez-Fernández, S. M. R. Wille, V. Hill and N. Samyn, Determination of Antidepressants in Hair via UHPLC–MS/MS as a Complementary Informative Tool for Clinical and Forensic Toxicological Assessments, *Ther. Drug Monit.*, 2016, **38**(6), 751–760.
35. M. Wikström, G. Thelander, I. Nyström and R. Kronstrand, Two fatal intoxications with the new designer drug methedrone (4-methoxymethcathinone), *J. Anal. Toxicol.*, 2010, **34**(9), 594–598.
36. R. Kronstrand, M. Forsman and M. Roman, Quantitative analysis of drugs in hair by UHPLC high resolution mass spectrometry, *Forensic Sci. Int.*, 2018, **283**, 9–15.
37. M. K. Nielsen, S. S. Johansen, P. W. Dalsgaard and K. Linnet, Simultaneous screening and quantification of 52 common pharmaceuticals and drugs of abuse in hair using UPLC-TOF-MS, *Forensic Sci. Int.*, 2010, **196**(1–3), 85–92.
38. X. Wang, S. S. Johansen, M. K. K. Nielsen and K. Linnet, Targeted analysis of 116 drugs in hair by UHPLC–MS/MS and its application to forensic cases, *Drug Test. Anal.*, 2017, **9**(8), 1137–1151.
39. X. Wang, Y. Zhuo, X. Tang, H. Qiang, W. Liu, H. Wu, P. Xiang, G. Duan and M. Shen, Segmental analysis of antidepressant and antipsychotic drugs in the hair of schizophrenic patients, *Drug Test. Anal.*, 2020, **12**(4), 472–484.
40. G. Skopp, A. Kniest, J. Haisser, K. Mann and D. Hermann, Buprenorphine and norbuprenorphine findings in hair during constant maintenance dosage, *Int. J. Leg. Med.*, 2011, **125**(2), 277–281.
41. M. N. Rabelo Alves, G. Zanchetti, A. Zanchetti, S. Tameni, B. Spinosa de Martinis and A. Polettini, Determination of cocaine and metabolites in hair by column-switching LC–MS–MS analysis, *Anal. Bioanal. Chem.*, 2013, **405**(19), 6299–6306.
42. O. Quintela, E. Lendoiro, A. Cruz, A. de Castro, A. Quevedo, C. Jurado and M. López-Rivadulla, Hydrophilic interaction liquid chromatography-tandem mass spectrometry (HILIC-MS/MS) determination of cocaine and

its metabolites benzoylecgonine, ecgonine methyl ester, and cocaethylene in hair samples, *Anal. Bioanal. Chem.*, 2010, **396**(5), 1703–1712.

43. F. Freni, S. Bianco, C. Vignali, A. Groppi, M. Moretti, A. Marco, M. Osculati and L. Morini, A multi-analyte LC–MS/MS method for screening and quantification of 16 synthetic cathinones in hair: Application to postmortem cases, *Forensic Sci. Int.*, 2019, **298**, 115–120.
44. P. Verri, C. Rustichelli, F. Palazzoli, D. Vandelli, F. Marchesi, A. Ferrari and M. Licata, Tramadol chronic abuse: an evidence from hair analysis by LC tandem MS, *J. Pharm. Biomed. Anal.*, 2015, **102**, 450–458.
45. Y. Hongkun, L. Jinlei, D. Lina, W. Yue, W. Junmei and A. Wurita, Detection and quantification of psychotropic drug etaqualone in human hair using GC–MS/MS, *Leg. Med.*, 2021, **53**, 101964.
46. I. Shu, A. Alexander, M. Jones, J. Jones and A. Negrusz, Determination of methamphetamine enantiomer composition in human hair by non-chiral liquid chromatography-tandem mass spectrometry method, *J. Chromatogr. B*, 2016, **1028**, 145–152.
47. N. Porpiglia, G. Musile, F. Bortolotti, E. F. De Palo and F. Tagliaro, Chiral separation and determination of ketamine and norketamine in hair by capillary electrophoresis, *Forensic Sci. Int.*, 2016, **266**, 304–310.
48. J. Rhee, J. Lee, J. Kim, M. Jang, I. Shin and S. Lee, Determination of Ketamine and Norketamine in Hair and Evaluation of Polydrug Use in Ketamine Abusers Using Hair Analysis in Korea, *J. Anal. Toxicol.*, 2021, **45**(9), 993–1005.
49. S. Lee, J. Kim, S. In, H. Choi, S. M. Oh, C. G. Jang and K. H. Chung, Development of a simultaneous analytical method for selected anorectics, methamphetamine, MDMA, and their metabolites in hair using LC–MS/MS to prove anorectics abuse, *Anal. Bioanal. Chem.*, 2012, **403**(5), 1385–1394.
50. E. Han, H. Yang, I. Seol, Y. Park, B. Lee and J. M. Song, Segmental hair analysis and estimation of methamphetamine use pattern, *Int. J. Leg. Med.*, 2013, **127**(2), 405–411.
51. R. Martin, J. Schürenkamp, A. Gasse, H. Pfeiffer and H. Köhler, Analysis of Psilocin, Bufotenine and LSD in Hair, *J. Anal. Toxicol.*, 2015, **39**(2), 126–129.
52. M. Nieddu, L. Burrai, M. P. Demontis, M. V. Varoni, E. Baralla, C. Trignano and G. Boatto, Simultaneous Determination of 11 Illicit Phenethylamines in Hair by LC–MS–MS: *In Vivo* Application, *J. Anal. Toxicol.*, 2015, **39**(7), 532–537.
53. E. Lendoiro, C. Jiménez-Morigosa, A. Cruz, M. Páramo, M. López-Rivadulla and A. de Castro, An LC–MS/MS methodological approach to the analysis of hair for amphetamine-type-stimulant (ATS) drugs, including selected synthetic cathinones and piperazines, *Drug Test. Anal.*, 2017, **9**(1), 96–105.
54. V. A. Boumba, M. Di Rago, M. Peka, O. H. Drummer and D. Gerostamoulos, The analysis of 132 novel psychoactive substances in human hair using a single step extraction by tandem LC/MS, *Forensic Sci. Int.*, 2017, **279**, 192–202.

55. Y. Huang, Q. Yang, K. Yoon, Y. Lei, R. Shi, W. Gee, E. T. Lin, R. M. Greenblatt and M Gandhi, Microanalysis of the antiretroviral nevirapine in human hair from HIV-infected patients by liquid chromatography-tandem mass spectrometry, *Anal. Bioanal. Chem.*, 2011, **401**(6), 1923–1933.
56. G. Frison, S. Frasson, F. Zancanaro, G. Tedeschi and L. Zamengo, Detection of 3-methylmethcathinone and its metabolites 3-methylephedrine and 3-methylnorephedrine in pubic hair samples by liquid chromatography-high resolution/high accuracy Orbitrap mass spectrometry, *Forensic Sci. Int.*, 2016, **265**, 131–137.
57. Y. J. Chang, M. R. Chao, S. C. Chen, C. H. Chen and Y. Z. Chang, A high-throughput method based on microwave-assisted extraction and liquid chromatography-tandem mass spectrometry for simultaneous analysis of amphetamines, ketamine, opiates, and their metabolites in hair, *Anal. Bioanal. Chem.*, 2014, **406**(9–10), 2445–2455.
58. M. J. Baptista, P. V. Monsanto, E. G. Pinho Marques, A. Bermejo, S. Avila, A. M. Castanheira, C. Margalho, M. Barroso and D. N. Vieira, Hair analysis for Δ^9-THC, Δ^9-THC-COOH, CBN and CBD, by GC/MS-EI. Comparison with GC/MS-NCI for Δ^9-THC-COOH, *Forensic Sci. Int.*, 2002, **128**(1–2), 66–78.
59. S. E. Breidi, J. Barker, A. Petróczi and D. P. Naughton, Enzymatic digestion and selective quantification of underivatised delta-9-tetrahydrocannabinol and cocaine in human hair using gas chromatography–mass spectrometry, *J. Anal. Methods Chem.*, 2012, **2012**, 907893.
60. M. Míguez-Framil, A. Moreda-Piñeiro, P. Bermejo-Barrera, P. López, M. J. Tabernero and A. M. Bermejo, Improvements on enzymatic hydrolysis of human hair for illicit drug determination by gas chromatography/mass spectrometry, *Anal. Chem.*, 2007, **79**(22), 8564–8570.
61. M. Míguez-Framil, P. Cabarcos, M. J. Tabernero, A. M. Bermejo, P. Bermejo-Barrera and A. Moreda-Piñeiro, Matrix solid phase dispersion assisted enzymatic hydrolysis as a novel approach for cocaine and opiates isolation from human hair, *J. Chromatogr. A*, 2013, **1316**, 15–22.
62. J. Al Jaber, D. Holt and A. Johnston, Method Development for the Detection of Basic/Weak Basic Drugs in Hair by LCMSMS: Comparison between Methanolic and Alkaline Extraction on Real Samples, *Pharmacol. Pharm.*, 2012, **3**, 263–274.
63. E. Han, H. Chung and J. M. Song, Segmental hair analysis for 11-nor-Δ^9-tetrahydrocannabinol-9-carboxylic acid and the patterns of cannabis use, *J. Anal. Toxicol.*, 2012, **36**(3), 195–200.
64. H. S. Cho, B. Cho, J. Sim, S. K. Baeck, S. In and E. Kim, Detection of 11-nor-9-carboxy-tetrahydrocannabinol in the hair of drug abusers by LC–MS/MS analysis, *Forensic Sci. Int.*, 2019, **295**, 219–225.
65. M. Minoli, I. Angeli, A. Ravelli, F. Gigli and F. Lodi, Detection and quantification of 11-nor-Δ9-tetrahydrocannabinol-9-carboxylic acid in hair by GC/MS/MS in Negative Chemical Ionization mode (NCI) with a simple and rapid liquid/liquid extraction, *Forensic Sci. Int.*, 2012, **218**(1–3), 49–52.

66. L. Mercolini, R. Mandrioli, M. Protti, M. Conti, G. Serpelloni and M. A. Raggia, Monitoring of chronic Cannabis abuse: an LC–MS/MS method for hair analysis, *J. Pharm. Biomed. Anal.*, 2013, **76**, 119–125.
67. M. Cobo-Golpe, A. de Castro-Ríos, A. Cruz, M. López-Rivadulla and E. Lendoiro, Determination and Distribution of Cannabinoids in Nail and Hair Samples, *J. Anal. Toxicol.*, 2021, **45**(9), 969–975.
68. Y. Shi, X. Cui, M. Shen and P. Xiang, Quantitative analysis of the endogenous GHB level in the hair of the Chinese population using GC/MS/MS, *J. Forensic Leg. Med.*, 2016, **39**, 10–15.
69. E. Bertol, A. Argo, P. Procaccianti, F. Vaiano, M. G. Di Milia, S. Furlanetto and F. Mari, Detection of gamma-hydroxybutyrate in hair: validation of GC–MS and LC–MS/MS methods and application to a real case, *J. Pharm. Biomed. Anal.*, 2012, **70**, 518–522.
70. M. Chèze, M. Deveaux, C. Martin, M. Lhermitte and G. Pépin, Simultaneous analysis of six amphetamines and analogues in hair, blood and urine by LC-ESI-MS/MS. Application to the determination of MDMA after low ecstasy intake, *Forensic Sci. Int.*, 2007, **170**(2–3), 100–104.
71. N. Deshmukha, I. Hussaina, J. Barker, A. Petroczi and D. P. Naughton, Analysis of anabolic steroids in human hair using LC–MS/MS, *Steroids*, 2010, **75**, 710–714.
72. M. Barroso, S. Costa, M. Dias, D. N. Vieira, J. A. Queiroz and M. López-Rivadulla, Analysis of phenylpiperazine-like stimulants in human hair as trimethylsilyl derivatives by gas chromatography–mass spectrometry, *J. Chromatogr. A*, 2010, **1217**(40), 6274–6280.
73. R. Wietecha-Posłuszny, A. Garbacik, M. Woźniakiewicz and P. Kościelniak, Microwave-assisted hydrolysis and extraction of tricyclic antidepressants from human hair, *Anal. Bioanal. Chem.*, 2011, **399**(9), 3233–3240.
74. R. Wietecha-Posłuszny, M. Woźniakiewicz, A. Garbacik, P. Chęsy and P. Kościelniak, Application of microwave irradiation to fast and efficient isolation of benzodiazepines from human hair, *J. Chromatogr. A*, 2013, **1278**, 22–28.
75. L. Sander, Pressurized Fluid Extraction, Journal of Research (NIST JRES), National Institute of Standards and Technology, Gaithersburg, MD, 2017, available from https://doi.org/10.6028/jres.122.007, accessed November 11, 2022.
76. F. Vincenti, C. Montesano, E. Oliva, F. Fanti, M. Vincenti, A. Salomone, D. Compagnone, R. Curini and M. Sergi, Accelerated Extraction and Analysis of Ethyl Glucuronide in Hair by Means of Pressurized Liquid Extraction Followed by Liquid Chromatography-Tandem Mass Spectrometry Determination, *J. Anal. Toxicol.*, 2021, **45**(9), 927–936.
77. C. Montesano, G. Vannutelli, M. Massa, M. C. Simeoni, A. Gregori, L. Ripani, D. Compagnone, R. Curini and M. Sergi, Multi-class analysis of new psychoactive substances and metabolites in hair by pressurized liquid extraction coupled to HPLC-HRMS, *Drug Test. Anal.*, 2017, **9**(5), 798–807.
78. M. Míguez-Framil, A. Moreda-Piñeiro, P. Bermejo-Barrera, I. Alvarez-Freire, M. J. Tabernero and A. M. Bermejo, Matrix solid-phase dispersion

on column clean-up/pre-concentration as a novel approach for fast isolation of abuse drugs from human hair, *J. Chromatogr. A*, 2010, **1217**(41), 6342–6349.

79. H. Miyaguchi, Determination of zolpidem in human hair by micropulverized extraction based on the evaluation of relative extraction efficiency of seven psychoactive drugs from an incurred human hair specimen, *J. Chromatogr. A*, 2013, **1293**, 28–35.
80. H. Miyaguchi and K. Kuwayama, Enantioselective determination of (R)-zopiclone and (S)-zopiclone (eszopiclone) in human hair by micropulverized extraction and chiral liquid chromatography/high resolution mass spectrometry, *J. Chromatogr. A*, 2017, **1519**, 55–63.
81. B. Flinders, E. Cuypers, H. Zeijlemaker, J. Tytgat and R. M. Heeren, Preparation of longitudinal sections of hair samples for the analysis of cocaine by MALDI-MS/MS and TOF-SIMS imaging, *Drug Test. Anal.*, 2015, 7(10), 859–865.
82. M. Poetzsch, M. R. Baumgartner, A. E. Steuer and T. Kraemer, Segmental hair analysis for differentiation of tilidine intake from external contamination using LC-ESI-MS/MS and MALDI-MS/MS imaging, *Drug Test. Anal.*, 2015, 7(2), 143–149.
83. H. Wang and Y. Wang, Matrix-assisted laser desorption/ionization mass spectrometric imaging for the rapid segmental analysis of methamphetamine in a single hair using umbelliferone as a matrix, *Anal. Chim. Acta*, 2017, **975**, 42–51.
84. H. Wang, Y. Wang, G. Wang and L. Hong, Matrix-assisted laser-desorption/ionization mass spectrometric imaging of olanzapine in a single hair using esculetin as a matrix, *J. Pharm. Biomed. Anal.*, 2017, **141**, 123–131.
85. A. Kernalléguen, C. Enjalbal, J. C. Alvarez, O. Belgacem, G. Léonetti, D. Lafitte and A. L. Pélissier-Alicot, Synthetic cannabinoid isomers characterization by MALDI-MS3 imaging: Application to single scalp hair, *Anal. Chim. Acta*, 2018, **1041**, 87–93.
86. F. Musshoff, T. Arrey and K. Strupat, Determination of cocaine, cocaine metabolites and cannabinoids in single hairs by MALDI Fourier transform mass spectrometry–preliminary results, *Drug Test. Anal.*, 2013, **5**(5), 361–365.
87. M. Shen, P. Xiang, Y. Shi, H. Pu, H. Yan and B. Shen, Mass imaging of ketamine in a single scalp hair by MALDI-FTMS, *Anal. Bioanal. Chem.*, 2014, **406**(19), 4611–4616.
88. T. Kamata, N. Shima, A. Miki, E. Matsuo, T. Yamamoto, H. Tsuchihashi, T. Sato, S. Shimma and M. Katagi, High Spatial-Resolution Matrix-Assisted Laser Desorption/Ionization-Ion Trap-Time-of-Flight Tandem Mass Spectrometry Imaging for Depicting Longitudinal and Transverse Distribution of Drugs Incorporated into Hair, *Anal. Chem.*, 2020, **92**(8), 5821–5829.
89. A. Miki, M. Katagi, T. Kamata, K. Zaitsu, M. Tatsuno, T. Nakanishi, H. Tsuchihashi, T. Takubo and K. Suzuki, MALDI-TOF and MALDI-FTICR imaging mass spectrometry of methamphetamine incorporated into hair, *J. Mass Spectrom.*, 2011, **46**(4), 411–416.

90. W. F. Duvivier, T. A. van Beek, E. J. Pennings and M. W. Nielen, Rapid analysis of Δ-9-tetrahydrocannabinol in hair using direct analysis in real time ambient ionization orbitrap mass spectrometry, *Rapid Commun. Mass Spectrom.*, 2014, **28**(7), 682–690.
91. W. F. Duvivier, M. R. van Putten, T. A. van Beek and M. W. Nielen, Un)targeted Scanning of Locks of Hair for Drugs of Abuse by Direct Analysis in Real Time-High-Resolution Mass Spectrometry, *Anal. Chem.*, 2016, **88**(4), 2489–2496.
92. J. Kim, J. Kim, H. Yum, M. Jang, J. Rhee, S. Lee and S. B. Han, Simultaneous determination of barbiturates, phenytoin and topiramate in hair by LC–MS/MS and application to real samples, *J. Pharmacol. Toxicol. Methods*, 2020, **106**, 106931.
93. M. Pujadas, S. Pichini, S. Poudevida, E. Menoyo, P. Zuccaro, M. Farré and R. de la Torre, Development and validation of a gas chromatography–mass spectrometry assay for hair analysis of amphetamine, methamphetamine and methylenedioxy derivatives, *J. Chromatogr. B*, 2003, **798**(2), 249–255.
94. Society of Hair Testing. Proficiency tests, Drugs of abuse results, available from https://www.soht.org/proficiency-test, accessed November 11, 2022.

CHAPTER 5

Recent Trends in Forensic and Clinical Applications of Hair Analysis

ISLAM AMINE LARABI*[a,b] AND JEAN-CLAUDE ALVAREZ[a,b]

[a] Department of Pharmacology and Toxicology, Raymond Poincaré Hospital, AP-HP, 92380 Garches, France; [b] Inserm U1018, Centre for Research in Epidemiology and Population Health (CESP), MOODS Team, MasSpecLab, UVSQ, Université Paris-Saclay, 78180 Montigny-le-Bretonneux, France
*Email: amine.larabi@uvsq.fr

5.1 Introduction

The extensive benefits of utilizing hair as a matrix for analysing drug exposure, as opposed to traditional matrices like blood or urine, are well-established. This recognition has resulted in the widespread adoption of hair analysis in various forensic contexts, including drug-related deaths, drug-facilitated crimes (DFCs), and child protection cases. Additionally, hair analysis finds valuable applications in clinical settings, such as drug rehabilitation programs and workplace drug testing. The unique characteristics of hair make it an invaluable tool in assessing and monitoring drug use across diverse scenarios. One of the key advantages of using hair as a specimen for analysis is its extensive detection window. Unlike other matrices, the detection period in hair is not limited to a few days or weeks but can extend for months or even years, depending on the length of the hair and assuming a uniform hair growth of 1 cm per month.[1] By analysing different segments of the hair, it becomes possible to establish a timeline of drug exposure and assess long-term patterns of substance use. The remarkable

Issues in Toxicology No. 47
Perspectives and Challenges of Hair Analysis
Edited by Pascal Kintz, Alberto Salomone and Marco Vincenti

Published by the Royal Society of Chemistry, www.rsc.org

temporal resolution offered by hair analysis significantly enhances the diagnostic capabilities in forensic and clinical investigations and provides valuable insights into an individual's drug history.

The majority of compounds are assimilated into the hair primarily through the bloodstream, where they remain stable and can be detected for varying durations.[2] Additionally, the process of sampling hair is non-invasive, straightforward, and does not necessitate any medical training. Another advantage of hair analysis is its ability to provide insights into the frequency and severity of drug consumption. While it may not be possible to determine the exact amount of drug intake through hair analysis, it can provide valuable information about the pattern and intensity of drug use. By analysing the distribution and concentration of drugs along the hair shaft, it becomes possible to distinguish between heavy, moderate, and light drug users.[3]

This information is particularly useful in assessing the level of drug consumption and understanding the habits and behaviours of individuals in forensic investigations and other relevant contexts.

Nonetheless, interpreting a positive hair test result can pose challenges due to the diverse routes through which substances can be incorporated into hair, including passive diffusion from blood stream, contamination by sebum, sweat, or external contamination. Additionally, various factors contribute to interindividual variability, such as genetic polymorphisms, hair color, metabolic disorders, dietary habits, and cosmetic use, all of which can influence the drug incorporation process and subsequent concentrations detected in hair. Consequently, the interpretation of hair concentration results can be intricate and nuanced.[4]

The objective of this chapter is to provide a summary of the recent trends in forensic and clinical applications of hair analysis.

5.2 Trends in Analytical Procedures

5.2.1 Single Hair and Micro-segmental Hair Analysis (MSA)

Segmental or sectional analysis is a method that involves dividing a hair shaft into smaller sections, typically centimetric, supra-centimetric, or sub-centimetric in length, to examine shorter periods of time. This approach allows for the retrospective determination of a timeline and trend of drug use. Comparing the concentrations of drugs in adjacent and successive segments provides valuable insights into the pattern and duration of drug use. However, factors like misalignment, variations in growth stages, and segmentation lengths limit the accuracy of time resolution. Fortunately, advancements in analytical methods, especially mass spectrometry (MS), have greatly improved the sensitivity of drug detection in hair, leading to a decrease in the number of hair strands required for analysis.[5,6]

In 1981, Smith *et al.* initially reported the radioimmunoassay analysis of phenobarbital using a mere 0.4 mg single hair sample.

Since then, the application of MS has made a substantial contribution to the reliable identification of various compounds. Several methods utilizing single hair analysis have been documented for substances such as amphetamine, methamphetamine, haloperidol, quinolone derivatives, methoxyphenamine, and chlorpheniramine.[7–11]

In 2009, a novel approach to single hair analysis emerged with the introduction of matrix-assisted laser desorption/ionization mass spectrometry (MALDI–MS), enabling the imaging of cocaine and its metabolites directly on a hair strand.[12]

In 2016, a ground-breaking technique called micro-segmental hair analysis (MSA) was introduced. This method revolutionizes the field by cutting a single hair strand at an incredibly precise interval of 0.4 mm, corresponding to approximately one day of hair growth assuming that hair strands grow at an average rate of $\approx$0.4 mm per day.

MSA is conducted on a set of five hair strands analysed separately, which can either be obtained during conventional hair analysis or specifically collected for this purpose. The method of acquiring the hair strands can be chosen based on the donor's preference, either through a haircut or hair plucking. Opting for a haircut is painless, but it must be postponed until the targeted hair region extends above the scalp, typically at least one month after the occurrence of drug exposure. On the other hand, hair plucking can be performed immediately following drug exposure. If hair plucking is chosen, it is crucial to select only those strands that cause discomfort when plucked, as they are more likely to be suitable specimens for drug testing. Hair strands that are easily plucked without pain may correspond to the catagen or telogen phase, where hair growth is arrested, and they will not provide an accurate reflection of recent substance consumption.

The recent review by Kuwayama *et al.* provides comprehensive details on MSA and the optimal approach for its execution.[7]

Through the utilization of MSA, it becomes feasible to mitigate numerous factors (*e.g.*, misalignment, variations in growth stages, segmentation lengths…) that might otherwise lead to inaccuracies, thereby considerably augmenting the precision of time resolution. Furthermore, this method facilitates a comprehensive examination of the mechanisms involved in the integration of drugs into hair and the estimation of the time of drug ingestion, making it invaluable in various contexts within clinical and forensic toxicology, including instances of drug facilitated sexual assault (DFSA),[13] or even death.[14]

While MSA holds promise as an approach that can potentially overcome several limitations of current hair testing methods, it is important to acknowledge that there is still room for improvement. It should be noted that interpreting the analytical results obtained from a single hair strand requires caution to avoid potential pitfalls in interpretation, such as:

- Hair strands are typically cut near the scalp using scissors, and the exact length of hair remaining on the scalp is unknown but estimated to be a

few millimetres. This length should be taken into account when estimating the day of drug ingestion. Analysing hair strands with the root can provide a more accurate estimation by considering the distance from the root end.
- Approximately 15–20% of hair strands selected for analysis may be in the telogen and catagen phases.[15] To ensure reproducibility of the analytical results, it is recommended to analyse two or preferably more hair strands independently.
- Even when collected from the same region of the head at the same time, individual hair strands can grow at different rates. Unless the actual growth rate of the analyzed hair strands is measured, this issue cannot be resolved.

These considerations are crucial to avoid potential inaccuracies and ensure reliable interpretations when utilizing this method for estimating the day of drug ingestion.[7]

5.2.2 Mass Spectrometry Imaging

Liquid chromatography coupled to tandem mass spectrometry (LC–MS/MS) is considered the gold standard technique in toxicology for analysing organic compounds. However, the year 2009 marked a significant breakthrough in the field of single hair analysis with the advent of MALDI–MS and its application for the detection of cocaine and its metabolites in hair.[12] This approach introduced the capability to directly visualize drugs on a hair strand through imaging, and has sparked new opportunities and advancements, leading to exciting developments in the field of forensic toxicology.[16–19] The underlying principle of this method involves placing a single hair on a MALDI support and applying a suitable matrix on top. Through a proton exchange between the matrix and the analyte, the analyte is ionized. This process enables the detection and tracing of the analyte along the hair strand, generating a comprehensive map of the desired mass distribution.

MALDI–MS has demonstrated successful applications in the analysis of various drugs, including cocaine,[12,20–25] methamphetamine,[26–28] ketamine,[29] methadone,[24] nicotine,[30] zolpidem,[24] methoxyphenamine,[28] tilidine,[31] and cannabis.[32]

5.2.3 High Resolution Mass Spectrometry

Untargeted liquid chromatography–high resolution mass spectrometry (LC–HRMS) techniques have become essential in systematic toxicological analysis.[33–35] These methods enable the investigation of a vast range of drugs during a single analytical cycle. However, the availability of shared mass spectra libraries and well-defined analytical procedures is still lacking, which poses a challenge in identifying newly introduced compounds.

In their study, Fabresse *et al.*,[36] focused on optimizing, validating, and applying an untargeted screening method for hair analysis using an ultra-high performance liquid chromatography system coupled with a benchtop Orbitrap mass spectrometer. The method incorporated data-dependent analysis (DDA) and utilized the MzCloud™ shared HRMS database. The developed method successfully detected 284 out of the 317 targeted compounds, including the identification of 72 new psychoactive substances (NPS). Notably, the method allowed for the first-time detection of AKB48-5F, a synthetic cannabinoid, in hair samples. This indicates the method's effectiveness in broadening the ability to detect a wide range of compounds, including newly emerging substances, in hair samples.

Several HRMS based methods applied to hair are now available in the literature.[37,38]

5.3 Forensic and Clinical Applications of Hair Analysis

Hair analysis plays a vital role in forensic and clinical toxicology, offering valuable insights in various scenarios. While detecting drugs in blood is crucial for determining the cause of overdose or death, it is also important to gather information about an individual's drug use leading up to their acute poisoning or demise. Hair analysis serves as an additional tool, complementing blood analysis, as it can detect drug use over an extended period. By examining hair samples, it is possible to confirm drug usage from weeks or months prior, even if the person was no longer under the influence of the substance at the time of blood sampling. This supplementary information can provide valuable insights into undisclosed medical conditions or a history of illicit drug use.

5.3.1 Acute Poisoning and Death

The primary focus in cases of overdose or death is determining the substances consumed shortly before the event and their potential role in causing or contributing to it.

Kuwayama *et al.*[39] introduced a novel method for estimating the time between drug ingestion and death using micro-segmental hair analysis. By incorporating internal temporal markers (ITMs) (which are ingested at a specific time interval before hair collection to mark timescales within individual hair strands), and measuring distances between drug segments and ITMs within individual hair strands, the day of drug ingestion and death can be estimated. The method was applied to a corpse with a documented history of lidocaine administration for surgery 57 days before the discovery, allowing for a more precise estimation of the day of death compared to conventional anatomical examinations.

Hair could be of great interest to objectify acute exposure. In that situation, the incorporation of the drug will not be through blood, but rather through contamination by sweat or sebum.

This phenomenon becomes particularly relevant when considering drugs from various classes that can impact the sweating response both centrally in the brain and in the peripheral nervous system. Examples of such drugs that cause hyperhidrosis include:

- Selective serotonin reuptake inhibitors (SSRIs) like citalopram.
- Tricyclic antidepressants such as amitriptyline.
- Antipsychotics like haloperidol and clozapine.
- Opioids including morphine, oxycodone, fentanyl, methadone, and tramadol.
- Non-steroidal anti-inflammatory medications (NSAIDs) like celecoxib and naproxen.
- Antimicrobial medications and antibiotics such as cephalosporins (*e.g.*, ceftriaxone or cefuroxime), quinolone antibiotics (*e.g.*, ciprofloxacin), and antivirals (*e.g.*, aciclovir or ribavirin).
- Medications that can affect endocrine function like systemic corticosteroids, thyroid medications, and diabetic medications like glipizide and insulin.
- Generalized hyperhidrosis can also occur during withdrawal from non-prescription drugs such as alcohol and caffeine.

In this context, Kintz *et al.*,[40] described a methodology to accurately document singular gamma-hydroxybutyrate (GHB) exposure resulting from sweat contamination. The approach involves detecting compounds excreted in sweat within the non-decontaminated proximal segment of hair, located near the scalp, where concentrations are significantly higher compared to distal segments due to sweat-induced contamination. This method enables discrimination between endogenous concentrations and exogenous exposure. Following a single exposure of 25 $mg\,kg^{-1}$, GHB hair concentration in the 0–1 cm proximal segment reached 1040 $ng\,mg^{-1}$, a substantial elevation compared to the normal population (GHB < 20 $ng\,mg^{-1}$) included in this study ($n = 12$).

This observation has already been reported by Nakahara *et al.*, in 1997 where hair roots were examined as a potential specimen for detecting acute methamphetamine (MA) poisonings using an animal model and fatal cases of MA intoxication.[41]

5.3.1.1 *Forensic Challenges: Decomposed or Mummified Cadavers*

Hair analysis has proven valuable in cases where other biological samples have decomposed or are unavailable. However, interpreting the results can be challenging due to the risk of body fluid contamination at the time of death or during the post-mortem period.[42]

Previous studies have shown that cocaine and its metabolites (benzoylecgonine BE, ecgonine methyl ester (EME), cocaethylene CE) and ethyl glucuronide (EtG) have been detected in the hair of ancient mummies dating from 2000 BCE to 1500 CE.[43–45]

In the case of an extensively decomposed body packer, toxicological investigations were conducted on hair and pellets. The analysis revealed the presence of methamphetamine, amiodarone, and its metabolite desethylamiodarone in the hair. The concentrations of methamphetamine in the hair were similar to regular consumers. However, interpreting the results was challenging due to potential contamination and the absence of the main metabolite of methamphetamine.[46]

These cases highlight the potential of hair analysis in revealing past drug consumption, even in historical or challenging forensic contexts.

5.3.1.2 *Tolerance Reversal: The Impact of Abstinence*

Hair analysis plays a crucial role in determining whether there was a recurrence of drug use prior to overdose or death. It is particularly important to consider the duration of drug use before the event when interpreting the presence of drugs in the bloodstream. When low levels of drugs are detected in this later, it is often presumed that the individual either had little experience with the drug or had abstained from drug use before the final dose. A reliable alternative is to conduct hair analysis. This information provides insights into the individual's tolerance level towards the detected drugs. Tolerance is especially significant for opioids,[42,47] but it also holds relevance for other psychoactive substances like benzodiazepines.[48]

Nevertheless, in a 2015 study conducted by Nielsen *et al.*, the focus was on methadone-related fatalities in Denmark and the associated risk factors. The findings contradicted the notion that concurrent use of central nervous system (CNS) depressants and reduced tolerance were significant factors in these deaths. Instead, the study emphasized the prevalence of co-ingested CNS depressants, indicating that drug–drug interactions played a more prominent role in fatal outcomes. Furthermore, the use of hair analysis provided valuable insights into poly-drug use, particularly highlighting the frequency of heroin abuse.[49]

5.3.1.3 *The Value of Hair Analysis in Cases with Inconclusive Traditional Analyses*

Hair analysis was conducted on a 30-year-old man found dead in his bed to determine the cause of death. Autopsy findings revealed cardiac hypertrophy and atherosclerotic coronary disease. The blood and urine analyses yielded negative results for alcohol and drugs of abuse. However, segmental hair analysis confirmed high levels of cocaine (range: 42 100–49 500 $pg\,mg^{-1}$), indicating heavy drug consumption, and confirming that the observed coronary abnormalities were directly linked to cocaine abuse.[42]

In another case involving a 44-year-old man with a history of alcoholism and drug abuse, the cause of death was inconclusive based on blood and urine analyses. However, pathological findings showed cirrhosis and gastric haemorrhage. Hair analysis, however, revealed the presence of EtG at 730 $pg\,mg^{-1}$, consistent with excessive alcohol consumption, suggesting it as a potential factor in the development of liver disease.[42]

Gheddar *et al.* recently investigated the use of keratinous matrices (head hair, body hair, and nail clippings) for analysing anabolic agents in a case involving a 29-year-old man found deceased at his residence.[50] Autopsy findings were unremarkable, but toxicological examination of blood, head hair, body hair, and nails revealed the presence of multiple anabolic steroids. Prolonged and repetitive use of these substances led to the conclusion that the cause of death was toxic, resulting from extensive abuse. Keratinous matrices proved more sensitive, detecting ten compounds compared to four in blood samples.

5.3.2 Hair Analysis in Drug-facilitated Crimes (DFC)

Drug facilitated crimes (DFC) refer to criminal acts that occur when a substance is administered to an individual with the intention of impairing their behavior, perceptions, or decision-making ability. It also encompasses situations where an individual takes advantage of someone who is incapacitated due to the voluntary consumption of a substance without their consent. These crimes involve the use of drugs or substances to facilitate the commission of an offense, such as sexual assault, robbery, or kidnapping, by exploiting the vulnerable state induced by the administered substance.[51]

The utilization of hair analysis in DFC cases has witnessed a notable upsurge, as evidenced by an extensive body of literature on the subject. This technique finds widespread application in cases where the assault is reported late, enabling the retrospective examination of hair samples. Hair analysis is also valuable in distinguishing between a single instance of drug exposure and chronic drug consumption, particularly in instances where false allegations may have been made.

The substances most frequently utilized in DFCs include benzodiazepines, Z-drugs (such as zopiclone, zolpidem, and zaleplon), and antihistamines. The consumption of drugs of abuse and alcohol can increase the vulnerability of victims, as they might have willingly ingested them, leading to disorientation or intoxication.[42]

The United Nations Office on Drugs and Crime (UNDOC) has provided a comprehensive list of compounds that should be targeted in DFC cases as part of the guidelines for forensic analysis of drugs involved in sexual assault and other criminal acts.[51] However, it is essential to customize this list according to the substances prevalent in each specific region or country, as the accessibility of the drug is a crucial factor in committing DFC.[52–55]

In their comprehensive review, Ping *et al.* examined the concentrations of drugs in hair and their significance in drug facilitated crimes. The study analyzed hair concentrations of 35 psychoactive drugs across 20 controlled dose studies and compared them to the 25 distinct drugs identified in reported case work.[56] This review contributes to a better understanding of drug concentrations in hair and their implications in forensic investigations related to drug-facilitated crimes.

5.3.2.1 *Drug-facilitated Sexual Assault*

Drug-facilitated sexual assaults (DFSA) are the most prevalent DFC, and involve non-consensual sexual acts perpetrated against a victim who is rendered incapacitated or unconscious by the effects of drugs, making them unable to resist or provide consent.[57] The intended effects involve inducing a state of partial consciousness or unconsciousness and causing anterograde amnesia.

Substances used in DFSA should ideally have low effective doses, with some exceptions like ethanol and gamma hydroxybutyric acid (GHB), and should produce rapid effects and memory impairment. These substances may include prescription medications that are commonly used for therapeutic purposes. They should also be soluble in aqueous mediums like beverages and devoid of taste. Commonly, DFSA substances are administered through drinks such as coffee, fruit juices, or more often alcoholic beverages.[42]

For instance, based on the latest national report in France, the most commonly identified compounds in DFSA cases are benzodiazepines (zolpidem, zopiclone, clonazepam, and alprazolam), antipsychotics (clozapine), antihistamines (doxylamine, loratadine), anticonvulsants (pregabalin), and opioids (tramadol, codeine, morphine). As for drugs of abuse, cannabis was found to be the most prevalent, followed by ketamine, MDMA, and cocaine.[58]

DFSA presents various challenges due to the administration of drugs in low doses, often as a single occurrence. Furthermore, there is often a significant delay between the assault and the victim reporting the incident or seeking medical attention, ranging from hours to days. This delay leads to late sample collection, making it challenging to detect the substances used. In such cases, hair analysis becomes a valuable tool.

Hair samples can be collected immediately after the assault, along with blood and urine samples. If the substance is present in the blood and/or urine but absent in the hair, it suggests a single instance of ingestion. Alternatively, delayed hair sampling (performed 4–6 weeks after the incident) is carried out when there is a substantial time gap between the assault and the report (few days). This involves collecting a hair strand closest to the scalp to identify the presence of drugs at the time of the incident.

Numerous case reports have documented the utilization of hair testing to underscore cases of DFSA.[59–65]

As an illustrative instance, we present a case involving an interim singer who was also an event host during a cruise. The individual became a suspected victim of DFSA. On the third day of the cruise, during a social gathering with friends and crew members, the victim's friend offered her three Ti'punch cocktails followed by a Long Island Iced Tea. The victim consumed a mixture of alcoholic beverages including vodka, gin, white rum, orange liqueur (Triple Sec, Cointreau, Grand Marnier), tequila, lemon juice, and Coca-Cola. Subsequently, she exhibited peculiar behavior and experienced memory loss after approximately two hours. She woke up in the bartender's cabin at 11:30 AM, discovering traces of vomit on her clothes. Upon returning to shore, she learned about the possibility of hair analysis and decided to file a complaint. Four weeks and three days after the alleged incident, the victim visited our laboratory for sample collection. Three hair strands, each measuring 15 cm, were collected, and four segments of 2 cm were analyzed from the root by LC/MS/MS to screen for commonly used substances in DFSA.

The analysis identified the presence of zolpidem and hydroxyzine in segment A at 1 pg mg^{-1} and 5 pg mg^{-1}, respectively (see Figure 5.1), corresponding to the time of the alleged assault. It also confirmed the absence

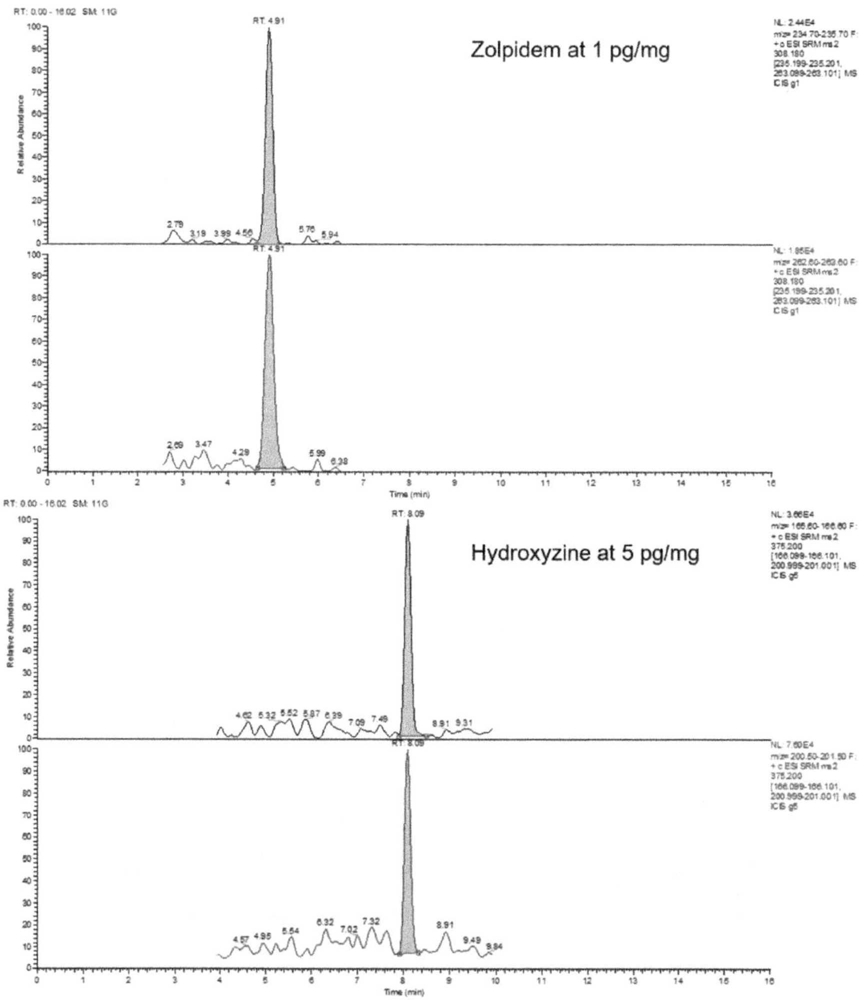

Figure 5.1 Extracted chromatogram from segment A of the patient supporting DFSA allegations.

of zolpidem and hydroxyzine in segments B, C, and D, except for hydroxyzine in segment B (found at 2.9 pg mg^{-1}), which may have been contaminated from the underlying segment A.

This finding indicates that the victim was not a habitual user of zolpidem and hydroxyzine, and a single administration of zolpidem and hydroxyzine occurred within two months prior to the sample collection.

Vomit analysis conducted on the victim's clothing provided further evidence supporting the alleged assault, as it confirmed the presence of zolpidem and hydroxyzine (see Figure 5.2).

While the majority of DFSA cases typically involve conventional DoA and pharmaceuticals, there is a growing concern regarding the emergence of new psychoactive substances (NPS) as a recent trend in these offenses.[66] With that regard, we described a case study focusing on the detection of 4-methylethcathinone (4-MEC), 3,4-methylenedioxypyrovalerone (MDPV), and doxylamine in hair samples.[57] The case involves a 44-year-old man who reported being assaulted after attending a party. Due to the delayed reporting, hair samples were collected for analysis 15 days later. Segmental hair analysis was conducted, and three hair segments (A, B, and C) were examined. The results indicated the presence of 4-MEC, MDPV, and doxylamine in the proximal hair segment (segment A) at 3 pg mg^{-1}, 5 pg mg^{-1}, and 9 pg mg^{-1}, respectively, indicating a single exposure to these substances within the month prior to the assault. This study emphasizes the significance of investigating NPS in cases of DFSA.

Figure 5.2 Vomit analysis on the victim's clothing confirmed the presence of zolpidem and hydroxyzine.

5.3.2.2 DFC in Children

DFCs in children serve three primary purposes: to subject them to sexual assault, to keep them quiet, or in cases of Munchausen syndrome by proxy.[67] The effectiveness of hair testing in evaluating the usage of incapacitating drugs in children with DFC has been validated by numerous studies.[68–71]

A retrospective study of suspected child abuse was conducted by Stauffer *et al.*, on 616 children (aged 0–17.5 years).[68] The study found that hair toxicology testing yielded positive results for at least one compound in 17.2% of cases, with unexplained drugs detected in 13.3% of cases. Multiple compounds, including parent drugs and/or metabolites, were identified in 48 cases. The most frequently detected substances included cocaine, benzoylecgonine, tetrahydrocannabinol, and methamphetamine. Environmental exposure was suggested in 68 cases where parent drugs were found without their metabolites.

However, interpreting the results of hair testing in children can be a complex task. The crucial aspect that demands clarification is discerning between isolated and repeated drug exposure using segmental hair analysis, as it carries significant implications both in court or in medical settings. Indeed, it is crucial to consider the potential diffusion along the hair shaft in children, as their hair tends to be more porous compared to adults.[72–74] When conducting segmental analysis to establish a single exposure, the optimal scenario is when the drug is detected solely in the segment that corresponds to the alleged time of the offense.[75] However, there are instances where the drug is found in several segments, which does not automatically indicate multiple exposures. Kintz *et al.*, suggests various potential explanations for the presence of multiple positive hair segments following a single drug exposure. These explanations encompass: (a) chronic consumption rather than a single instance of drug use, (b) variances in hair growth rate and the efficacy of sample collection, (c) external contamination, (d) incorporation of the drug and its metabolites through sweat, and (e) hair quality (damage, cosmetic treatments…), porosity, and so forth.[76]

In a study by Alvarez *et al.*, the efficacy of hair analysis in differentiating between chronic and acute drug use in children was investigated.[67] Hair samples were collected from 18 children, ranging from 1 day to 15 years old, who had received drug treatment or experienced intoxication. The samples were analyzed using LC/MS/MS. In the youngest children (1 day to 29 months), drugs administered in hospitals or causing intoxication were detected in all hair segments regardless of the sampling time. The concentrations remained consistent along the hair shaft, suggesting the diffusion and incorporation of analytes from sebum or sweat into the hair. Concentrations were highest shortly after administration and decreased over time. Hair analysis successfully detected intoxicating compounds even when they were undetectable in blood or urine, but it was not suitable for determining long-term drug exposure. In children aged 34 months to 15 years, drugs used in hospitals or causing intoxication were not detected in hair segments taken 1–5 days after drug administration, indicating the absence of drug incorporation from sebum or sweat into the hair. Hair analysis proved effective in identifying chronic drug administration in children over 34 months, similar to adults.

5.3.2.3 DFC in the Elderly

Elder abuse is a globally acknowledged issue that is on the rise. It is particularly prevalent within nursing homes and within the family setting.[42]

Alvarez *et al.* conducted a study to investigate cases of mistreatment among older individuals in a geriatric long-term care unit.[77] The disappearance of specific psychotropic drugs from the unit's pharmacy raised suspicions of misuse by an unidentified nurse. Blood samples were discreetly collected from seven patients with frequent agitation crises, and hair testing was performed on five of them. Strengthened safety measures were implemented, and a second round of blood testing was conducted on 18 patients from adjacent geriatric units. Hair testing was also performed. The initial blood analysis revealed one patient testing positive for non-prescribed midazolam. Hair testing of the selected patients identified non-prescribed midazolam in three of them, along with other substances. In the subsequent blood analysis, three patients tested positive for non-prescribed medicines. The second hair analysis showed the disappearance of benzodiazepines and "Z drugs" except for clobazam. New medications and decreased concentrations were observed. The findings led to the suspension of a suspected nurse, and subsequent hair analysis showed the absence of non-prescribed psychoactive substances.

These results supported the investigation and highlighted the role of hair testing in detecting drug misuse and ensuring the secure delivery of medications in geriatric departments.

5.3.3 Drug Abuse

5.3.3.1 Hair as an Epidemiological-based Tool for Monitoring Drug Use

To effectively address public health challenges related to drug use, it is crucial to monitor patterns and trends in drug consumption. A range of methodologies is employed to gather this essential information. Traditional methods include surveys and web-based data collection, which provide insights into drug use patterns, risk perceptions, social and health consequences, and overall impact. However, these methods can be limited by reporting biases and may not capture a comprehensive picture of drug use. In recent years, innovative techniques such as wastewater and syringe residue analysis have been utilized to monitor drug use. These approaches provide indirect indication of drug consumption by measuring drug residues in sewage or analysing residual content in used syringes among injection drug users. While valuable, they do not provide direct information on individual drug exposure or allow for population-level analysis. In this context, hair analysis has emerged as a promising tool in epidemiology. Hair analysis allows for the direct assessment of drug exposure in individuals or populations by analysing their own samples. This method overcomes some of the limitations associated with traditional approaches, providing more accurate and objective data on drug use.[3]

We conducted a six-year study using hair testing to assess the prevalence of NPS and conventional drugs of abuse (DOA) among high-risk populations in Paris and its suburbs.[3] Among 480 patients, 29% tested positive for NPS, with amphetamines, cocaine, and opioids showing prevalence rates of 32%, 38.5%, and 52% respectively. The study identified 27 different NPS, with cathinones like 4-MEC and mephedrone being the most commonly detected. Synthetic cannabinoids were also found in hair samples for the first time. Combining multiple substances was common among NPS users. Hair concentrations varied widely, indicating sporadic rather than chronic use in most cases.

In a 10-year study by Dugues *et al.*, hair samples from 939 individuals in addiction or intoxication settings were analyzed.[78] NPS were detected in 31% of cases, with 1551 instances identified among 288 individuals. Amphetamines, cocaine, and opioids had higher prevalence rates than NPS. 51 different NPS were found, with ketamine, X-MMC, and dextromethorphan being frequently detected. Most NPS users were male, with a median age of 30 years.

Many studies have also highlighted the importance of hair analysis in monitoring drug use patterns and trends, including both traditional drugs and NPS.[79–85]

The European Monitoring Centre for Drugs and Drug Addiction (EMCDDA) launched the European Hair Drug Analysis Network (EHDAN) project in 2022 to evaluate the effectiveness of hair testing as an indicator of drug use in the general population.[86] This collaborative initiative involves experts from multiple countries and aims to advance scientific understanding of hair analysis for epidemiological studies. The project's findings contribute to evidence-based policies and interventions to address drug-related challenges. Hair samples were collected at drug-checking services and music festivals in France, Italy, and Portugal. A study in France analyzed hair samples from 100 participants, revealing a diverse range of NPS. Figure 5.3 visually represents the variety of NPS identified through hair analysis of the first 39 samples.

These preliminary results allowed first reported concentrations of 2-CB: 8 pg mg^{-1}; 3-HO-PCP: range: 1–31 pg mg^{-1} ($n=6$); 5-MAPB: 4–480 pg mg^{-1} ($n=4$); MeO-PCE: 4–39 pg mg^{-1} ($n=3$); threo-isopropylphenidate (IPPH): 12–35 pg mg^{-1} ($n=2$); and bentazepam: $<$1–80 pg mg^{-1} ($n=5$) (Figure 5.4) to be described.

The complete dataset from all 100 participants will be analyzed and compared with self-reported information to classify the population based on drug use frequency. This classification will establish reference concentrations for light, medium, and heavy users. The study aims to gain a comprehensive understanding of drug use patterns and behaviors among

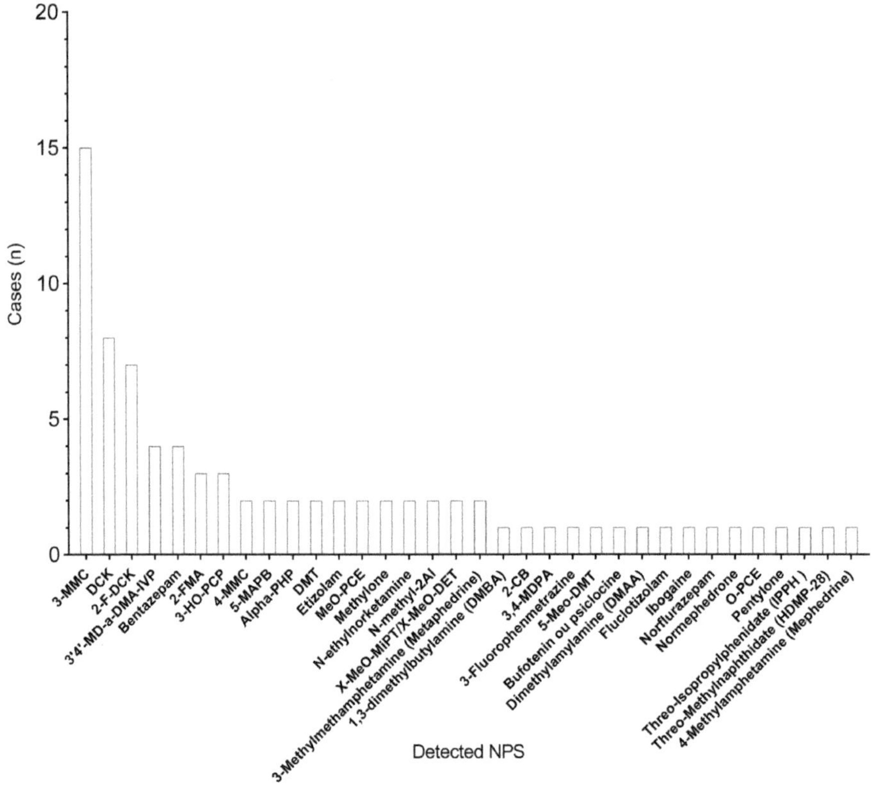

Figure 5.3 NPS identified in hair samples of the EHDAN project in France ($n = 39$).

attendees of these venues. By doing so, it intends to propose harm reduction programs customized for individuals affected by drug use disorders.

5.3.3.2 *Recreational and Chronic Drug Use*

Hair analysis is a crucial tool in evaluating instances of sporadic or persistent substance use. It enables the examination of various segments of the hair to establish a chronological account of drug use and track changes in substance levels over a period of time. The existing body of literature is filled with articles that describe the utilization of hair testing to document the recreational use of drugs of abuse.[87–96]

In a study, we showcased the efficacy of hair testing in determining exposure patterns to different fentanyls.[97,98] The research focused on a 43-year-old male who experienced an unintentional overdose after using a powder labeled as butyrylfentanyl. Hair samples were collected one month and one year later. The results revealed repetitive exposure to 3-fluorofentanyl, with higher concentrations during the overdose period. The powder was confirmed to contain 3-fluorofentanyl, while butyrylfentanyl was not detected. Furanylfentanyl and fentanyl were also found in the hair samples. In the

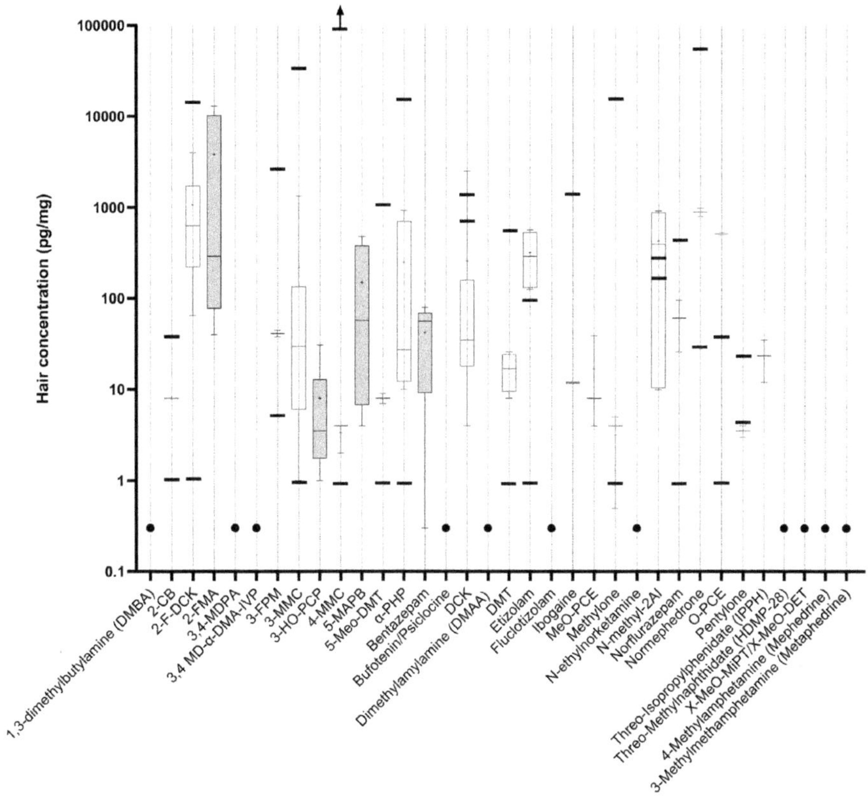

Figure 5.4 The graph depicts the 25th, 50th (median), and 75th percentiles of the data using bars, while the whiskers denote the minimum and maximum concentrations. The plus sign (+) represents the mean concentration. Additionally, horizontal thick black lines indicate reported concentrations from existing literature, where applicable. Concentrations reported for the first time are highlighted with gray filled bars. Notably, black filled circles at a concentration of 1 pg mg^{-1} signify the detection of 11 compounds utilizing LC–HRMS with a shared database, for which reference standards were unavailable. The upward arrow concerning 4-MMC indicates concentrations higher than 10 000 pg mg^{-1}.

second round of analysis conducted one year later, the patient showed the use of three additional fentanyls: methoxyacetylfentanyl, carfentanil, and acetyl fentanyl. The findings indicated repetitive exposure to methoxyacetylfentanyl and sporadic use of carfentanil and acetyl fentanyl. The concentration of 3-fluorofentanyl decreased over time, while furanylfentanyl and fentanyl concentrations increased despite the initiation of methadone treatment. Interestingly, the patient denied using three of the six detected fentanyls. This study is significant as it demonstrates the presence of multiple fentanyls in hair samples from an individual with confirmed drug abuse. The findings contribute to understanding fentanyl exposure and highlight the usefulness of hair testing in identifying various fentanyls in cases of overdose and drug abuse.

The consumption of synthetic cannabinoids (SC) in the Paris area was investigated by Dugues *et al.*, focusing on a drug user who unknowingly consumed nine SC.[99] The patient, a 42-year-old male with addiction to cannabis, alcohol, and occasional use of cocaine, codeine, and heroin, experienced withdrawal symptoms and intense craving. Hair analysis was conducted to assess his drug use. Three segments (2 cm each) of hair were analyzed, confirming exposure to cannabis, cocaine, MDMA, and codeine. Additionally, nine undisclosed SC were identified and quantified in the hair segments (A/B/C, $pg\,mg^{-1}$): JWH-019 (50/27/12), JWH-200 (20/13/5), JWH-250 (30/42/40), AM-2201 (10/32/3), BB-22 (10/10/10), 5F-PB-22 (14/64/10), 5F-APINACA (20/<10/<3), ADB-PINACA (43/20/10), AB-FUBINACA (36/10/<3). These concentrations indicated regular and increasing consumption in the six months leading up to the hair sampling. Notably, the concentrations of BB-22 in the hair were the first reported data in the literature.

These findings underscore the importance of systematic screening for NPS, and highlight the significant role of hair analysis in detecting and monitoring their consumption.

5.3.3.3 *Rehabilitation Programs: Applications and Pitfalls*

The long-term use of drugs of abuse can have various detrimental effects on both physical and mental health. Treatment programs, counselling, and support groups can aid in recovery and improving overall health outcomes. In certain court proceedings, it may be necessary to evaluate drug abstinence following a prolonged period of drug consumption or assess the effectiveness of detoxification programs.[42]

These situations typically arise in legal contexts where drug use or addiction may have legal implications, such as child custody cases, probation or parole hearings, and drug-related offenses, including driving under the influence of drugs (DUID).[100–102]

To evaluate drug abstinence, urine testing is commonly used due to its non-invasiveness, cost-effectiveness, and ability to provide recent drug use information. Blood testing may be utilized in situations where a more immediate assessment of drug presence is required. Hair testing, on the other hand, provides a longer retrospective record of drug use and can assess abstinence over a more extended period.[42]

Segmental hair analysis enables the retrospective determination of drug use patterns by comparing drug concentrations in different segments of the hair. The proximal segment, closer to the root, reflects the most recent drug exposure, allowing for the assessment of recent drug use or verification of drug abstinence during that specific timeframe. Comparing concentrations in the proximal segment to those in the distal segment, representing the furthest period from hair collection, provides insights into changes in drug use over time. When using segmental hair analysis to verify abstinence, negative results are expected in the segments nearest to the root, indicating recent abstinence.

However, segmental hair analysis has limitations. Drug concentrations in hair decrease over time due to natural wash and various factors such as hair treatments and exposure to sunlight. This can result in decreasing concentrations of a drug from the proximal to the distal segment of the hair, potentially leading to the interpretation of increasing drug use or non-compliance with abstinence programs. The hair growth cycle, further complicates the interpretation of drug use patterns. Latent or dormant hair in the telogen phase (resting phase) that previously incorporated drugs can contribute to the positivity of newly grown hair that does not contain the drug. Hair growth rate is influenced by multiple factors, adding another layer of complexity to hair test interpretation. External contamination from sweat and axial diffusion along the hair shaft can also introduce complications in hair testing interpretation.[15]

In our recent study, we investigated the duration of ketamine detection in hair after treatment discontinuation and conducted a literature review on the topic. We focused on a 22-year-old female who underwent ketamine treatment for major depression and collected a 26 cm hair lock for analysis 17 weeks after treatment. Using LC–MS/MS, we found that ketamine remained detectable in hair even after 4 months of treatment cessation, with an estimated elimination half-life of 0.88 months. This suggests that it may take up to 7 months for ketamine to become undetectable in hair. Our study provides the first published data on ketamine disappearance from hair. Previous studies have shown varying detection durations for different substances, such as heroin, cocaine, tramadol, promazine, amphetamine, methamphetamine, and THC-COOH (Table 5.1).[15,103–109]

The findings contribute to the overall understanding of drug detection in hair, and have the potential to improve the interpretation of hair testing results.

Table 5.1 Timeframe (months) for achieving negative hair results after abstinence: summary of findings from existing literature.

Compounds	Delay to obtain a negative hair result after abstinence (months)	Ref.
Amphetamine	4–5	109 and 110
Cocaine	3–4	105 and 106
Heroin	3	104
Ketamine	6–7	103
Methamphetamine	4–5	109 and 110
Promazine	3–4	108
THC-COOH	6–7	107
Tramadol	3–4	107

5.3.4 Hair Testing for Clinical Diagnosis and Monitoring

5.3.4.1 *Cortisol and Chronic Stress*

Hair analysis has long been utilized to identify exposure to xenobiotics, showcasing its clinical potential. However, the scope of its application has significantly widened with the ability to assess endogenous molecules in hair, enabling the detection of endocrinological disruptions.

A growing body of evidence suggests that elevated hair cortisol levels are associated with chronic stress related diseases such as mental health disorders (*i.e.*, depression, post-traumatic stress disorder PTSD), sleep disorder, cardiovascular disease (CVD), obesity, and has been linked with established metabolic risk factors including high blood pressure, diabetes, and adiposity.[110–118]

A recent review conducted by Decker Soares *et al.*[119] focused on the measurement of cortisol in hair using LC–MS/MS. The lack of a standardized procedure for quantifying cortisol in hair prompted the exploration of various techniques for sample preparation, chromatographic separation, and mass spectrometric detection. Among the commonly employed analytical approaches were simple methanolic extraction, reversed-phase separation, and MRM detection in negative ion mode. The reported assays were found to possess sufficient sensitivity for clinical applications.

5.3.4.2 *Therapeutic Drug Monitoring*

The use of hair for therapeutic drug monitoring (TDM) holds promise, but also presents challenges. While some studies have demonstrated a relationship between drug dose and hair concentration or between serum and hair concentrations of parent drug and/or metabolites, others have shown a lack of association.[120–122]

Metcafe *et al.* found evidence that higher anti-tuberculosis (TB) drug concentrations in hair predict extensively-drug resistant-TB clinical outcomes in a high-burden setting.[123]

In a study conducted by Capron *et al.*,[124] the median concentrations of Imatinib (IM), a commonly prescribed drug for chronic myeloid leukemia (CML), were compared between patients with therapeutic failure and those with partial or optimal response. The results showed that patients with therapeutic failure had significantly lower median IM concentrations in both hair (4.63 *vs.* 7.93, $p = 0.040$) and plasma (629.5 *vs.* 1084.8, $p = 0.009$) compared to patients with a positive response. A hair IM concentration below 5.8 ng mg^{-1} demonstrated 83% sensitivity and 70% specificity in identifying patients with therapeutic failure.

Zakrzewska *et al.*[125] conducted a study to assess the utility of measuring levetiracetam concentration in hair by LC/MS–MS as an indicator of dosing schedule. They used 12 rats divided into different treatment groups and a control group. The results showed that pulverized hair had a stronger correlation between levetiracetam concentration and the number of drug doses. However, in a clinical setting with 47 epilepsy patients from another study, levetiracetam hair concentrations showed only a slight correlation with plasma concentrations ($r = 0.36$), and no significant correlation was observed with the daily dose ($\mathrm{mg\,kg^{-1}}$).[126]

In the study of Sun *et al.*, no significant correlation was shown between hair and serum concentrations of the total fraction of risperidone and its main active metabolite 9-hydroxyrisperidone in 34 treated patients for schizophrenia.[127] In another population of schizophrenic patients ($n = 46$), Wang *et al.* showed a significant relationship between the administered dose and the concentration in the proximal hair segment was found only for clozapine, norclozapine, and chlorpromazine among the 23 investigated analytes, and conclude that the idea of using hair concentrations to monitor a daily dose is inapplicable.[128]

However, in terms of assessing patients' compliance with treatment, hair testing has emerged as a valuable tool in the era of monitoring, as several studies have demonstrated.[129–136]

5.3.4.3 Gestational Drug Exposure and Paediatric Hair Analysis

In addition to the aforementioned concerns regarding drug-facilitated crimes (DFC) in children, the identification of substance use disorder during pregnancy through toxicological analysis provides indirect evidence of underlying mental health issues in the mother.[137,138] Moreover, it poses developmental risks for the newborn. Two well-documented scenarios are as follows: (1) foetal alcohol spectrum disorder (FASD), characterized by significant impairments such as facial dysmorphology, growth retardation, and neurodevelopmental abnormalities, and (2) neonatal abstinence syndrome (NAS), which manifests as various severe clinical outcomes, primarily associated with chronic opioid exposure but also with other drug classes such as selective serotonin reuptake inhibitors (SSRIs).[139,140]

Segmenting a maternal hair sample collected at birth into three equal segments enables the evaluation and comparison of drug or ethanol usage patterns during different stages of gestation.[120,138]

The review of Wang and Drummer examines the interpretation of drug presence in children's hair for clinical and forensic purposes.[141] The analysis includes three categories of publications: newborns with *in utero* or short-term exposure, children passively exposed through their environment, and those exposed through accidental ingestion or deliberate administration. Cocaine, codeine, 6-AM, and morphine generally exhibit higher concentrations in hair from *in utero* exposure compared to passive exposure, although overlap exists. The concentration ranges ($\mathrm{ng\,mg^{-1}}$) observed *in utero* compared to in passive exposure are as follows:

- cocaine [0–257] *in utero vs.* [0–75] in passive exposure;
- codeine [0.4–41] *in utero vs.* [0.006–0.44] in passive exposure;

- 6-MAM [<2–29] *in utero vs.* [0.01–11] in passive exposure;
- morphine [1.7–35] *in utero vs.* [0.01–1.3] in passive exposure.

Methamphetamine shows no significant difference between passive exposure [0.1–131] ng mg^{-1} and *in utero* exposure [0–23] ng mg^{-1}, but data is limited.

Codeine and methadone concentrations do not differ between passive and active exposure:

- codeine [0.006–0.44] in passive exposure *vs.* [0.09–5] in active exposure;
- methadone [0.007–2.2] in passive exposure *vs.* [0.05–21] in active exposure.

However, data on other drugs and comparisons is insufficient. The comparison of hair concentrations is complicated by variations in extraction and washing techniques, as well as the absence of relevant contextual information. These findings emphasize the challenges of interpreting hair concentrations, particularly in children, without considering the surrounding circumstances.

Kintz *et al.* studied several cases of children exposed to methadone and discussed the limitations of child hair testing.[142] According to the author, limited available hair from children for analysis, along with the finer and more porous nature of their hair compared to adults, increases the risk of contamination. The timing of drug incorporation into hair is complex due to asynchronous hair growth and the potential for prenatal drug exposure. Identifying the source of drug incorporation, whether systemic or external, is crucial for assessing the actual health risk to children. However, distinguishing between the two modes of exposure is challenging, as drugs can be ingested or inhaled, as well as externally deposited from smoke, dust, or contaminated surfaces.

These limitations should be considered when interpreting hair results of children, and communicated to child protection authorities and courts to ensure a realistic assessment of the children health risk.

5.3.4.4 Alcohol Use Disorder

The direct analysis of ethanol in hair is impossible due to its volatility and the potential for external absorption. However, ethanol metabolites such as ethyl glucuronide (EtG) and fatty acid ethyl esters (FAEE) (*i.e.*, ethyl palmitate, EtPa) can be detected in hair, and are established biomarkers that effectively assess excessive alcohol consumption. The Society of Hair Testing (SoHT) has defined cut-off values for evaluating both abstinence and heavy drinking (>60 g ethanol per day) using these biomarkers.[143] The substantial body of literature surrounding hair analysis in monitoring alcohol use disorder (AUD) serves as strong evidence of its credibility.[144–146]

A dedicated chapter of this book will comprehensively explore and provide updates on all aspects related to the subject.

5.3.5 Workplace Drug Testing

Workplace drug testing utilizes a range of sample types, such as blood, sweat, urine, oral fluid (OF), and hair, each serving distinct purposes. For immediate detection of drug use during working hours, urine and OF tests are commonly employed, with blood samples reserved for accidents. These tests provide insight into recent drug use, typically within 24 to 72 hours, depending on the substance. They are suitable for "random drug testing" or "for cause" situations where assessing impairment is crucial. In contrast, hair testing offers a long-term historical view, making it valuable for hiring decisions and revealing consistent drug use patterns following a positive urine or oral fluid test.[147–149]

To ensure legally defensible workplace drug testing, the European Workplace Drug Testing Society (EWDTS) has developed and regularly updated guidelines. These guidelines aim to establish best practices for drug testing in the workplace, while accommodating the specific customs and legislation of individual countries.[150]

In a recent systematic review conducted in 2022 by Morse *et al.*, the focus was on assessing the efficacy, effectiveness, and cost-effectiveness of workplace-based interventions aimed at preventing and treating problematic substance use.[151] The review acknowledges the significant impact that employee alcohol and drug use can have on the workplace, leading to issues such as absenteeism, reduced productivity, high turnover, and safety concerns. The analysis included a total of 39 articles, consisting of 28 primary research articles and 11 reviews. However, the synthesis of findings was somewhat limited due to variations in the designs of interventions and evaluations across the studies. Nevertheless, there were promising outcomes observed for several types of workplace-based interventions, particularly in the area of substance use screening, which demonstrated potential effectiveness in addressing problematic substance use.

A dedicated chapter within this book will cover this topic, providing a detailed analysis and comprehensive exploration.

5.3.6 Doping

This section will offer a succinct overview of the subject matter, as the book will comprehensively explore the utilization of hair testing in doping cases through a dedicated chapter.

Anti-doping laboratories accredited by the World Anti-Doping Agency (WADA) primarily rely on urine and blood as their primary matrices for routine analysis. When adverse results occur, the responsibility to prove innocence falls on the athlete, who has the right to request additional investigations, such as hair analysis, to provide supplementary evidence.[152] The Richard Gasquet case set a precedent for the acceptance of hair analysis in certain doping cases.[153] In 1999, the Society of Hair Testing (SoHT) published guidelines that outlined recommended practices and standards for conducting hair analysis in the context of doping.[154]

Hair testing can be a valuable tool to document instances of dietary or environmental contamination in specific cases.[155–157] The Court of Arbitration for Sport (CAS) played a significant role in acknowledging hair testing as evidence in contamination scenarios, notably with the introduction of the "kissing defense" theory.[158] This groundbreaking ruling has led to an increased acceptance and utilization of hair analysis in doping cases.[159,160] However, it is important to recognize the limitations and advantages of hair analysis, and it should be considered as a supplementary tool rather than a standalone alternative, as emphasized in the regulations set by WADA.[161,162]

5.3.7 Biomonitoring of Hazardous Compounds

Over the years, a significant number of scientific papers have underscored the practicality of hair testing as a monitoring tool for assessing exposure to organic pollutants[163–175] and metals.[176–182] Recognizing the importance of these topics, this book will feature dedicated chapters that specifically delve into these areas, providing comprehensive coverage and insights into the utilization of hair analysis as an effective means of monitoring organic pollutants and metals.

5.3.8 A Perspective on Hair Analysis: Investigating Xenobiotic Metabolism

The mechanisms underlying the incorporation of xenobiotics into hair, as well as the factors influencing their stability, remain insufficiently understood. Three established pathways for drug integration into hair include direct absorption from the bloodstream, uptake through sebum and sweat, and external contamination. Among the proposed incorporation models, the "biochemical concept" offers a tentative explanation for the lower metabolite to parent drug ratio (MPR) observed due to high lipophilicity of parent drugs, highlighting the dependency on drug properties such as drug pK_a, structure, size, lipophilicity, protein binding capacity, and melanin affinity. Moreover, basic drugs demonstrate a greater propensity for hair incorporation compared to neutral or acidic drugs, leading to higher concentrations of these substances in hair. This limitation arises from the absence of an electrical bond between the negatively charged melanin present in hair and the acid molecules, which exist as anions at the acidic pH of hair (pH 5.0). Furthermore, acidic molecules encounter challenges in crossing the membrane of the blood capillary, owing to the pH gradient between the blood and the hair follicle.[2]

It is important to acknowledge that not all drugs conform to the criteria set forth by the previously mentioned model. In certain cases, dealkylated metabolites (such as nortriptyline, norclomipramine, norclobazam, nordiazepam, norflurazepam) have shown a greater propensity for integration into the hair shaft compared to their parent drugs, despite having lower lipophilicity.[183] This observation aligns with their respective blood MPR.[184]

The aforementioned observation raises intriguing questions regarding the mechanisms underlying the incorporation of drugs into hair. It also prompts further exploration into the intrinsic metabolic activity of the hair follicle (HF) and the potential occurrence of *in situ* formation of specific metabolites within the follicle.

Previous studies have provided evidence of significant enzymatic activity within hair follicles, encompassing both phase I (oxidation, involving cytochromes) and phase II (conjugation, involving UDP glucuronosyltransferases or UGT) metabolic processes.[185–188]

In line with these findings, our recent experimental investigations focused on the metabolism of NPS belonging to ketamine derivatives, such as O-PCE, 3-HO-PCE, and 2F-DCK.[189–191] By comparing the metabolic profiles generated through *in vitro* models using human liver microsomes with those obtained from authentic biological samples, including hair, we were able to identify a majority of the metabolites in the hair matrix. Intriguingly, some of these metabolites were exclusively present in hair samples, raising questions regarding the possibility of *in situ* metabolism occurring within the hair follicle itself.

These observations emphasize the need for further exploration of the unique metabolic processes taking place within the hair follicle and their contribution to drug metabolism.[192]

References

1. P. Kintz, *Hair analysis in clinical and forensic toxicology*, Elsevier, Boston, MA, 1st edn, 2015.
2. G. A. A. Cooper, in *Hair Analysis in Clinical and Forensic Toxicology*, Elsevier, 2015, pp. 1–22.
3. I. A. Larabi, N. Fabresse, I. Etting, L. Nadour, G. Pfau, J. H. Raphalen, P. Philippe, Y. Edel and J. C. Alvarez, *Drug Alcohol Depend.*, 2019, **204**, 107508.
4. I. Benhalima, E. Castex, G. Dumont, A. Gish, F. Hakim, D. Allorge and J.-M. Gaulier, *Leg. Med. Tokyo Jpn.*, 2022, **56**, 102032.
5. N. Allibe, P. Kintz, A. Faure, F. Paysant, A.-P. Michard-Lenoir, F. Stanke-Labesque, V. Scolan and H. Eysseric-Guerin, *Curr. Pharm. Des.*, 2017, **23**(36), 5502–5510.
6. C. Wiedfeld, G. Skopp and F. Musshoff, *Drug Test. Anal.*, 2021, **13**, 817–832.
7. K. Kuwayama, H. Miyaguchi, T. Kanamori, K. Tsujikawa, T. Yamamuro, H. Segawa, Y. Okada and Y. T. Iwata, *Forensic Toxicol.*, 2022, **40**, 215–233.
8. O. Suzuki, H. Hattori and M. Asano, *J. Forensic Sci.*, 1984, **29**, 11712J.

9. H. Matsuno, T. Uematsu and M. Nakashima, *Br. J. Clin. Pharmacol.*, 1990, **29**, 187–194.
10. T. Uematsu, M. Nakano, H. Akiyama and M. Nakashima, *Br. J. Clin. Pharmacol.*, 1993, **35**, 199–203.
11. T. Kamata, N. Shima, K. Sasaki, S. Matsuta, S. Takei, M. Katagi, A. Miki, K. Zaitsu, T. Nakanishi, T. Sato, K. Suzuki and H. Tsuchihashi, *Anal. Chem.*, 2015, **87**, 5476–5481.
12. S. Vogliardi, D. Favretto, G. Frison, S. D. Ferrara, R. Seraglia and P. Traldi, *J. Mass Spectrom. JMS*, 2009, **44**, 18–24.
13. D. Xu, J. Ji, P. Xiang, H. Yan and M. Shen, *Forensic Toxicol.*, 2022, **40**, 374–382.
14. K. Kuwayama, M. Nariai, H. Miyaguchi, Y. T. Iwata, T. Kanamori, K. Tsujikawa, T. Yamamuro, H. Segawa, H. Abe, H. Iwase and H. Inoue, *Int. J. Legal Med.*, 2019, **133**, 117–122.
15. I.-A. Larabi, I. Etting and J.-C. Alvarez, *Drug Test. Anal.*, 2023, **15**(9), 971–979.
16. T. Porta, C. Grivet, T. Kraemer, E. Varesio and G. Hopfgartner, *Anal. Chem.*, 2011, **83**, 4266–4272.
17. A. Miki, M. Katagi, T. Kamata, K. Zaitsu, M. Tatsuno, T. Nakanishi, H. Tsuchihashi, T. Takubo and K. Suzuki, *J. Mass Spectrom.*, 2011, **46**, 411–416.
18. A. Miki, M. Katagi, N. Shima, H. Kamata, M. Tatsuno, T. Nakanishi, H. Tsuchihashi, T. Takubo and K. Suzuki, *Forensic Toxicol.*, 2011, **29**, 111–116.
19. A. Kernalléguen, P. Kintz, G. Leonetti, D. Lafitte and A.-L. Pélissier-Alicot, *J. Anal. Toxicol.*, 2023, **47**, 417–428.
20. S. Vogliardi, D. Favretto, G. Frison, S. Maietti, G. Viel, R. Seraglia, P. Traldi and S. D. Ferrara, *Anal. Bioanal. Chem.*, 2010, **396**, 2435–2440.
21. T. Porta, C. Grivet, T. Kraemer, E. Varesio and G. Hopfgartner, *Anal. Chem.*, 2011, **83**, 4266–4272.
22. F. Musshoff, T. Arrey and K. Strupat, *Drug Test. Anal.*, 2013, **5**, 361–365.
23. E. Cuypers, B. Flinders, I. J. Bosman, K. J. Lusthof, A. C. Van Asten, J. Tytgat and R. M. A. Heeren, *Forensic Sci. Int.*, 2014, **242**, 103–110.
24. M. Poetzsch, A. E. Steuer, A. T. Roemmelt, M. R. Baumgartner and T. Kraemer, *Anal. Chem.*, 2014, **86**, 11758–11765.
25. A. Kernalléguen, R. Steinhoff, S. Bachler, P. S. Dittrich, F. Saint-Marcoux, S. El Bakhi, F. Vorspan, G. Léonetti, D. Lafitte, A.-L. Pélissier-Alicot and R. Zenobi, *Anal. Chem.*, 2018, **90**, 2302–2309.
26. A. Miki, M. Katagi, T. Kamata, K. Zaitsu, M. Tatsuno, T. Nakanishi, H. Tsuchihashi, T. Takubo and K. Suzuki, *J. Mass Spectrom. JMS*, 2011, **46**, 411–416.
27. K. Kuwayama, K. Tsujikawa, H. Miyaguchi, T. Kanamori, Y. T. Iwata and H. Inoue, *Anal. Bioanal. Chem.*, 2012, **404**, 1823–1830.
28. T. Kamata, N. Shima, K. Sasaki, S. Matsuta, S. Takei, M. Katagi, A. Miki, K. Zaitsu, T. Nakanishi, T. Sato, K. Suzuki and H. Tsuchihashi, *Anal. Chem.*, 2015, **87**, 5476–5481.

29. M. Shen, P. Xiang, Y. Shi, H. Pu, H. Yan and B. Shen, *Anal. Bioanal. Chem.*, 2014, **406**, 4611–4616.
30. T. Nakanishi, T. Nirasawa and T. Takubo, *J. Anal. Toxicol.*, 2014, **38**, 349–353.
31. M. Poetzsch, M. R. Baumgartner, A. E. Steuer and T. Kraemer, *Drug Test. Anal.*, 2015, **7**, 143–149.
32. E. Beasley, S. Francese and T. Bassindale, *Anal. Chem.*, 2016, **88**, 10328–10334.
33. E. Zarrouk, M. Lenski, C. Bruno, V. Thibert, P. Contreras, K. Privat, A. Ameline and N. Fabresse, *Toxicol. Anal. Clin.*, 2022, **34**, 3–18.
34. L. Boland, L. Carrara, L. Chanu and T. Willeman, *Toxicol. Anal. Clin.*, 2022, **34**, 19–28.
35. S. S. Wille, A. A. Boland, V. V. D. Fazio and N. N. Samyn, *Toxicol. Anal. Clin.*, 2022, **34**, S54–S55.
36. N. Fabresse, I. A. Larabi, T. Stratton, R. Mistrik, G. Pfau, G. Lorin de la Grandmaison, I. Etting, S. Grassin Delyle and J.-C. Alvarez, *Drug Test. Anal.*, 2019, **11**, 697–708.
37. C. Montesano, G. Vannutelli, M. Massa, M. C. Simeoni, A. Gregori, L. Ripani, D. Compagnone, R. Curini and M. Sergi, *Drug Test. Anal.*, 2017, **9**, 798–807.
38. E. Marchei, M. C. Rotolo, G. Mannocchi, A. Capomassi, L.-M. Gómez-Ruiz, A. Acosta-López, R.-Y. Ramos-Gutiérrez, M.-B. Varela-Busaka, S. Pichini and Ó. García-Algar, *J. Pharm. Biomed. Anal.*, 2022, **211**, 114607.
39. K. Kuwayama, M. Nariai, H. Miyaguchi, Y. T. Iwata, T. Kanamori, K. Tsujikawa, T. Yamamuro, H. Segawa, H. Abe, H. Iwase and H. Inoue, *Int. J. Legal Med.*, 2019, **133**, 117–122.
40. P. Kintz, *J. Anal. Toxicol.*, 2016, **40**, 563–564.
41. Y. Nakahara, R. Kikura, M. Yasuhara and T. Mukai, *Forensic Sci. Int.*, 1997, **84**, 157–164.
42. C. Jurado, in *Hair Analysis in Clinical and Forensic Toxicology*, ed. P. Kintz, A. Salomone and M. Vincenti, Academic Press, Boston, 2015, pp. 241–273.
43. L. W. Cartmell, A. Aufderhide and C. Weems, *J. Okla. State Med. Assoc.*, 1991, **84**, 11–12.
44. A. C. Springfield, L. W. Cartmell, A. C. Aufderheide, J. Buikstra and J. Ho, *Forensic Sci. Int.*, 1993, **63**, 269–275.
45. F. Musshoff, C. Brockmann, B. Madea, W. Rosendahl and D. Piombino-Mascali, *Forensic Sci. Int.*, 2013, **232**, 213–217.
46. F. Hakim, S. Nassibou, A. Gish, B. Lima, J.-F. Wiart, C. Richeval, J. Outreville, V. Quétard, D. Allorge and J.-M. Gaulier, *J. Anal. Toxicol.*, 2022, **46**, e60–e64.
47. H. Druid, J. J. Strandberg, K. Alkass, I. Nyström, F. C. Kugelberg and R. Kronstrand, *Forensic Sci. Int.*, 2007, **168**, 223–226.
48. G. Høiseth, M. Arnestad, R. Karinen, L. Morini, S. Rogde, C. Sempio, V. Vindenes and Å. M. L. Øiestad, *J. Anal. Toxicol.*, 2018, **42**, 49–54.

49. M. K. K. Nielsen, S. S. Johansen and K. Linnet, *Forensic Sci. Int.*, 2015, **248**, 134–139.
50. L. Gheddar, A. Blanchot, A. Ameline, N. Arbouche, J.-S. Raul and P. Kintz, *J. Anal. Toxicol.*, 2023, **46**, e307–e313.
51. UNODC, Guidelines for the forensic analysis of drugs facilitating sexual assault and other criminal acts https://www.unodc.org/documents/scientific/forensic_analys_of_drugs_facilitating_sexual_assault_and_other_criminal_acts.pdf (accessed 18 June 2023).
52. S. Djezzar, N. Richard and M. Deveaux, in *Toxicological Aspects of Drug-Facilitated Crimes*, Elsevier, 2014, pp. 11–46.
53. V. Dumestre-Toulet and H. Eysseric-Guérin, in *Toxicological Aspects of Drug-Facilitated Crimes*, Elsevier, 2014, pp. 223–254.
54. A.-S. Lemaire-Hurtel and J.-C. Alvarez, in *Toxicological Aspects of Drug-Facilitated Crimes*, Elsevier, 2014, pp. 47–91.
55. M. Chèze and J.-M. Gaulier, in *Toxicological Aspects of Drug-Facilitated Crimes*, Elsevier, 2014, pp. 181–222.
56. P. Xiang, M. Shen and O. H. Drummer, *J. Forensic Leg. Med.*, 2015, **36**, 126–135.
57. I. A. Larabi, M. Martin, I. Etting, P. Penot, N. Fabresse and J. C. Alvarez, *Drug Test. Anal.*, DOI: 10.1002/dta.2377.
58. ANSM, Soumission Chimique-Résultats de l'enquète 2020 https://ansm.sante.fr/page/resultats-denquetes-pharmacodependance-addictovigilance (accessed 20 June 2023).
59. F. P. Busardò, F. Vaiano, G. Mannocchi, E. Bertol, S. Zaami and E. Marinelli, *Drug Test. Anal.*, 2017, **9**, 953–956.
60. K. Kuwayama, M. Nariai, H. Miyaguchi, Y. T. Iwata, T. Kanamori, K. Tsujikawa, T. Yamamuro, H. Segawa, H. Abe, H. Iwase and H. Inoue, *Forensic Sci. Int.*, 2018, **288**, 23–28.
61. K. Sasaki, N. Shima, T. Kamata, A. Ishikawa, A. Nitta, M. Wada, S. Nakano-Fujii, H. Kakehashi, T. Sato and M. Katagi, *Forensic Sci. Int.*, 2021, **325**, 110881.
62. A. Carfora, C. P. Campobasso, P. Cassandro, R. Petrella and R. Borriello, *J. Anal. Toxicol.*, 2022, **46**, e16–e20.
63. D. Xu, J. Ji, P. Xiang, H. Yan and M. Shen, *Forensic Toxicol.*, 2022, **40**, 374–382.
64. S. E. Melchior, M. K. K. Nielsen, A. R. Oropeza, J. Banner and S. S. Johansen, *Forensic Sci. Int.*, 2023, **347**, 111678.
65. A.-L. Pélissier-Alicot, P. Kintz, A. Ameline, C. Sastre, V. Baillif-Couniou and G. Leonetti, *Toxicol. Anal. Clin.*, 2022, **34**, 127–131.
66. K. S. Hagan and L. Reidy, *Forensic Sci. Int.*, 2015, **257**, 71–75.
67. J. C. Alvarez, L. Lasne, I. Etting, G. Chéron, V. Abadie, N. Fabresse and I. A. Larabi, *Int. J. Legal Med.*, 2018, **132**, 165–172.
68. S. L. Stauffer, S. M. Wood and M. D. Krasowski, *J. Forensic Leg. Med.*, 2015, **33**, 61–67.
69. L.-M. Mehling, S. S. Johansen, X. Wang, E. Doberentz, B. Madea and C. Hess, *Forensic Sci. Int.*, 2016, **259**, e25–e31.

70. X. Wang, S. S. Johansen, M. K. K. Nielsen and K. Linnet, *Forensic Sci. Int.*, 2018, **285**, e1–e12.
71. C. Wiedfeld, G. Skopp, U. Kuepper and F. Musshoff, *Drug Test. Anal.*, 2021, **13**, 1580–1584.
72. M. R. Harkey, *Forensic Sci. Int.*, 1993, **63**, 9–18.
73. B. Buffoli, F. Rinaldi, M. Labanca, E. Sorbellini, A. Trink, E. Guanziroli, R. Rezzani and L. F. Rodella, *Int. J. Dermatol.*, 2014, **53**, 331–341.
74. F.-C. Yang, Y. Zhang and M. C. Rheinstädter, *PeerJ*, 2014, **2**, e619.
75. C. Rey-Salmon and G. Pépin, *Arch. Pediatr. Organe Off. Soc. Francaise Pediatr.*, 2007, **14**, 1318–1320.
76. P. Kintz, M. Villain, V. Dumestre-Toulet and B. Ludes, *J. Clin. Forensic Med.*, 2005, **12**, 36–41.
77. J.-C. Alvarez, J.-L. Le Quintrec, M. Gourdon, I. Etting, H. Michelon and A. Larabi, *Toxicol. Anal. Clin.*, 2019, **31**, S17–S18.
78. P. Dugues, M. Martin, E. Abe, I. Etting, Y. Edel, J. C. Alvarez and A. Larabi, *Toxicol. Anal. Clin.*, 2022, **34**, S80.
79. J.-C. Alvarez, N. Fabresse and I. A. Larabi, *Curr. Pharm. Des.*, 2017, **23**, 5487–5495.
80. X. Wang, J. Cui, Y. Zhuo, B. Shen, S. Zhang, W. Liu, M. Shen and P. Xiang, *J. Anal. Toxicol.*, 2020, **44**, 482–489.
81. J. Chas, R. Bauer, I. A. Larabi, G. Peytavin, P. Roux, E. Cua, L. Cotte, A. Pasquet, C. Capitant, L. Meyer, F. Raffi, B. Spire, G. Pialoux, J.-M. Molina and J.-C. Alvarez, *J. Acquired Immune Defic. Syndr.*, 2021, **86**, 552–561.
82. J. M. Matey, G. Montalvo, C. García-Ruiz, F. Zapata, A. López-Fernández and M. A. Martínez, *Forensic Sci. Int.*, 2021, **329**, 111080.
83. L.-M. Gómez-Ruiz, E. Marchei, M. C. Rotolo, P. Brunetti, G. Mannocchi, A. Acosta-López, R.-Y. Ramos-Gutiérrez, M.-B. Varela-Busaka, S. Pichini and O. Garcia-Algar, *Pharmaceuticals*, 2022, **15**, 382.
84. A. Niebel, F. Pragst, F. Krumbiegel and S. Hartwig, *Forensic Sci. Int.*, 2022, **331**, 111148.
85. A. Niebel, L. Westendorf, F. Krumbiegel, S. Hartwig, M. K. Parr and M. Tsokos, *Drug Test. Anal.*, 2022, **14**, 110–121.
86. EMCDDA, European Hair Drug Analysis Network (EHDAN), https://www.emcdda.europa.eu/activities/hair-testing_en (accessed 24 June 2023).
87. M. Martin, J. F. Muller, K. Turner, M. Duez and V. Cirimele, *Forensic Sci. Int.*, 2012, **218**, 44–48.
88. A. Natekar, I. Motok, P. Walasek, C. Rao, G. Clare-Fasullo and G. Koren, *J. Popul. Ther. Clin. Pharmacol. J. Thérapeutique Popul. Pharamcologie Clin.*, 2012, **19**, e466–e472.
89. Y. P. Gaillard, A.-C. Cuquel, A. Boucher, L. Romeuf, F. Bevalot, J.-M. Prevosto and J.-M. Menard, *J. Forensic Sci.*, 2013, **58**, 263–269.
90. J. Kim, S. In, Y. Park, M. Park, E. Kim and S. Lee, *Anal. Bioanal. Chem.*, 2013, **405**, 9769–9778.
91. A. Namera, K. Konuma, T. Saito, S. Ota, H. Oikawa, S. Miyazaki, S. Urabe, H. Shiraishi and M. Nagao, *J. Chromatogr. B: Anal. Technol. Biomed. Life Sci.*, 2013, **942–943**, 15–20.

92. J. C. Alvarez, I. Etting, E. Abe, A. Villa and N. Fabresse, *Forensic Sci. Int.*, 2016, **270**, 39–45.
93. J. C. Alvarez, N. Fabresse, A. Knapp and I. El Hajj Sleiman, *Toxicol. Anal. Clin.*, 2016, **29**, 64–70.
94. J. Sim, H. S. Cho, J. Lee, S. In and E. Kim, *J. Pharm. Biomed. Anal.*, 2017, **140**, 162–168.
95. N. Arbouche, J.-S. Raul, D. Garnier, P. Kintz and A. Ameline, *Drug Test. Anal.*, 2019, **11**, 610–616.
96. S. Mestria, S. Odoardi, G. Biosa, V. Valentini, G. Di Masi, F. Cittadini and S. Strano-Rossi, *Forensic Sci. Int.*, 2021, **323**, 110817.
97. I. A. Larabi, M. Martin, N. Fabresse, I. Etting, Y. Edel, G. Pfau and J. C. Alvarez, *Forensic Toxicol.*, 2020, **38**, 277–286.
98. I. A. Larabi, M. Martin, I. Etting, G. Pfau, Y. Edel and J. C. Alvarez, *Drug Test. Anal.*, 2020, dta.2868.
99. P. Dugues, E. Abe, I. Etting, A. H. Nguyen, Y. Edel, J.-C. Alvarez and I. A. Larabi, *Toxicol. Anal. Clin.*, 2023, **35**, 78–84.
100. Consensus de la Société Française de Toxicologie Analytique (SFTA), *Ann. Toxicol. Anal.*, 2004, **16**, 229–230.
101. E. Lendoiro, A. de Castro, C. Jiménez-Morigosa, X. A. Gomez-Fraguela, M. López-Rivadulla and A. Cruz, *Forensic Sci. Int.*, 2018, **286**, 239–244.
102. E. J. D. Ogden, J. C. Verster, A. C. Hayley, L. A. Downey, B. Hocking, C. K. Stough, A. B. Scholey and Y. Bonomo, *Intern. Med. J.*, 2018, **48**, 908–915.
103. M. Shen, P. Xiang, Y. Sun and B. Shen, *Forensic Sci. Int.*, 2013, **227**, 64–68.
104. M. Felli, S. Martello, R. Marsili and M. Chiarotti, *Forensic Sci. Int.*, 2005, **154**, 96–98.
105. F. Garcia-Bournissen, M. Moller, M. Nesterenko, T. Karaskov and G. Koren, *Forensic Sci. Int.*, 2009, **189**, 24–27.
106. P. Kintz, A. Ameline and J.-S. Raul, *J. Anal. Toxicol.*, 2020, **44**(1), 65–68.
107. P. Kintz, A. Ameline and J.-S. Raul, *J. Anal. Toxicol.*, 2022, **46**, 803–806.
108. N. Suwannachom, T. Thananchai, A. Junkuy, T. E. O'Brien and P. Sribanditmongkol, *Forensic Sci. Int.*, 2015, **254**, 80–86.
109. T. Wang, B. Shen, H. Wu, J. Hu, H. Xu, M. Shen and P. Xiang, *Forensic Sci. Int.*, 2018, **284**, 153–160.
110. E. Iob and A. Steptoe, *Curr. Cardiol. Rep.*, 2019, **21**, 116.
111. N. El Mlili, H. Ahabrach and O. Cauli, *Life*, 2021, **11**, 81.
112. E. Koumantarou Malisiova, I. Mourikis, C. Darviri, N. C. Nicolaides, I. M. Zervas, C. Papageorgiou and G. P. Chrousos, *Physiol. Behav.*, 2021, **229**, 113244.
113. S. Noushad, S. Ahmed, B. Ansari, U.-H. Mustafa, Y. Saleem and H. Hazrat, *Int. J. Health Sci.*, 2021, **15**, 46–59.
114. E. E. Psarraki, I. Kokka, F. Bacopoulou, G. P. Chrousos, A. Artemiadis and C. Darviri, *Psychoneuroendocrinology*, 2021, **124**, 105098.
115. F. G. Schaafsma, G. Hulsegge, M. A. de Jong, J. Overvliet, E. F. C. van Rossum and K. Nieuwenhuijsen, *J. Occup. Health*, 2021, **63**, e12189.

116. R. A. Kitani, K. Letsou, I. Kokka, C. Kanaka-Gantenbein and F. Bacopoulou, *Children*, 2022, **9**, 715.
117. Y. Li, W. Jia, N. Yan, Y. Hua, T. Han, J. Yang, L. Ma and L. Ma, *J. Affective Disord.*, 2023, **329**, 438–447.
118. L. M. Pageau, T. J. Ng, J. Ling, B. A. Given, L. B. Robbins, P. Deka and E. C. Schlegel, *J. Hypertens.*, 2023, **41**, 875–887.
119. D. R. Decker Soares, M. V. Antunes and R. Linden, *Bioanalysis*, 2021, **13**, 1145–1155.
120. P. Kintz, E. Russell, M. Baber and S. Pichini, in *Hair Analysis in Clinical and Forensic Toxicology*, Elsevier, 2015, pp. 141–159.
121. V. Avataneo, A. D'Avolio, J. Cusato, M. Cantù and A. De Nicolò, *J. Pharm. Biomed. Anal.*, 2019, **166**, 40–51.
122. J.-C. Alvarez, E. Abe, A. Knapp, C. Mayer, I. Etting and I. A. Larabi, *Toxicol. Anal. Clin.*, 2021, **33**, S37–S38.
123. J. Metcalfe, P. Bacchetti, R. Gerona, A. Esmail, K. Dheda and M. Gandhi, *ERJ Open Res.*, 2019, **5**, 00046–02019.
124. A. Capron, M. V. Antunes, S. C. Wagner, V. S. Mattevi, N. Vieira, R. Leite, F. Reginato, M. Capra, L. Fogliatto, P. Wallemacq and R. Linden, *Clin. Chim. Acta Int. J. Clin. Chem.*, 2016, **453**, 42–47.
125. A. Zakrzewska, M. Dermanowski, P. Mierzejewski and J. Kuczyńska, *Basic Clin. Pharmacol. Toxicol.*, 2021, **128**, 419–429.
126. K. Karaś-Ruszczyk, J. Kuczyńska, H. Sienkiewicz-Jarosz, I. Kurkowska-Jastrzębska, P. Bienkowski, M. Restel, J. Samochowiec and P. Mierzejewski, *Ther. Drug Monit.*, 2017, **39**, 263–268.
127. X. Sun, L. Wang, F. Yang, J. Ren, P. Jiang, H. Liu, H. Li, C. Li and C. Zhang, *Gen. Psychiatry*, 2019, **32**, e100042.
128. X. Wang, Y. Zhuo, X. Tang, H. Qiang, W. Liu, H. Wu, P. Xiang, G. Duan and M. Shen, *Drug Test. Anal.*, 2020, **12**, 472–484.
129. A. Ferrari, M. Licata, C. Rustichelli, C. Baraldi, D. Vandelli, F. Marchesi, F. Palazzoli, P. Verri and E. Silingardi, *Eur. J. Clin. Pharmacol.*, 2017, **73**, 197–203.
130. J. R. Castillo-Mancilla and J. E. Haberer, *Curr. HIV/AIDS Rep.*, 2018, **15**, 49–59.
131. D. Hassall, N. Brealey, W. Wright, S. Hughes, A. West, R. Ravindranath, F. Warren and P. Daley-Yates, *Pulm. Pharmacol. Ther.*, 2018, **51**, 59–64.
132. M. Haedener, W. Weinmann, D. Eich, M. Liebrenz, T. Wuethrich and A. Buadze, *PLoS One*, 2021, **16**, e0248747.
133. F. Taus, R. Gottardo, M. Ballotari, C. Utzeri and F. Tagliaro, *Drug Test. Anal.*, 2022, **14**, 1773–1778.
134. I. A. Larabi, N. Fabresse, A. Knapp, M. Forcet, F. J. Baud, G. L. Grandmaison and J. C. Alvarez, *Drug Test. Anal.*, 2018, **10**(4), 694–700.
135. I. A. Larabi, E. Abe, I. Etting and J.-C. Alvarez, *Toxicol. Anal. Clin.*, 2016, **28**, 303–310.
136. S. Georgin-Lavialle, E. Abe, A. Larabi, L. Savey, S. Ducharme-Bénard, V. Hentgen, G. Grateau and J.-C. Alvarez, *Rheumatol. Oxf. Engl.*, 2021, **60**(3), 1563–1564.

137. L. J. Langman, A. M. Rushton, D. Thomas, P. Colbourne, I. Seiden-Long, M. M. Brun, D. Colantonio and P. J. Jannetto, *Clin. Biochem.*, 2023, **111**, 1–10.
138. A. Bertaso, R. Gottardo, M. Murari, M. Mazzola, N. M. Porpiglia, F. Taus, R. Beghini, F. Gandini and F. Bortolotti, *Drug Test. Anal.*, 2023, **15**(9), 980–986.
139. B. T. McPhail, C. Emoto, D. Butler, T. Fukuda, H. Akinbi and A. A. Vinks, *J. Clin. Pharmacol.*, 2021, **61**, 857–870.
140. J. Wang and F. Cosci, *Psychother. Psychosom.*, 2021, **90**, 299–307.
141. X. Wang and O. H. Drummer, *Forensic Sci. Int.*, 2015, **257**, 458–472.
142. P. Kintz, A. Farrugia, A. Ameline, A. Eibel and J.-S. Raul, *Forensic Sci. Int.*, 2017, **280**, 176–180.
143. SoHT, 2019 Consensus for the use of alcohol markers in hair for supporting the assessment of abstinence and chronic alcohol consumption https://www.soht.org/images/pdf/Revision_2019_Alcoholmarkers.pdf (accessed 23 June 2023).
144. B. M. R. Appenzeller, R. Agirman, P. Neuberg, M. Yegles and R. Wennig, *Forensic Sci. Int.*, 2007, **173**, 87–92.
145. A. Biondi, F. Freni, C. Carelli, M. Moretti and L. Morini, *Forensic Sci. Int.*, 2019, **300**, 106–119.
146. F. Pragst, in *Hair Analysis in Clinical and Forensic Toxicology*, Elsevier, 2015, pp. 71–139.
147. L. Tsanaclis and J. Wicks, in *Hair Analysis in Clinical and Forensic Toxicology*, Elsevier, 2015, pp. 197–239.
148. G. N. Stowe, R. B. Paulsen, V. A. Hill and M. I. Schaffer, *J. Anal. Toxicol.*, 2019, **43**, 553–563.
149. M. D. M. Ramírez Fernández, S. M. R. Wille, V. di Fazio, N. Kummer, V. Hill and N. Samyn, *Ther. Drug Monit.*, 2015, **37**, 600–608.
150. A. Salomone, L. Tsanaclis, R. Agius, P. Kintz and M. R. Baumgartner, *Drug Test. Anal.*, 2016, **8**, 996–1004.
151. A. K. Morse, M. Askovic, J. Sercombe, K. Dean, A. Fisher, C. Marel, M.-L. Chatterton, F. Kay-Lambkin, E. Barrett, M. Sunderland, L. Harvey, N. Peach, M. Teesson and K. L. Mills, *Front. Public Health*, 2022, **10**, 1051119.
152. World Anti-Doping CODE 2021 https://www.wada-ama.org/sites/default/files/resources/files/2021_wada_code.pdf (accessed 22 June 2023).
153. Court of Arbitration for Sport, Arbitration CAS 2009/A/1926 International Tennis Federation (ITF) v. Richard Gasquet & CAS 2009/A/1930 World Anti-Doping Agency (WADA) v. ITF & Richard Gasquet, award of 17 December 2009 https://jurisprudence.tas-cas.org/Shared%20Documents/1926,%201930.pdf (accessed 22 June 2023).
154. SoHT, Consensus of the Society of Hair Testing on hair testing for doping agents, https://www.soht.org/images/pdf/Consensus_Doping_Agents.pdf (accessed 22 June 2023).
155. M. Thevis, L. Geyer, H. Geyer, S. Guddat, J. Dvorak, A. Butch, S. S. Sterk and W. Schänzer, *Drug Test. Anal.*, 2013, **5**, 372–376.

156. U. Mareck, H. Geyer, T. Schertel, S. Petring, O. Krug and M. Thevis, *Drug Test. Anal.*, 2021, **13**, 1911–1914.
157. J.-C. Alvarez, I. Etting, L. Juillard, Z. Massy and I.-A. Larabi, *Clin. Chim. Acta Int. J. Clin. Chem.*, 2023, **546**, 117395.
158. P. Kintz, *J. Anal. Toxicol.*, 2021, **45**, e3–e5.
159. N. Fabresse, S. Grassin-Delyle, I. Etting and J.-C. Alvarez, *Int. J. Legal Med.*, 2017, **131**, 989–999.
160. L. Gheddar, J.-S. Raul and P. Kintz, *J. Chromatogr. B: Anal. Technol. Biomed. Life Sci.*, 2021, **1187**, 123048.
161. M. Thevis, H. Geyer, L. Tretzel and W. Schänzer, *J. Pharm. Biomed. Anal.*, 2016, **130**, 220–230.
162. P. Kintz, L. Gheddar, A. Ameline, N. Arbouche and J.-S. Raul, *Drug Test. Anal.*, 2020, **12**, 316–322.
163. S. Zhang, X. Yan, B. Tang, W. Luo, S. Chen, X. Luo, J. Zheng, B. Mai and Y. Yu, *Sci. Total Environ*, 2023, **892**, 164341.
164. F.-J. Peng, C. Emond, E. M. Hardy, N. Sauvageot, A. Alkerwi, M.-L. Lair, B. M. R. Appenzeller and NESCAV project group for the Grand Duchy of Luxembourg, *Environ. Int.*, 2021, **153**, 106526.
165. A. Pena, S. Duarte, A. M. P. T. Pereira, L. J. G. Silva, C. S. M. Laranjeiro, M. Oliveira, C. Lino and S. Morais, *Molecules*, 2021, **27**, 242.
166. E. Park, J. Lee, J. Lee, J. Lee, H. S. Lee, Y. Shin and J.-H. Kim, *Chemosphere*, 2021, **277**, 130215.
167. A. Iglesias-González, E. M. Hardy and B. M. R. Appenzeller, *Environ. Int.*, 2020, **134**, 105332.
168. L. Li, S. Yang, T. Chen, L. Han and G. Lian, *J. Pharm. Sci.*, 2018, **107**, 1122–1130.
169. X. González-Gómez, N. Cambeiro-Pérez, E. Martínez-Carballo and J. Simal-Gándara, *Sci. Total Environ*, 2018, **625**, 311–319.
170. D. A. Haines, G. Saravanabhavan, K. Werry and C. Khoury, *Int. J. Hyg. Environ. Health*, 2017, **220**, 13–28.
171. V. Yusa, M. Millet, C. Coscolla, O. Pardo and M. Roca, *Chemosphere*, 2015, **139**, 91–108.
172. V. Yusa, M. Millet, C. Coscolla and M. Roca, *Anal. Chim. Acta*, 2015, **891**, 15–31.
173. E. M. Hardy, R. C. Duca, G. Salquebre and B. M. R. Appenzeller, *Forensic Sci. Int.*, 2015, **249**, 6–19.
174. J.-C. Alvarez, I. Etting and I. A. Larabi, *Biomed. Chromatogr. BMC*, 2022, **36**, e5391.
175. P. Dugues, N. Fabresse and J.-C. Alvarez, *Toxicol. Anal. Clin.*, 2021, **33**, S24–S25.
176. C. Jota Baptista, F. Seixas, J. M. Gonzalo-Orden and P. A. Oliveira, *Environ. Sci. Pollut. Res. Int.*, 2022, **29**, 18398–18407.
177. S. Grassin-Delyle, M. Martin, O. Hamzaoui, E. Lamy, C. Jayle, E. Sage, I. Etting, P. Devillier and J.-C. Alvarez, *Talanta*, 2019, **199**, 228–237.
178. J.-P. Goullé, E. Saussereau, L. Mahieu and M. Guerbet, *Bioanalysis*, 2014, **6**, 2245–2259.

179. J.-P. Goullé, E. Saussereau, L. Mahieu, D. Bouige, M. Guerbet and C. Lacroix, *Ann. Biol. Clin.*, 2010, **68**, 429–440.
180. J.-P. Goullé, E. Saussereau, L. Mahieu, D. Bouige, M. Guerbet and C. Lacroix, *Bull. Acad. Natl. Med.*, 2008, **192**, 555–565; discussion 565–567.
181. J.-P. Goullé, L. Mahieu, J. Castermant, N. Neveu, L. Bonneau, G. Lainé, D. Bouige and C. Lacroix, *Forensic Sci. Int.*, 2005, **153**, 39–44.
182. P. Kintz, J.-P. Goullé, P. Fornes and B. Ludes, *J. Anal. Toxicol.*, 2002, **26**, 584–585.
183. K. Rygaard, K. Linnet and S. S. Johansen, *Metabolites*, 2021, **11**, 686.
184. C. Hiemke, N. Bergemann, H. W. Clement, A. Conca, J. Deckert, K. Domschke, G. Eckermann, K. Egberts, M. Gerlach, C. Greiner, G. Gründer, E. Haen, U. Havemann-Reinecke, G. Hefner, R. Helmer, G. Janssen, E. Jaquenoud, G. Laux, T. Messer, R. Mössner, M. J. Müller, M. Paulzen, B. Pfuhlmann, P. Riederer, A. Saria, B. Schoppek, G. Schoretsanitis, M. Schwarz, M. S. Gracia, B. Stegmann, W. Steimer, J. C. Stingl, M. Uhr, S. Ulrich, S. Unterecker, R. Waschgler, G. Zernig, G. Zurek and P. Baumann, *Pharmacopsychiatry*, 2018, **51**, 9–62.
185. M. J. Finnen and S. Shuster, *Biochem. Pharmacol.*, 1985, **34**, 3571–3575.
186. H. Merk, *Z. Hautkr.*, 1985, **60**, 11–18, 21.
187. H. F. Merk, H. Mukhtar, B. Schutte, I. Kaufmann, M. Das and D. R. Bickers, *Biochem. Biophys. Res. Commun.*, 1987, **148**, 755–761.
188. H. F. Merk, H. Mukhtar, I. Kaufmann, M. Das and D. R. Bickers, *J. Invest. Dermatol.*, 1987, **88**, 71–76.
189. I. A. Larabi, F. Zerizer, A. Ameline, I. Etting, D. Joseph, P. Kintz and J. C. Alvarez, *Drug Test. Anal.*, 2021, **13**, 1108–1117.
190. I. A. Larabi, D. Joseph, C. Lesueur and J.-C. Alvarez, *Metabolites*, 2023, **13**, 432.
191. D. Joseph, C. Lesueur, F. Zerizer, A. Fenot, J. C. Alvarez and I. A. Larabi, *J. Anal. Toxicol.*, 2023, bkad030.
192. S. Kazem, E. C. Linssen and S. Gibbs, *Drug Discovery Today*, 2019, **24**, 1899–1910.

CHAPTER 6

Multi-analyte Methods and High-resolution Mass Spectrometry for Hair Analysis

MARCO VINCENTI

Department of Chemistry, University of Turin, Via Pietro Giuria 7, 10125 Torino, Italy
Email: marco.vincenti@unito.it

6.1 Evolution of Mass Spectrometry Instrumentation

This chapter focuses on discussing the mass-spectrometric instrumentation most commonly used in hair analysis, together with its evolution following both technological advancements and new analytical challenges that involve hair testing. As a matter of fact, parallel progress of hair testing queries' complexity and the performance of mass-spectrometric instrumentation is observed. In turn, these analytical performances are strictly connected with technological innovations and original devices concerning mass spectrometers. For this reason, only the most recent studies (*i.e.*, from around 2016) have been considered, the former being covered in a preceding book chapter.[1]

For nearly all applications involving complex mixtures such as hair extracts, a chromatographic separation of the mixture components precedes the mass-spectrometric (MS) characterization of the targeted analytes. In particular, gas- and liquid-chromatography (LC) are competitively used, the latter finding increasing acknowledgement in recent years. The progressive detachment from the gas-chromatographic (GC) technique in hair analysis occurred despite the fact that GC–MS assembly can take advantage of

Issues in Toxicology No. 47
Perspectives and Challenges of Hair Analysis
Edited by Pascal Kintz, Alberto Salomone and Marco Vincenti

Published by the Royal Society of Chemistry, www.rsc.org

electron-impact (EI) as the ionization technique which offers a highly-specific and fully-informative spectroscopic characterization of the analytes, not exploitable in LC–MS instrumentation.[2] However, EI delivers a relatively inefficient ionization of the analytes and this low ionization yield is only partially compensated by the relatively high ion-transmission efficiency observed in GC–MS coupling. Another disadvantage of the use of GC as the separation technique is its mismatch with polar and highly-hydrophilic analytes, unless prior chemical derivatization is accomplished, which in turn results in increased complexity, sample-processing time, and uncertainty in quantitation.[2] Despite these drawbacks, GC–MS arrangements are still widely used for both targeted and untargeted hair analysis whenever extreme sensitivity conditions are not required. In the latter case, GC–MS/MS configurations may significantly increase the signal/noise ratio filling the gap with LC–MS/MS instrumentation.

Beyond the multifaceted considerations expressed in the preceding paragraph, it is undeniable that LC–MS/MS instrumentation is increasingly becoming the gold standard for the large majority of toxicological determinations conducted by hair analysis.[1] The matrix itself and its monthly-lasting exposition to environmental and climate agents makes the incorporation of volatile substances less favorable than those with high molecular weight and/or polar character, not directly manageable by GC.

Several factors have contributed to the success of LC–MS/MS, both in the low- and high-resolution (HRMS) versions. The most important technological advancement in toxicological laboratories has been the generalized transition from HPLC to UHPLC (ultra-high-pressure LC). In comparison with HPLC, UHPLC guarantees significant improvement in at least one of the following parameters: chromatographic resolution, sensitivity, productivity, the latter possibly being the most decisive.[3] A very large number of samples per day can now be processed at equal or improved sensitivity and resolution with respect to the past and also in comparison with GC-based methods of analysis, resulting in crucial cost reduction.

The second fundamental improvement of LC–MS instrumentation concerns their interfacing. In LC–MS, the analytes charging is induced by either electrospray (ESI) or – less frequently – atmospheric pressure chemical ionization (APCI), both occurring at ambient pressure. The ionization efficiency is high, in comparison with EI, but the ion transmission is relatively low, due to both the large pressure gap to be maintained between the ionization chamber and ultra-high vacuum of the analyzer housing and the large number of ion-molecule collisions occurring along the ion path.[2] The radical improvement of ion-focusing devices (multipole ion guides, ion-mobility electrodynamic focusing, *etc.*), some of which based on innovative principles, improved the ion-transmission efficiency by orders-of-magnitude, resulting in unprecedented sensitivity and detection limits in the femto- and atto-moles range, crucial in hair analysis where the available matrix is very limited.[4]

Unlike EI, ESI and APCI produce mass spectra with very few ion signals, typically the protonated (positive ions) or deprotonated (negative ions) molecular ion, with no or very few of the fragment ions that are essential for the identification of the analytes and their structural characterization. To compensate for this deficiency, it is necessary to induce the fragmentation of molecular ions in the post-ionization stage, either by collisional activation or other methods that generate vibrational excitation of the protonated molecular ions. The fragment ions generated by collisional activation of a precursor ion, mass-selected by a first analyzer, should then be determined by a second mass analyzer. Thus, LC separation habitually implies the need of a tandem mass spectrometer to fulfill a wide range of analytical determinations.[2]

The most common tandem mass spectrometers commercially available have a quadrupole as the first mass analyzer and either a low resolution (another quadrupole) or a high resolution (typically a time-of-flight or an Orbitrap®) mass analyzer as the second one. In between, a collision cell designed as a quadrupole or a multipole ion-guide is generally present to induce the fragmentation of precursor ions and focus the generated product ions into the second mass analyzer.[4] Notably, also the initial quadrupole can be operated as an ion-guide by zeroing its continuous potential, so as to convey all ions produced in the ionization chamber into the second mass analyzer, without preliminary mass selection. The availability of a high-resolution mass analyzer allows the determination of the so-called "exact mass" – that includes 3–4 decimal places and in turn provides both a reasonable clue of the elemental composition of precursor and product ions – and a considerable improvement of the corresponding ion-signal selectivity.[2]

Besides the supplementary information that a high-resolution mass analyzer can offer, another important difference between a quadrupole (Q) and both Orbitrap® and time-of-flight (TOF) mass analyzers is that the first one works sequentially (one mass at a time) while the latter typically collect full mass spectra (almost simultaneous detection of all ions).[5] The sequential *vs.* simultaneous operating mode of mass analyzers differentiates their application areas, particularly in hair analysis, and the queries that a specific MS experiment is asked to investigate. In hair testing, the extremely limited amount of sample and analytes constantly available makes the high-sensitivity requirement mandatory, whatever MS experiment is conceived to address a specific question.[1] Therefore, tandem quadrupole instruments are typically used for targeted analyses under selected reaction monitoring (SRM) operating conditions, whereas HRMS analyzers are also consistently used for untargeted analysis.

Ion-trap (IT) devices deserve further specific considerations. Initially developed as inexpensive 3D-analyzers operating in the full-scan mode (non-sequential), they evolved in the linear (2D) version as devices capable of multi-stage mass analysis (for example MS^3), so assuring improved specificity in the analytes' detection.[2] Moreover, IT devices are increasingly adopted to collect ions during variable time intervals before their pulsed transmission to either a further mass analyzer, such as TOF or Orbitrap®, or a reaction region where charge exchange or ion-molecule reactions may take place.[4] Altogether, ion-traps are extremely adaptable devices that, in their different shapes and functions, may be used as either a low-resolution mass analyzer or ion collecting and focusing equipment.

In untargeted analysis, hybrid MS/HRMS instruments are typically employed, such as Q-TOF or Q-Orbitrap®. Due to the complexity of the task that relies on the large size and quality of the collected data, a variety of experimental strategies are adopted, including polarity-switching, alternate HRMS and MS/HRMS data collection, combination with ion-mobility devices. In particular, the so-called "data-dependent" and "data-independent" acquisition modes are mostly used.[5] In the first operating mode, a preliminary full-scan high-resolution spectrum is acquired, on which the most abundant molecular ion signals are recognized by the software. Immediately after, a sequence of MS/HRMS experiments are automatically executed using the molecular ions previously identified as precursor ions. This alternate sequence is repeated cyclically with high frequency.[5] The final outcome is that both precursor and product ions are determined in high resolution for all the most abundant species eluted in any moment of the chromatographic run. The data-independent acquisition mode uses the *Q* analyzer to select mass intervals (for example, 15–20 Da wide) in rapid progression; the molecular ions falling in each interval undergo simultaneous collisional activation and fragmentation; the resulting product ions are detected in high resolution and possibly attributed to the corresponding molecular ion.[5]

6.2 Evolution of Hair Testing Purposes

Most of the recurring scopes of hair testing are still the same after decades and are mainly focused in the toxicological area of investigations. Typical target analytes are drugs of abuse, pharmaceutical drugs, markers of alcohol consumption, environmental and workplace contaminants, and doping agents.[6] However, the progressive diffusion of a growing number of synthetic drugs and new psychoactive substances (NPS) in a rising population percentage has made the instrumentation requisites needed to fulfill hair testing more demanding.[1] On the other hand, the ever-increasing performance of analytical instrumentation has driven hair testing toward more challenging tasks, which is apparent from the most recent literature. In particular, HRMS instruments were not frequently present in toxicology laboratories ten years ago, but became quite common in recent years. The literature review about the current objectives trends of hair testing pursued by means of modern mass spectrometers includes the general tasks listed below.

(a) Increase of sample throughput. As in any other context involving economical constrains, toxicology laboratories have been asked to increase their productivity and reduce costs. In hair analysis, these goals have been pursued by optimizing and simplifying the sample treatment procedures,[7] particularly for hair extraction, implementing laboratory automation as much as possible, and shortening the instrumental processing time. The latter scope was accomplished by a combination of increased detection specificity (by means of HRMS and MS/MS) and reduction of chromatographic run time (by means of UHPLC and fast-GC). In particular, increased detection specificity made the analyte separation conditions less stringent, allowing the mobile phase gradient in LC and the oven temperature ramp in GC[8] to be drastically accelerated. Enactment of such strategies in toxicology laboratories nowadays allows processing of up to approximately 100 hair samples per day on each instrument, reducing the number of instruments needed and the corresponding investment and maintenance costs. At the same time, a reduction of waste is achieved.

(b) Wide multi-analyte targeting. The primary and obvious advantage of using a single analytical method to verify the presence and quantify all the targeted analytes in a hair strand is again to reduce the time, costs, and waste for each sample to be analyzed. This benefit is particularly gained in routine analysis targeted to traditional drugs. However, other important applications encourage the development and validation of analytical methods addressed to the concurrent detection of tens or hundreds of targeted substances in hair samples. These applications include both the search for a large range of NPS and other synthetic drugs for screening purposes (for example, within epidemiological studies) or the detection of potential intoxicating agents in the habitual *vs.* incidental consumption of drugs in the course of post-mortem investigations.

(c) Targeted and untargeted metabolomic studies. Metabolomic studies used to discover distinctive biomarkers suggestive of specific pathologies or the effects of a drug administration can be accomplished on any kind of biological matrix, but hair offers the advantage of leveling off the short-term influence of environmental and dietary factors, making this matrix increasingly investigated by means of HRMS untargeted analysis. More studies are also conducted on hair to detect unknown metabolites of drugs that undergo rapid biochemical transformation.

(d) Extreme sensitivity and specificity. Both single hair[9] and segmental analysis typically require ultimate sensitivity and specificity, due to the extremely limited amount of sample available for the analysis. While these objectives are not new,[1] the improvement of the performance of mass spectrometers is progressively bringing these applications out of the proof-of-concept stage and into customary practice. On the other hand, the ability to work on shorter hair segments is likely to offer a more detailed description of the investigated substance intake chronology.

(e) Comprehensive screening for postponed reexamination. Targeted hair analysis generally offers adequate sensitivity for the detection and quantification of a complete range of pre-established substances. However, forensic investigations frequently evolve with new circumstantial evidence being brought to light after the analytical and toxicological work has been completed. In such cases, hair specimen should be reexamined, provided that a leftover aliquot remained after the first analysis. Just like full-scan GC–EIMS, the current high-resolution mass spectrometers offer a variety of techniques to acquire LC–HRMS and LC–MS/HRMS data without preliminary definition of the targeted substances, namely under untargeted conditions, allowing retrospective inspection of results possibly collected years earlier.

6.3 Multi-targeted Methods for Pharmaceuticals and Traditional Drugs of Abuse

Given that hair analysis is consistently used for detecting the exposure to certain substances of the tested subjects during extended periods of time, the applications that combine hair testing and targeted multi-analyte methods are generally addressed to verify the potential and often unpredictable exposure to a variety of molecules, typically drugs of abuse and/or abused pharmaceuticals, occurring accidentally or with continuity throughout months. The main purpose is qualitative, to confirm or screen-out exposure, but a quantitative evaluation is needed as soon as any of the targeted analytes is detected in the keratin matrix, so as to check the frequency and intensity of the intake. Within this general frame, the posed questions that the analysis is asked to reply are variables and include drive ability assessment, compliance to abstinence, toxicological overview, epidemiological studies, chronic and acute poisoning investigations, and many others.

The most critical aspect of wide multi-targeted methods for hair analysis is the expected variability of the analyte extraction yields from a complex matrix such as keratin. The difficult predictability of extraction efficiencies depends on the hair features (color, thickness, previous cosmetic treatments) as well as the number and heterogeneity of the targeted substances. For example, it makes a significant difference if the screening is addressed to many substances of the same class or the targeted analytes have radically different structures and physical and chemical properties.[10] The use of non-specific extraction conditions inevitably leads to reduced yields, even if several up-to-date studies have taken on the task to develop conditions that allow acceptable extraction of extremely large analytes panels. For example, a recent study compared the extraction efficiencies obtained from seven combinations of solvents within one- or two-step extraction strategies.[11] Two-step procedures with methanol – actually an acidic water/MeOH mixture in the second stage – proved to provide the highest efficiencies on an authentic hair pool.[11] Similar conclusions about the superiority of methanol as an extraction solvent were drawn in a subsequent study.[12]

Comprehensive overviews have been published in recent years to illustrate the different approaches followed by toxicological laboratories throughout the world to treat hair specimens before the instrumental determination of multi-targeted drugs.[1,7] More recently, further analytical methods have been proposed to screen extremely large panels of pharmaceuticals and drugs of abuse in hair samples.[11–17] Krumbiegel *et al.* developed a method aimed to detect and quantitate 76 substances in nails and compared the resulting data from post-mortem nail samples with those arising from hair collected from the same corpses.[13] Wang *et al.* targeted 116 drugs, including hypnotics, anxiolytics, analgesics, benzodiazepines, antihistamines, antidepressants, antipsychotics, and anticonvulsants, with LOD values in the low- to fractional $pg\,mg^{-1}$ range, to investigate forensic cases and post-mortem specimens.[14] Notably, the high method's sensitivity is ideally sufficient to detect the intake of a single drug dose by the deceased subject.[14] An analogous array of substances was aimed for by an analytical method developed by Shin *et al.*, which reached again LOD values in the low $pg\,mg^{-1}$ range for 75 medicaments and drugs of abuse.[12]

All the methods previously cited[11–14] used various models of triple-quadrupole MS/MS instrumentation, constantly operated in the SRM mode. In SRM, the identification criteria to be fulfilled for positive testing typically include the coincidence of the chromatographic retention time, together with the co-presence and correct intensity ratio of two precursor-product ion transitions. In the same period, a multi-targeted method based on a single stage HRMS instrument was developed by Odoardi *et al.* and proposed as an alternative to the SRM approach.[16] In this method, different identification criteria were used, including the chromatographic retention time, exact mass of monoisotopic protonated molecular ion, and its isotopic pattern, which all needed to match with the corresponding parameters of the pure standard for positive identification. In the case of positive screening, the presence of the detected substance was confirmed by inducing the molecular ion fragmentation by in-source collision-induced dissociation and monitoring the corresponding product ions and their relative abundance.[16] The clear advantage of this approach is that the number of targeted analytes is potentially unlimited, as long as the corresponding pure standards are available for matching the identification criteria and verifying the validation parameters. While HRMS instrumentation definitely provides increased flexibility for multi-targeted analysis with respect to triple-quadrupole MS/MS instruments, the large variety of different TOF and Orbitrap® mass spectrometers commercially available does not allow updated conclusions to be drawn about their comparative performances, in terms of sensitivity and accuracy. However, the combination of a low-resolution quadrupole with a HRMS analyzer, nowadays offered in the most expensive MS/MS configurations, is likely to provide wide flexibility and superior performances at the same time.

Beside the technical aspects of hair sample treatments and instrumental determinations, multi-targeted procedures involve a variety of hidden dangers and subtle features to be considered in both their quality assessment and thorough interpretation of the resulting data. In particular, the validation of analytical methods for hair analysis is typically conducted under conditions radically different from those found in authentically positive hair samples. The latter consistently contain only one or few of the targeted substances, which are firmly incorporated into the keratin matrix, whereas validation of the analytical methods is habitually operated on blank hair samples externally spiked with tens or hundreds of target analytes at low and even at high concentration, plausibly interfering with each other. Thus, extraction yields, matrix effects, and calibration curves are possibly affected by uncontrolled bias, in turn making the accuracy of the declared results unreliable. A recent method addressed to the simultaneous detection of 116 targeted analytes within a single sample treatment and analysis, including opioids, stimulants, benzodiazepines, z-drugs, antidepressants, and neuroleptics, took up the challenge of verifying the possible occurrence of substance-specific or class-specific sources of bias and proposed different solutions to compensate for them.[15] In particular, different calibration ranges were set up depending on the expected concentrations in hair for each drug, in turn leading to calibration curves mostly quadratic and with different weighting factors.[15] In multi-targeted methods, homogeneously linear calibration curves with identical weighting factors, as frequently found in the scientific literature, are not correct and reveal a rough and over-optimistic validation process.[16,17] Other sources of bias that were investigated in the cited study[15] included the presence of specific interferences from the matrix or co-eluting analytes, the consequent signal suppression that may arise from them, and the occasionally low and scarcely reproducible extraction yield that may affect certain drugs. A subsequent study from the same group extended this careful approach to cannabinoids and endogenous steroids.[18] Since this extension implied some modifications of the hair treatment procedure previously optimized, the coherence of the quantitative results obtained from the two procedures on authentic hair samples was positively verified.[18]

Multi-targeted methods in hair analysis may sometimes investigate more restricted sets of analytes, namely those belonging to the same class of drugs. In particular, homologous substances and/or metabolites of the drug primarily targeted are used to discriminate direct drug consumption from external contamination or to distinguish the intake of legal *vs.* illegal formulations. A typical case of this sort is represented by phytocannabinoids, whose determination in conventional and alternative matrices, including hair, has been recently reviewed.[19] A second, recently published review, confirms that phytocannabinoids and their metabolites represent a widely investigated subject of analytical toxicology in which multi-targeting is restricted to a specific class of substances and applied to various biological matrices, including hair.[20]

Whenever detected in hair, the metabolites of the ingested substances generally show very low concentration, due to their limited presence in blood and the low affinity of hydrophilic compounds with the keratin substrate. The incorporation of antipsychotic drugs into keratinized matrices (nails and hair) represents the focus of a recent study from Cobo-Golpe *et al.*, which demonstrated substantial agreement between the outcome of the two matrices and higher concentration in hair than in toe- and fingernails.[21] The positive detection of phase-II opiate glucuronide metabolites in hair has been demonstrated for the first time in a recent study conducted by Grabenauer *et al.*[22] The analytical method was validated for nine opiates and six glucuronide metabolites yielding LOD values in the 0.2–1.0 pg mg^{-1} range. The method was applied to 46 authentic hair samples collected from known drug users, revealing that the relative hair concentration of the four detected glucuronide metabolites with respect to the parent opiate concentrations ranged from 0.59% up to 2.33%.[22] A third example of class-specific targeted method of analysis is the one published by Narváez *et al.*, reporting the determination of multiple mycotoxins in human hair by LC–MS/HRMS with Q-TOF configuration.[23]

A different field of interest where wide multi-targeted methods of hair analysis are applied is that concerning doping controls in horse racing. Unlike in human doping investigations where hair testing is typically used to provide an answer to specific case questions (see the specific chapter in this book), horse hair is increasingly used for wide screening of doping substances with the purpose of covering the control over extended periods of time. Notably, the horse mane has an expected growth of 2.0–2.5 cm per month. Indeed, hair testing provides the opportunity for longitudinal monitoring of drug exposure for both humans and horses, but ethical and commercial constraints are less stringent in the latter case. A practical application of equine hair testing implicates the possible illegal administration of anabolic and androgenic steroids, which – for pharmacokinetic reasons – are typically administered as synthetic ester derivatives. Unlike natural steroids, the corresponding esters are not produced by any endogenic process nor are present in cattle feed. Therefore, their detection in horse hair, even at trace levels, provides direct proof of doping. In a 2017 study, Kwok *et al.* developed and validated for qualitative identification purposes a LC–MS/HRMS method for the detection of 48 anabolic steroids and several of their esters, namely those commonly available in pharmacological preparations.[24] The same laboratory improved the analytical method further to extend the analysis to 72 androgenic steroids and their esters, with estimated LOD values in the sub to low pg mg^{-1} level.[25] Besides steroids, many other pharmaceutical drugs are prohibited in horse racing, making them all the target analytes in anti-doping controls. To test for the presence of these drugs in mane hair, the group mentioned above developed a different analytical method aimed at the detection of 121 prohibited substances, but in this case a low resolution triple-quadrupole was employed.[26] Also in this study, low LOD values were obtained (range: 0.25–10 pg mg^{-1}) despite the fact that considerable ion suppression was observed for most analytes, which was classified as a matrix effect.[26]

An alternative explanation is that the observed ion suppression may arise from the reciprocal interference of 121 substances eluting within a 4 min interval of retention times.

6.4 Multi-targeted Methods for Fentanyl and Other Synthetic Opioids

The substances generally referred to as synthetic opioids were progressively synthetized and registered as prescription drugs to medically treat acute and chronic pain in the most extreme expressions, such as for example terminal cancer. However, the increasing misuse and gradual transition from the legal to the illegal market made these drugs the subject of serious concern, up to the point to become the cause of epidemic overdose deaths in the U.S. and around the world. Nowadays, quite a large number of different synthetic opioids are present in the illegal market, raising the need to develop specific multi-targeted methods to determine their presence in both traditional and alternative biological matrices, including hair. The assortment of the targeted substances embraces both the original pharmaceutical products and several new analogs, specifically synthetized as drugs of abuse, which can also be classified among the NPS.

The choice of the mass spectrometric method used to detect these substances largely depends on the underlying purpose of the analysis. As for other classes of substances, hair testing is adopted to verify misuse during extended periods of time *vs.* episodic intake, and its frequency. The cited widespread misuse stimulated several epidemiological studies devoted to investigating the extension, distribution, and specific features of this phenomenon. For example, hair from a population of heroin-addicted subjects was screened to assess whether intentional or unintentional simultaneous exposure to synthetic opioids was evident.[27] Due to the pharmaceutical potency of fentanyl analogs, they are common heroin adulterants that expose the unaware users to highly increased overdose risks. From the initial drug array (seven substances), the LC–MS/MS method was extended to 13 opioids[28] and applied to authentic hair testing of an increasing population of heroin users ($n = 293$).[29] In these studies, a triple-quad instrument was used, operating in the SRM mode. Notably, LOD values as low as 0.1 pg mg^{-1} were obtained for most targeted analytes and extremely large calibration intervals[28] proved to be necessary to quantify opioid concentrations ranging from the LOQ (0.3 pg mg^{-1}) up to several thousand pg mg^{-1} (8 ng mg^{-1} for fentanyl, 35 ng mg^{-1} for tramadol), depending on whether the detected opioid was accidentally present in a heroin dose or was the substance most frequently abused.[29] In a subsequent study, the same group developed a different analytical technique for opioid detection in hair, based on UHPLC–MS/HRMS, that combined a quadrupole and a TOF mass analyzer.[30] The objective of the novel method was to detect opioid metabolites and unexpected opioids, besides the previously targeted ones, using an untargeted approach. To this purpose, data acquisition cyclically included a TOF high-resolution full MS scan followed by multiple MS/HRMS scans (according to the SWATH™ technique) during a cycle interval of 0.55 s.[30] Even if the LOD values were slightly higher (from 0.2 to 1.2 pg mg^{-1}) than those obtained with the SRM technique previously cited,[28] the analysis of 100 authentic hair samples revealed the sporadic presence of two unexpected fentanyl analogs, not included in the original targeted method. Indeed, the opportunity of retrospective re-examination of previously processed hair samples represents a valuable benefit of untargeted methods, especially in post-mortem investigations.

In the same period (2019–2022), further targeted methods for hair testing were developed in various laboratories to tackle the fentanyl epidemic.[31–36] All these methods used triple-quad MS/MS instruments operating in the SRM mode to detect and quantify up to 27 fentanyl analogs and related opioids.[35] Figures of merit were similar for all methods, the differences mostly depending on the hair extract clean-up procedure and the specific instrument model used for the analysis. A comparison of these differences was summarized in a paper from Platosz *et al.*[35] In particular, significantly higher sensitivity was observed when the method was focused on fentanyl and other synthetic analogs, excluding the classic prescription opioids (morphine, codeine, tramadol, oxycodone, hydrocodone) but including several metabolites of fentanyl analogs.[31] Notably, two studies reported to have carried out segmental hair analysis.[33,36] In the study from Wei and Su, the quantitative results obtained on hair samples spiked by soaking from the LC–MS/MS method are compared with those obtained from a GC–MS/MS method developed in parallel.[36] No significant differences in the mean concentrations of the analytes were observed and both methods were considered suitable for authentic hair sample analysis addressed to fentanyl and analogs.

6.5 Multi-targeted Methods for New Psychoactive Substances

The appearance of new drugs of synthetic origin, collectively called NPS, started in the first decade of the millennium with the onset of the wave of synthetic cannabinoids. Soon afterwards, several other classes of NPS emerged into the black/internet market, followed by an immediate upsurge and rapid turnover of hundreds of new drugs. While the earlier analytical methods for hair testing were focused on cannabinoids,[37–39] it soon became evident that multi-targeted methods should theoretically cover an extremely large number of substances, with widely different chemical structures and physical/chemical properties. However, some analytical methods kept targeting single classes of relatively uniform drugs, depending on the specific populations or applications the method was developed for. A characteristic example is represented by cathinones and other stimulants, entactogens and psychedelic drugs,[40–42] typically abused by the juvenile population at discotheques and musical events.[43,44] The structural homogeneity of the targeted substances makes the analytical methods more robust in terms of extraction, purification, and quantification reproducibility. Most of the latter methods used triple quadrupole instruments in the SRM operating mode, with the exception of one,[42] which exploited full scan and targeted data-dependent MS/HRMS scan acquisitions with an Orbitrap® analyzer for screening and quantification, respectively.

Other examples of analytical methods addressed to specific NPS categories were typically developed to investigate repeated cases of acute or fatal intoxication, which are often associated with the rapid diffusion of these NPS in a certain country or geographical area. Examples of this sort include tryptamines,[45] synthetic benzodiazepines,[46] and cannabinoids.[47] Notably, the latter method included as many as 41 metabolites among the targeted substances, besides 18 cannabinoids, in order to provide further support to positive testing in authentic hair analysis.

Among the analytical procedures for hair and nail testing published in recent years, much more abundant were those targeting very large sets of analytes belonging to different NPS classes.[48–57] Quite clearly, considerable effort was devoted in most of these studies to finding the most appropriate experimental conditions for the extraction, handling, selective recognition, and quantification of disparate substances with highly variable molecular weight, polarity, proton affinity, spectral features, and stability. In 2017, Boumba *et al.* had already proposed a screening method for 132 NPS based on a single-step hair extraction with acidified methanol followed by LC–MS/MS (SRM) detection.[48] Such a wide screening was intended to provide initial clues in medico-legal cases where NPS use was suspected. More recently, Rubicondo *et al.* applied the same extraction method and similar instrumental conditions and purposes, but extended the detection to 49 conventional drugs, besides 120 NPS.[49] Another screening method for 98 NPS was proposed by Nzekoue *et al.*, but in this case M3® reagent was used for hair extraction, instead of acidified methanol.[50] A recent note with limited details claims the simultaneous determination of a 513 NPS record, even if it is unclear how the validation parameters could be maintained under control in the routine work for such a huge number of analytes.[51]

An interesting alternative to conventional hair extraction with methanol was suggested by Vincenti *et al.*, who used a pressurized liquid extractor to perform both the decontamination and extraction steps, followed by a clean-up and concentration step achieved by dispersive liquid/liquid microextraction.[52] An Orbitrap® high-resolution mass analyzer was employed for detection, operating in the parallel reaction monitoring mode. Similarly, high-resolution data were obtained in two further epidemiological studies devoted to the detection of NPS in the hair of selected populations.[53,54] In the first paper, NPS were checked in the hair samples that previously tested positive to ketamine, on the assumption that poly-consumption of ketamine and NPS is expected in the population of young poly-abusers.[53] In the second study, a wide panel of traditional drugs and NPS was checked in the hair of pregnant Mexican women.[54] In both studies, the data-dependent MS/HRMS acquisition mode with full scan of the product ions proved particularly suited for performing a feasible hair screening.

Besides hair, the alternative keratin matrix providing similar chronological data is that of nails. This matrix, by itself or in combination with hair, has also been the biological material on which the presence of NPS and other drugs has been checked. In particular, Mannocchi *et al.* described a LC–MS/MS (SRM) method for the determination of 87 NPS and 32 traditional drugs of abuse on both hair and nails,[55] while Liu *et al.* published an original procedure for determining more than 100 NPS in nail samples.[56] In the latter method, cryogenic grinding of the nails followed by further high-speed fine grinding in the extraction solvent achieved high efficiency and analyte recovery in a short time, allowing its application on a large scale, as in the presented 294 authentic samples.[56] Both methods used triple quadrupoles in the SRM operating mode.

Quite clearly, the analyte's multiplication and differentiation emphasize the uncertainty factors typically affecting hair analysis, that include the lack of knowledge about the real extraction and recovery yields, the variability of hair color, thickness, and permeability, the possible multiplicity of the analyte's incorporation mechanisms, together with a large variety of individual features (phenotypic, behavioral, physio-pathological). In general, these uncontrolled elements make it virtually impossible to correlate the taken drug dose with its hair concentration, with the exception of few widely studied substances, but it is even difficult to guarantee that the detected concentration corresponds to that present in the hair. Analyte multiplication and differentiation adds a further dimension of complexity, namely the practical impossibility to maintain the analytical procedure under strict control for extended periods of time and under the declared validation conditions obtained when the method was developed. As a matter of fact, the extract from an authentic hair sample in which one or two targeted analytes are present is radically different from a validation sample prepared by extracting a blank hair portion spiked with hundreds of substances at low, intermediate, and even high concentration, possibly interfering with one another (at least in the matrix effect assessment).

Knowing all these elements of uncertainty, most of the wide multi-targeted methods were developed for screening purposes only, designating the quantitation to dedicated procedures, possibly committed to a more restricted number of analytes of the same class. In this sense, even the methods in which reliable calibration curves were obtained and accuracy data were determined should be regarded as essentially qualitative or "semi-quantitative", provided that such a definition can make any sense. Positive detection of a drug above a certain cut-off level offers reliable proof of exposure, and guarantees a good level of specificity.

6.6 Untargeted Methods: Looking for Metabolites

The application of hair analysis to queries of forensic toxicology often involves the need for identifying the presence of metabolites of the ingested/injected/inhaled substance together with the parent one, in order to exclude that detection of the latter may arise from external contamination, instead of real intake.[57] Several other reasons occasionally occur to focus the analytical investigation toward the search of metabolites, besides the original drug. However, the hair concentration of these metabolites – whenever present – is generally very low, due to both their higher hydrophilic character, which disfavor their interaction with hair melanin, and the various hair incorporation mechanisms occurring from matrices (blood, sweat, sebum) where the parent substance is prevalent. These circumstances generally make the detection of metabolites troublesome and the instrumental requirements for their detection highly demanding.

While the main metabolites of the traditional drugs have been known for a long time and the analytical methods for hair testing often include them among the targeted analytes, the increasing number of NPS successfully penetrating the black market and their rapid turnover prevent the acquisition of a stable reference framework of metabolites to be looked for. These difficulties do not reduce the importance of gaining knowledge on the number and nature of the metabolites formed from each NPS and incorporated into the hair shaft, for which an increasing number of studies are addressing this task.[58] Besides NPS, also for pharmaceutical drugs, poisons, and pollutants, the simultaneous determination of the parent substance and their metabolites in the hair matrix is creating increasing interest.

Tandem mass spectrometry has been used for the search of unexpected metabolites for a long time. In particular, triple quadrupole instruments were exploited in the so-called "precursor ion scan" and "neutral loss scan" operating modes, in which a structural feature characteristic for both the parent substance and their metabolites yields – by collisional activation – a specific product ion or neutral loss whose corresponding precursor ions are investigated by full MS scanning. In practice, these techniques are "targeted" with respect to the product ion (or neutral loss), and "untargeted" with respect to the precursor ions, but the sequential mode in which quadrupoles scan the mass range results in a significantly reduced sensitivity.

As long as high-resolution mass analyzers became more widely available, the strategies for metabolites' detection evolved significantly, because the opportunity to determine the exact mass and the corresponding elemental composition of precursor and/or product ions in single or tandem mass analysis introduced a crucial improvement in the analytical sensitivity. Since the main primary metabolic processes are basically the same for all drugs, the exact mass of several expected metabolites can be predicted. For example, if an oxidation process introduces a hydroxyl group into the parent drug structure, several isomeric metabolites can be generated, but all having the same elemental composition (and exact mass). Therefore, a simple LC–HRMS experiment conducted in the untargeted operation mode (*i.e.*, detecting all ions at once by means of a TOF or an Orbitrap® mass analyzer) yields a dataset that can subsequently be checked in search of the protonated or deprotonated molecular ions of the expected metabolites. Subsequently, a targeted LC–MS/HRMS experiment may select the metabolites' precursor ions with the first mass analyzer and detect all the highly-resolved product ions after collisional-activated fragmentation. These MS/MS spectra can then be stored to form shared databases, containing continuously growing information.[59,60] Other "at once" experiments involving either data-dependent or data-independent MS acquisition modes may possibly provide comparable information, at the expense of some specificity or sensitivity. It is worth noting that all these strategies have a partially-targeted nature, in that they are all addressed to investigate the metabolic evolution of a certain substance, even when the specific metabolites are not known in advance.

6.7 Untargeted Methods: Metabolomics

Unlike the search for metabolites, what is meant by "metabolomics" or "global metabolomics" corresponds to a truly untargeted investigation. In general, metabolomics investigates the "effect" produced by a certain event on the whole metabolic system of the implicated organisms, resulting in a modification of the physiological spectrum of the substances present in the tested biological specimen (blood, urine, hair, breath, oral fluid, *etc.*). The metabolites with widely increased or decreased concentrations, together with the unexpected ones, become the "biomarkers" of the event that hit the organisms. In clinical biochemistry, this event is generally a pathology or a dysfunction and the biomarkers are used to diagnose this disease or discriminate among different pathologies.[61,62] In forensic toxicology, the event under investigation is commonly the exposure to a certain substance. Within the latter context, hair testing may be used to investigate the long-term effects of the chronical or repeated exposure to certain substances (drugs, poisons, environmental or workplace contaminants).[61] Thus, hair metabolomics looks for any substance whose concentration in hair is significantly and consistently increased (or decreased) by the effect of exposure. These are identified as exposure biomarkers and may, or may not, include direct metabolites of the substance the investigated subjects were exposed to.

In the preceding paragraph, the term "any substance" was underlined just to highlight the strictly untargeted nature of metabolomics investigations. In order to detect the best biomarkers, all the chemical species present in hair should theoretically be identified and at least roughly quantified, on comparative basis (*i.e.*, exposed subjects *vs.* unexposed subjects). This is practically impossible, as long as innumerable substances are present in any hair sample, at highly variable concentration levels. Even the most sophisticated strategies and methods of multi-dimensional chromatography and mass spectrometry are forced to restrict the array of detected substances to few thousand, typically the most abundant ones, excluding the minor metabolites. The resulting big-data files should then be preliminarily handled to align, normalize, and filter the chromatographic profiles and lastly be interpreted by means of machine learning algorithms. Dedicated software is provided by all instrumental producers to undertake all these highly complex tasks, even if the final outcome is frequently beyond the control of the experimenter.

The clear advantage of using hair for metabolomic studies is that the composition of hair is relatively stable over time, not depending on the wide daily fluctuations observed in biological fluids as an effect of diet, circadian rhythm, physical activity, stress.[63,64] Individual daily fluctuations represent a major source of noise when metabolomic profiles of groups of subjects are compared. On the other side, hair metabolomics is unable to give prominence to the short-term effects of a single acute event, due to the unselective character of the approach.

All the substances forming the metabolomic profile are those captured in the inner part of the hair shaft, with the exclusion of surface contaminants. Therefore, the initial subject of investigation is to find the most appropriate decontaminating conditions to clean the hair surface from any impurity and the most effective extraction conditions to recover all the substances from its inner part.[61,63,64] A different approach gave value to both solutions of hair wash and hair extract to highlight short-term and long-term exposures, respectively.[65] In any of these solutions, about 10 000 substances were initially characterized by their MS features, then about 3000 were filtered, 274–276 were fully identified, and 645–89 were recognized as significant, respectively.[65] These numbers are cited as roughly representative of what is expected from a hair metabolomic study conducted by a LC–MS/HRMS instrument in either data-dependent of data-independent operating modes.

Other potential sources of bias in metabolomic investigations are hair color,[66] hair length,[66] and cosmetic treatments.[67] It is well known that the melanin content in hair influences both its color and its affinity toward certain substances, making their incorporation increase with the eumelanin concentration.[1] In particular, lipophilic and basic substances are supposed to exhibit the highest discrimination between dark and blond hair, being captured more extensively by dark hair,[66] even if no systematic study has so far been published on the correlations occurring among hair color, molecular properties, and differential hair incorporation. In metabolomic studies that compare different cohorts of subjects to highlight a specific difference between them (*e.g.*, the exposure to a pollutant or the occurrence of a pathological condition), the hair color represents a confounding factor to be considered, likewise hair aging and frequent exposure to solar irradiation.[66] Also, cosmetic treatments, such as hair bleaching and perming, modify to various extent the metabolome.[67] In particular, Eisenbeiss *et al.* applied untargeted metabolomics to identify undeclared treatments addressed to alter the toxicological results from hair testing.[67]

Several applications of hair metabolomics for toxicological purposes are present in the recent literature, even if only few of them study the long-term global effect of drug abuse.[68–70] Jeong, Lee and coworkers investigated the metabolome modifications associated to the mental disorders affecting chronic methamphetamine abusers.[68,69] Likewise, Xie *et al.* studied the hair metabolomic profile of heroin abusers to verify the biochemical modifications induced by chronic opiate consumption.[70] Other published papers reported the use of untargeted MS analysis to identify the metabolites produced after some NPS consumption. Larabi *et al.* took advantage of data-dependent acquisition to identify nine metabolites of deschloro-*N*-ethyl-ketamine in the hair of a deceased abuser.[71] The acquisition cycles included one full HRMS scan and up to five data-dependent MS/HRMS product ions scans. Other untargeted methods for NPS metabolites identification were applied to *in vitro* cultures and were supported by *in silico* predictions; the presence of expected metabolites was subsequently confirmed in hair following a targeted approach.[72,73]

The clinical applications of metabolomics are by far the most commonly recurring in the scientific literature, but hair has been rarely selected as the matrix of choice for these applications. Nevertheless, hair metabolomics may provide unparalleled insight in a variety of clinical purposes: diagnosis, pathology evolution over time and monitoring, prognosis, and evaluation therapy efficacy.[62] Pioneering studies on hair metabolomics used GC–MS instrumentation for sample analysis,[74,75] since EI ionization provides an extremely valuable tool for metabolite identification, even if it restricts the array of detectable substances to the volatile ones. The cited paper investigated the pregnancy complications determined by fetal growth restriction.[74] The pregnancy complications were the subject of a following review, suggesting hair as an ideal matrix for metabolomics studies, even if all the literature cited therein only targeted specific analytes, indirectly demonstrating that hair metabolomics was still in the immature stage in 2018.[76] The subsequent hair metabolomics studies on the effects of air pollution on gestational diabetes mellitus[77] and cervical cancer[78] substantially improved the analytical approach and data analysis strategy, but still relied on GC–MS instrumentation.

The third application area of hair metabolomics has been environmental toxicology.[79,80] In a 2017 study, tetrahydroxylated polycyclic aromatic hydrocarbons (PAH) were determined in hair collected from PAH-exposed rats.[79] These metabolites were suggested as possible biomarkers for the formation of their DNA-adducts, which after all were experimentally identified in diverse biological matrices. Although the whole analytical procedure was explicitly targeted, the spectrum of different PAHs and isomers was very broad and the sensitivity requirements very strict. The application of an appropriate metabolites' derivatization, followed by electron-capture negative ionization combined with GC–MS/MS, operating in the SRM mode, provided adequate sensitivity to detect as many as 23 tetrahydroxylated PAHs.[79] In a different study, the general metabolomic patterns obtained from urine and hair of rats exposed to di-(2-propylheptyl)-phthalate were compared by means of several machine-learning techniques.[80] Full scan HRMS spectra were collected by a TOF mass analyzer after UHPLC separation and the chemical structure of the exposure metabolite biomarkers was tentatively identified by subsequent MS/MS experiments. Indeed, this is one of the few published studies in which true general hair metabolomics is achieved, resulting in an extended list of 62 useful metabolites, whose concentration differences between treated and untreated animals proved statistically significant.[80]

6.8 Post-mortem and Drug-facilitated Crimes

Further classes of forensic inquiries in which both large multi-analyte targeting and broad untargeted methods find valuable application are those related to drug-facilitated crimes and post-mortem investigations. When the administration of an intoxicating substance is alleged but its chemical identity is totally unknown nor can a hypothesis be made on its pharmacological classification, then the toxicological investigation should theoretically adopt an untargeted method of analysis, possibly including a data-dependent or data-independent UHPLC–MS/HRMS procedure. However, the number and variety of drugs and poisons actually acquirable by an ordinary criminal is not limitless, making the list of candidate drugs/poisons certainly wide yet countable. In any circumstance in which an acute intoxication had occurred, hair is not considered the matrix of choice, unless a considerable time had elapsed between administration and biological fluid sampling. In particular, segmental hair analysis is considered to be of unique interest when a single drug administration needs to be chronologically pinpointed, but in this case high sensitivity requirements are needed. The combination of the two cited factors (limited number of candidate analytes and high sensitivity requirement) makes the use of wide multi-analyte targeted methods more practical than untargeted methods of analysis.

In 2022, Niebel *et al.* published a retrospective study in which hair samples from over 1200 post-mortem cases occurring between 2008 and 2020 were re-analyzed in search of possible previously undetected NPS.[81] The updated LC–MS/MS (SRM) procedure targeted 58 relatively common NPS, revealing a relatively widespread use of most of these new synthetic drugs, as was demonstrated by 381 positive testing (31.6%) and 48 detected substances in at least one case.[81]

A recent review compares the published analytical methods used to investigate the forensic cases of drug-facilitated sexual assaults, confirming that targeted LC–MS/MS (SRM) methods are by far the most widely employed ones, due to their sensitivity, selectivity, and wide applicability.[82] While most of these procedures are implemented on triple quadrupole instruments,[82] high resolution mass analyzers are increasingly used, especially when confirmation methods for specific classes of drugs are required, such as for example synthetic ketamines[83] and barbiturates.[84]

6.9 Conclusions

The appearance and sudden upsurge of the NPS problem in the second decade of the millennium, with the intoxication cases related to them, captured most of the recent research efforts dedicated to hair testing. The new theme, together with the synthetic opioid epidemics, attracted considerable interest and required innovative analytical approaches because the number of candidate target analytes and the corresponding metabolites abruptly grew by one order-of-magnitude, at least. While the subject of new synthetic drugs (identification and quantification in hair, metabolite characterization, *etc.*) is still far from coming to an end, some repetitiveness is perceptible in the dedicated scientific literature. This is likely to direct really innovative research toward unprecedented objectives and applications of hair analysis.

As was mentioned in Section 6.2, innovative research in hair testing is continuously empowered by the technological improvements of analytical instrumentation and related software of data management. In particular, parallel progress of hair testing ambitions with mass spectrometric devices, performances, and MS data interpretation is expected. In the subject matter of multi-analyte methods, a very significant step has already been made by the sharp improvement of the sensitivity, speed, and resolution of mass analyzers in combination with ion-focusing devices. If anything, the simultaneous detection of hundreds of target analytes has created new puzzles, inasmuch as the conditions found in real hair extracts appear to be progressively farthest from the spiked surrogates used for analytical method development and validation. New protocols should be elaborated to reduce this gap.

More striking innovation is expected by the continuous development and diffusion of high-resolution mass analyzers, which are manufactured by a multiplicity of highly competing producers. The combination of extremely effective chromatographic systems, that generate fast-evolving eluate composition profiles, with ultra-high-resolution mass analyzers, characterized by high-frequency cycle scanning, creates an enormous flow of data requiring fast storage and subsequent powerful mining. Indeed, MS manufacturers nowadays hire more software developers than hardware specialists to support effective data mining programs. Within this context, hair analysis is likely to take considerable advantage from HRMS evolution. In particular, the research devoted to hair metabolomics will probably find a substantial growth in the near future, taking into account that hair metabolomics delivers a profiling for the tested subjects largely different and independent from those arising from blood, urine, and breath. The ability of hair metabolomics to generate profiles free from short-term random modifications will likely bring this discipline out of the current early-stage of development, since it provides a unique synopsis of chronic exposure. The drawback of the extremely limited abundance of most metabolites present in the hair matrix has already been overcome by the sensitivity and resolution performances guaranteed by the current HRMS instrumentation. The consolidation of technological advancements and data-mining software improvements will prospectively support further incentive to hair metabolomics studies.

References

1. P. Kintz, A. Salomone and M. Vincenti, *Hair Analysis in Clinical and Forensic Toxicology*, Academic Press, London, 2015.
2. J. T. Watson and O. D. Sparkman, *Introduction of Mass Spectrometry: Instrumentation, Applications, and Strategies for Data Interpretation*, John Wiley & Sons, Inc., Chichester, United Kingdom, 4th edn, 2007.
3. M. W. Dong, *HPLC and UHPLC for Practicing Scientists*, John Wiley & Sons, Inc., Hoboken, New Jersey, 2nd edn, 2019.
4. J. H. Gross, *Mass Spectrometry: A Textbook*, Springer International Publishing, Cham, Switzerland, 3rd edn, 2017.
5. J. Guo and T. Huan, *Anal. Chem.*, 2020, **92**, 8072.
6. P. Kintz, *Analytical and Practical Aspects of Drug Testing in Hair*, CRC Press, Boca Raton, FL, USA, 2007.
7. T. Baciu, F. Borrull, C. Aguilar and M. Calull, *Anal. Chim. Acta*, 2015, **856**, 1.
8. F. Franz, H. Jechle, V. Angerer, M. Pegoro, V. Auwärter and M. A. Neukamm, *Anal. Chim. Acta*, 2018, **1006**, 61.
9. H. C. Chen, P. D. Lee and Y. Z. Chang, *J. Pharm. Biomed. Anal.*, 2022, **213**, 114678.
10. H. Khajuria, B. P. Nayak and A. Badiye, *Med., Sci. Law*, 2018, **58**, 137.
11. M. M. Madry, T. Kraemer and M. R. Baumgartner, *Forensic Sci. Int.*, 2018, **282**, 137.
12. Y. Shin, T. Y. Kong, J. C. Cheong, J. Y. Kim, J. I. Lee and H. S. Lee, *Biomed. Chrom.*, 2019, **9**, 4600.
13. F. Krumbiegel, M. Hastedt, L. Westendorf, A. Niebel, M. Methling, M. K. Parr and M. Tsokos, *Forensic Sci. Med. Pathol.*, 2016, **12**, 416.
14. X. Wang, S. S. Johansen, M. K. Klose Nielsen and K. Linnet, *Drug Test. Anal.*, 2017, **9**, 1137.
15. C. Scholz, J. Cabalzar, T. Kraemer and M. R. Baumgartner, *J. Anal. Toxicol.*, 2021, **45**, 701.
16. E. Alladio, E. Amante, C. Bozzolino, F. Seganti, A. Salomone, M. Vincenti and B. Desharnais, *Talanta*, 2020, **215**, 120867.
17. W. L. Cheng, C. Markus, C. Y. Lim, R. Z. Tan, S. K. Sethi and T. P. Loh, *Ann. Lab. Med.*, 2023, **43**, 5.
18. C. Scholz, M. R. Baumgartner, T. Kraemer and T. M. Binz, *Anal. Methods*, 2022, **14**, 4583.
19. A. G. Nicolaou, M. C. Christodoulou, I. J. Stavrou and C. P. Kapnissi-Christodoulou, *J. Chromatogr. A*, 2021, **1651**, 462277.
20. W. H. Abd-Elsalam, M. A. Alsherbiny, J. Y. Kung, D. W. Pate and R. Löbenberg, *Talanta*, 2019, **204**, 846.

21. M. Cobo Golpe, A. de Castro Ríos, A. Cruz, M. Páramo, M. López Rivadulla and E. Lendoiro, *J. Pharm. Biomed. Anal.*, 2020, **189**, 113443.
22. M. Grabenauer, N. D. Bynum, K. N. Moore, R. M. White, J. M. Mitchell, E. D. Hayes and R. Flegel, *J. Anal. Toxicol.*, 2018, **42**, 115.
23. A. Narváez, Y. Rodríguez-Carrasco, A. Ritieni and J. Mañes, *J. Chromatogr. B: Anal. Technol. Biomed. Life Sci.*, 2022, **1191**, 123117.
24. K. Y. Kwok, T. L. S. Choi, W. H. Kwok, J. K. Y. Wong and T. S. M. Wan, *J. Chromatogr. A*, 2017, **1493**, 76.
25. T. L. S. Choi, K. Y. Kwok, W. H. Kwok, Y. Y. K. Tsoi, J. K. Y. Wong and T. S. M. Wan, *J. Chromatogr. A*, 2018, **1566**, 51.
26. J. K. Y. Wong, T. L. S. Choi, K. Y. Kwok, E. N. Y. Lei and T. S. M. Wan, *J. Pharm. Biomed. Anal.*, 2018, **158**, 189.
27. J. J. Palamar, A. Salomone, R. Bigiarini, M. Vincenti, P. Acosta and B. Tofighi, *Am. J. Drug Alcohol Abuse*, 2019, **45**, 90.
28. A. Salomone, J. J. Palamar, R. Bigiarini, E. Gerace, D. Di Corcia and M. Vincenti, *J. Anal. Toxicol.*, 2019, **43**, 259.
29. A. Salomone, R. Bigiarini, J. J. Palamar, C. McKnight, L. Vinsick, E. Amante, D. Di Corcia and M. Vincenti, *Anal. Toxicol.*, 2020, **44**, 362.
30. A. Salomone, D. Di Corcia, P. Negri, M. Kolia, E. Amante, E. Gerace and M. Vincenti, *Anal. Bioanal. Chem.*, 2021, **413**, 225.
31. F. P. Busardò, J. Carlier, R. Giorgetti, A. Tagliabracci, R. Pacifici, M. Gottardi and S. Pichini, *Front. Chem.*, 2019, **7**, 184.
32. I. A. Larabi, M. Martin, I. Etting, G. Pfau, Y. Edel and J. C. Alvarez, *Drug Test. Anal.*, 2020, **12**, 1298.
33. M. M. Ramírez Fernández, S. M. R. Wille, D. Jankowski, V. Hill and N. Samyn, *Forensic Sci. Int.*, 2020, **307**, 110.
34. F. Freni, M. Moretti, D. Radaelli, C. Carelli, A. M. M. Osculati, L. Tronconi, C. Vignali and L. Morini, *J. Pharm. Biomed. Anal.*, 2020, **189**, 113476.
35. N. A. Platosz, T. M. Binz, M. R. Baumgartner, E. Lendoiro, A. de Castro and M. Concheiro, *J. Anal. Toxicol.*, 2021, **45**, 943.
36. Q. Wei and F. Hai Su, *ACS Omega*, 2022, **7**, 19176.
37. J. Znaleziona, P. Ginterová, J. Petr, P. Ondra, I. Válka, J. Ševčík, J. Chrastina and V. Maier, *Anal. Chim. Acta*, 2015, **874**, 11.
38. F. Franz, H. Jechle, V. Angerer, M. Pegoro, V. Auwärter and M. A. Neukamm, *Anal. Chim. Acta*, 2018, **1006**, 61.
39. A. G. Nicolaou, M. C. Christodoulou, I. J. Stavrou and C. P. Kapnissi-Christodoulou, *J. Chromatogr. A*, 2021, **1651**, 462277.
40. A. Salomone, G. Gazzilli, D. Di Corcia, E. Gerace and M. Vincenti, *Anal. Bioanal. Chem.*, 2016, **408**, 2035.
41. E. Lendoiro, C. Jiménez-Morigosa, A. Cruz, M. Páramo, M. López-Rivadullaa and A. de Castro, *Drug Test. Anal.*, 2017, **9**, 96.
42. N. La Maida, G. Mannocchi, S. Pichini, G. Basile, A. Di Giorgi, F. P. Busardò and E. Marchei, *Eur. Rev. Med. Pharmacol. Sci.*, 2022, **26**, 5033.
43. J. J. Palamar, A. Salomone, M. Vincenti and C. M. Cleland, *Drug Alcohol Depend.*, 2016, **161**, 200.

44. J. J. Palamar, A. Salomone, E. Gerace, D. Di Corcia, M. Vincenti and C. M. Cleland, *Int. J. Drug Policy*, 2017, **48**, 91.
45. Y. Shi, R. Wang, S. Yuan, H. Qiang, M. Shen, B. Shen, O. H. Drummer, Z. Yu, Y. Zhao and P. Xiang, *J. Chromatogr. B: Anal. Technol. Biomed. Life Sci.*, 2020, **1159**, 122392.
46. B. Cho, H. S. Cho, J. Kim, J. Sim, I. Seol, S. K. Baeck, S. In, D. H. Shin and E. Kim, *Forensic Sci. Int.*, 2020, **306**, 110058.
47. L. DeFreitas, A. M. Fonseca Pego, R. Kronstrand, E. Lendoiro, A. de Castro-Ríos and M. Concheiro, *J. Anal. Toxicol.*, 2022, **46**, 852.
48. V. A. Boumba, M. Di Rago, M. Peka, O. H. Drummer and D. Gerostamoulos, *Forensic Sci. Int.*, 2017, **279**, 192.
49. J. Rubicondo, L. Scuffi, L. Pietrosemoli, M. Mineo, F. Terranova, M. Bartucca, C. Trignano, E. Bertol and F. Vaiano, *J. Anal. Toxicol.*, 2022, **46**, e262.
50. F. K. Nzekoue, M. Agostini, M. Verboni, C. Renzoni, L. Alfieri, S. Barocci, M. Ricciutelli, G. Caprioli and S. Lucarini, *J. Pharm. Biomed. Anal.*, 2021, **205**, 114310.
51. J. Musiał, J. Powierska-Czarny, J. Czarny, M. Raczkowski, N. Galant, B. Buszewski and R. Gadzała-Kopciuch, *Arch. Toxicol.*, 2022, **96**, 2927.
52. F. Vincenti, C. Montesano, L. Cellucci, A. Gregori, F. Fanti, D. Compagnone, R. Curini and M. Sergi, *J. Chromatogr. A*, 2019, **1605**, 360348.
53. J. M. Matey, G. Montalvo, C. García-Ruiz, F. Zapata, A. López-Fernández and M. A. Martínez, *Forensic Sci. Int.*, 2021, **329**, 111080.
54. E. Marchei, M. C. Rotolo, G. Mannocchi, A. Capomassi, L. M. Gómez-Ruiz, A. Acosta-López, R. Y. Ramos-Gutiérrez, M. B. Varela-Busaka, S. Pichini and Ó. García-Algar, *J. Pharm. Biomed. Anal.*, 2022, **211**, 114607.
55. G. Mannocchi, A. Di Trana, A. Tini, S. Zaami, M. Gottardi, S. Pichini and F. P. Busardò, *Anal. Bioanal. Chem.*, 2020, **412**, 5125.
56. P. Liu, W. Liu, H. Qiao, S. Jiang, Y. Wang, J. Chen, M. Su and B. Di, *Anal. Chim. Acta*, 2022, **1226**, 340170.
57. O.H. Drummer, D. Gerostamoulos, M.A. LeBeau and F. Pragst, *J. Anal. Toxicol.*, 2020, **44**, e6.
58. J. M. Matey, A. López-Fernández, C. García-Ruiz, G. Montalvo, M. D. Moreno and M. A. Martínez, *J. Anal. Toxicol.*, 2022, **46**, e1.
59. N. Fabresse, I. A. Larabi, T. Stratton, R. Mistrik, G. Pfau, G. L. de la Grandmaison, I. Etting, S. Grassin-Delyle and J.-C. Alvarez, *Drug Test. Anal.*, 2019, **11**, 697.
60. M. Mardal, P. W. Dalsgaard, B. S. Rasmussen, K. Linnet and C. B. Mollerup, *Anal. Chem.*, 2023, **95**, 4592.
61. Y. R. Lee, J. Hong and B. C. Chung, *Mass Spectrom. Lett.*, 2022, **13**, 2.
62. W.-J. Jang, J. Y. Choi, B. Park, J. H. Seo, Y. H. Seo, S. Lee, C.-H. Jeong and S. Lee, *Molecules*, 2019, **24**, 2195.
63. L. Eisenbeiss, A. E. Steuer, T. M. Binz, M. R. Baumgartner and T. Kraemer, *Anal. Bioanal. Chem.*, 2019, **411**, 3963.

64. W. C.-W. Chang, P.-H. Wang, C.-W. Chang, Y.-C. Chen and P.-C. Liao, *Talanta*, 2021, **223**, 121708.
65. Y. Chen, J. Guo, S. Xing, H. Yu and T. Huan, *Front. Chem.*, 2021, **9**, 674265.
66. L. Eisenbeiss, T. M. Binz, M. R. Baumgartner, T. Kraemer and A. E. Steuer, *Metabolites*, 2020, **10**, 381.
67. L. Eisenbeiss, T. M. Binz, M. R. Baumgartner, T. Kraemer and A. E. Steuer, *Analyst*, 2020, **145**, 6586.
68. S. Kim, W.-J. Jang, H. Yu, J. Kim, S.-K. Lee, C.-H. Jeong and S. Lee, *Int. J. Mol. Sci.*, 2020, **21**, 6041.
69. M. J. Seo, S.-H. Song, S. Kim, W. J. Jang, C.-H. Jeong and S. Lee, *Arch. Pharm. Res.*, 2021, **44**, 890.
70. P. Xie, T.-J. Wang, G. Yin, Y. Yan, L.-H Xiao, Q. Li and K.-S. Bi, *J. Mol. Neurosci.*, 2016, **58**, 93.
71. I. A. Larabi, F. Zerizer, A. Ameline, I. Etting, D. Joseph, P. Kintz and J.-C. Alvarez, *Drug Test. Anal.*, 2021, **13**, 1108.
72. S. Mestria, S. Odoardi, S. Federici, S. Bilel, M. Tirri, M. Marti and S. Strano Rossi, *J. Anal. Toxicol.*, 2021, **45**, 475.
73. R. Goncalves, N. Castaing, C. Richeval, D. Ducint, K. Titier, E. Morvan, A. Grélard, A. Loquet and M. Molimard, *Forensic Sci. Int.*, 2022, **333**, 111215.
74. K. Sulek, T.-L. Han, S. Granato Villas-Boas, D. S. Wishart, S.-E. Soh, K. Kwek, P. D. Gluckman, Y.-S. Chong, L. C. Kenny and P. N. Baker, *Theranostics*, 2014, **4**, 953.
75. J. Yang, Y. Wei, H. Qi, N. Yin, Y. Yang, Z. Li, L. Xu, X. Wang, P. Yuan, L. Li, T.-L. Han and Y. Zhao, *Mol. Med.*, 2020, **26**, 37.
76. T. D. J. Delplancke, Y. Wu, T.-L. Han, L. R. Joncer, H. Qi, C. Tong and P. N. Baker, *BioMed Res. Int.*, 2018, 2815439.
77. X. Chen, X. Zhao, M. B. Jones, A. Harper, J. V. de Seymour, Y. Yang, Y. Xia, T. Zhang, H. Qi, J. Gulliver, R. D. Cannon, R. Saffery, H. Zhang, T.-L. Han and P. N. Baker, *Front. Endocrinol.*, 2022, **13**, 1060309.
78. R. Ran, X. Zhong, Y. Yang, X. Tang, M. Shi, X. Jiang, A. Lin, X. Gan, T. Yu, L. Hu, X. Dong and T.-L. Han, *Med. Oncol.*, 2023, **40**, 75.
79. N. Grova, J.-P. Antignac, E. M. Hardy, F. Monteau, K. Pouponneau, B. Le Bizec and B. M. R. Appenzeller, *Anal. Chim. Acta*, 2017, **995**, 65.
80. C.-L. Shih, H.-Y. Wu, P.-M. Liao, J.-Y. Hsu, C.-Y. Tsao, V. G. Zgoda and P.-C. Liao, *Anal. Chim. Acta*, 2019, **1052**, 84.
81. A. Niebel, L. Westendorf, F. Krumbiegel, S. Hartwig, M. K. Parr and M. Tsokos, *Drug Test. Anal.*, 2022, **14**, 110.
82. N. Almofti, A. Ballesteros-Gómez, S. Rubio and E. Girela-López, *Talanta*, 2022, **250**, 123713.
83. S. Mestria, S. Odoardi, G. Biosa, V. Valentini, G. Di Masi, F. Cittadini and S. Strano-Rossi, *Forensic Sci. Int.*, 2021, **323**, 110817.
84. D. Wen, Y. Shi, X. Zhang, B. Xie, W. Liu, F. Yu, P. Xiang, B. Cong and C. Ma, *Forensic Sci. Res.*, 2022, **7**, 78.

CHAPTER 7

Markers of Alcohol Use

L. MORINI,* M. MORETTI AND C. CARELLI

Department of Public Health, Experimental and Forensic Medicine, University of Pavia, Italy
*Email: luca.morini@unipv.it

7.1 Introduction

Alcohol abuse and misuse constitute a growing problem with relevant clinical, social, economic, and legal implications.[1] Epidemiological studies consistently report worldwide increases in the percentage of alcohol users, in the amount consumed per capita and in the extent of alcohol-related mortality and disability, along with a decrease in the age of first use. In 2015, the World Health Organization (WHO) stated that almost 6% of worldwide deaths are linked to excessive alcohol use, with the situation worsening significantly (25%) among young people (age group 20–39).[2]

According to the global status report on alcohol and health of 2018, released by the WHO, a constant increase of total alcohol per capita consumption was observed between 2005 and 2016, among people over 15 years of age. In particular, the report highlighted that the highest amounts of alcohol consumption are recorded in the European Region, though a declining trend was observed recently.[3]

Depending on the monitored substance, hair analysis could provide important data regarding the past and/or continuative alcohol consumption. In particular, WHO defines heavy episodic drinking (HED) as the consumption of 60 or more grams of pure alcohol on at least one occasion at least once per month. Hence, the sensitivity and specificity of an alcohol biomarker in diagnosing chronic excessive consumption will depend on its ability to discriminate an ethanol (EtOH) daily intake above 60 g.[3]

Issues in Toxicology No. 47
Perspectives and Challenges of Hair Analysis
Edited by Pascal Kintz, Alberto Salomone and Marco Vincenti

Published by the Royal Society of Chemistry, www.rsc.org

It is also important to report the definitions of alcohol abuse and dependence in accordance with the Diagnostic and Statistical Manual of mental disorders (DSM IV and V). The DSM-IV (published in 1994) likewise had no "alcoholism" diagnosis but instead described two distinct disorders—alcohol abuse and alcohol dependence—with specific criteria for each diagnosis. The DSM-5 combines those two disorders into one alcohol use disorder with sub-classifications of severity.[4]

7.1.1 Alcohol Dependence (DSM IV)

A pathological pattern of substance use that leads to clinically significant impairment or discomfort as manifested by three or more of the following conditions occurring at any time during the same 12-month period:

(1) tolerance
  (a) need for significantly higher doses of the substance to achieve intoxication or desired effect
  (b) a significantly diminished effect with continued use of the same amount of the substance
(2) withdrawal, as manifested by each of the following:
  (a) characteristic withdrawal syndrome
  (b) the same substance is taken to lessen or avoid withdrawal symptoms
(3) the substance is often taken in larger amounts or for longer periods than expected by the subject
(4) persistent desire or unsuccessful attempts to reduce or control substance use
(5) large amounts of time are spent on activities necessary to obtain the substance, to take it, to recover from its effects
(6) interruption or reduction of important social, work, or recreational activities due to the use of the substance
(7) continued use of the substance despite awareness of having persistent problem recurrent physical or psychological problem, likely caused or exacerbated by alcohol use.

7.1.2 Alcohol Abuse (DSM IV)

A pathological pattern of alcohol use that brings recurrent clinically significant impairment or distress within a 12-month period manifested by one or more of the following conditions:

(1) continued substance use with the inability to perform major tasks related to work, home, school (absences, neglect)
(2) recurrent use of alcohol in physically hazardous situations (driving, operating machinery...)
(3) legal problems related to substance use (arrest for fighting, resisting...)

(4) continued use of alcohol despite persistent social or interpersonal problems caused or exacerbated by alcohol use (family arguments over consequences of intoxication, physical confrontations).

Although there is a lot of overlap between the criteria (list of symptoms) used by the DSM-IV and the DSM-5, there are two significant changes. The DSM-5 eliminates having legal problems as a result of drinking as a criterion for diagnosis but adds craving for alcohol as a criterion.

7.2 Alcohol Biomarkers – Ethyl Glucuronide

Effective strategies to contain the increasing diffusion of risky drinking behaviours crucially rely on the availability of objective diagnostic tests. Quantification of ethyl glucuronide (EtG) in hair (hEtG) has established itself, over the past decade, as the most reliable indicator of long-term drinking patterns in both clinical and forensic settings.[5–11] Examples of clinical uses include: identification of relapses in drinking;[12] screening for alcohol-use disorders;[13,14] screening for fetal alcohol exposure;[15–19] motivating changes in drinking behaviour in at-risk individuals;[20] evaluating effectiveness of interventions for alcohol-related problems;[21,22] assessing liver transplant eligibility.[23] Examples of forensic uses include: post-mortem cases;[24] child-custody cases;[25] driving offenses and re-instatement of driving license;[20,26–28] firearm license eligibility; establishing alcohol use after detoxification; establishing alcohol use after detoxification or during probation; loss of employment; workplace screening.[29,30] Furthermore, use of hEtG in epidemiological studies to investigate drinking patterns and better define 'at-risk drinking' and its correlations to chronic disease and all-cause mortality is also gaining popularity.[20]

Alcohol is mostly (90–98%) metabolized in the liver by alcohol dehydrogenase, while 2–8% excreted unchanged mainly through breath and urine and about 1.5% goes through a non-oxidative metabolism to β-D-ethyl glucuronide (EtG), ethyl sulphate (EtS), ethyl phosphate (EtP), phosphatidyl ethanol (PEth), and fatty acid ethyl esters (FAEEs).[31–33]

EtG is a minor non-oxidative ethanol metabolite ($\leq$1% of ingested ethanol) resulting from enzymatic ethanol (EtOH) glucuronidation in the liver. This small, polar, slightly acidic, and relatively stable molecule was found in hair in 1997 by Sachs[34] and later by Skopp *et al.*[5] EtG in hair has been studied as a potential biomarker for monitoring past and continuative alcohol ingestion.[8,35–42] It can be detected in several body fluids and tissues for variable intervals of time after ethanol ingestion: blood and urine EtG levels are commonly used to assess short-term consumption (up to a few days), whereas keratinized matrices (mainly hair) are increasingly employed to monitor abstinence and chronic abuse (over months); detection of EtG in meconium, on the other hand, is of great interest – along with maternal and neonatal PEth analysis in whole blood or dried blood spots (DBS) – as a reliable diagnostic tool of gestational drinking and fetal alcohol exposure.[16,43–49] Reliable quantification of EtG in both traditional (*i.e.*, blood and derivatives, urine) and non-traditional (*e.g.*, hair, nails, meconium) matrices is typically carried out using chromatographic (GC or LC) methods coupled to single or tandem mass spectrometry (MS); given EtG's physicochemical properties, combined with the much simpler sample preparation compared to GC methods, LC–MS/MS in negative ion mode has emerged as the most suitable method for hEtG analysis and is currently the most commonly used technique for investigation of this alcohol-use marker, although no standardized protocol(s) has been devised so far.

7.2.1 Advancements in hEtG Analysis

An overview of published GC–MS and LC–MS/MS methods for hEtG analysis is provided in Tables 7.1 and 7.2. The first attempt to systematically measure EtG in hair was made in 2000 by Skopp *et al.*,[5] who developed and validated a GC–EI–MS method employing methyl glucuronide as the internal standard, obtaining a limit of detection (LOD) = 2200 pg mg^{-1} and a lower limit of quantification (LLOQ) = 5000 pg mg^{-1} (using 50 mg of hair), which in practice only allowed detection of EtG positivity in some heavy drinkers, rather than reliable quantification. In 2002, Janda *et al.*[50] developed the first LC–ESI–MS/MS method, introducing an isotopically labeled internal standard (EtG-D5) and obtaining LOD = 51 pg mg^{-1} and LLOQ = 102 pg mg^{-1} (using 100 mg of hair). Post-column addition of acetonitrile was adopted to improve the ionization process without influencing the chromatographic separation, hence overcoming the sensitivity problems caused by EtG's highly polar nature. In 2004, two novel GC–MS methods were published by Jurado *et al.*[51] and Yegles *et al.*,[8] respectively obtaining LLOQ = 50 pg mg^{-1} (using 100 mg of hair) and LLOQ = 6 pg mg^{-1} (using 30 mg of hair). Both methods utilized penta-deuterated EtG (EtG-D5) as the internal standard, ultrasonication in water as the extraction technique and PFPA as the derivatizing agent, however, they differed in terms of pulverization technique (respectively, manually cut and ball milled), of absence[51] or presence[8] of solid phase extraction (SPE) clean-up and of ionization mode (respectively, electron impact (EI) and negative chemical ionization (NCI)).

Differences in sample preparation among the above-mentioned methods also involved the decontamination step, as they all comprised two subsequent washings but with different solvents (see Tables 7.1 and 7.2). In 2006, Morini *et al.*[9] developed and fully validated a highly sensitive (LOD = 2 pg mg^{-1} and LLOQ = 3 pg mg^{-1} using 100 mg of hair) LC–ESI–MS/MS method, employing EtG-D5 as internal standard, post-column addition of acetonitrile and, for the first time, dichloromethane and methanol for the washing procedure and multiple reaction monitoring (MRM) mode to operate the triple-quadrupole MS. Most of the subsequently-proposed LC–MS/MS

Table 7.1 Overview of the available GC–MS methods for determination of hETG.[a]

Authors, year [ref.]	Amount of hair (mg)	Decontamination	Homogenization	Extraction	Clean-up	Derivatization	Analytical technique	Monitored ions/ transitions (m/z)	LLoQ ($pg\,mg^{-1}$)
Skopp *et al.*, 2000[5]	50	5 ml ether, 5 ml acetone	Pulverized	0.25 ml deionized water, 1 ml methanol (5 h incubation, 3 h sonication)	—	MSTFA	GC–EI–MS	261; 292; 375; 405	5000.0
Jurado *et al.*, 2004[51]	100	Deionized water, acetone	Cut	2 ml deionized water (2 h sonication, overnight incubation)	—	PFPA	GC–EI–MS	333; 234; 495	50.0
Yegles *et al.*, 2004[8]	30	Deionized water, heptane	Pulverized	2 ml deionized water (2 h sonication)	Isolute NH_2 SPE	PFPA	GC–NCI–MS	496; 347	6.0
Paul *et al.*, 2008[14]	10	Methanol	Cut	1 ml deionized water (overnight sonication)	Oasis®MAX SPE	BSTFA	GC–EI–MS/MS	261 → 143	20.0
Paul *et al.*, 2011[56]	20								10.0
Pragst *et al.*, 2008[57] Kerekes *et al.*, 2009[58]	30	Deionized water, acetone	Pulverized	2 ml deionized water (2 h sonication)	Oasis®MAX SPE	PFPA	GC–NCI–MS	496; 349	2.3
Kharbouche *et al.*, 2009[59]	30	Deionized water, acetone	Pulverized	1 ml deionized water (2 h sonication)	Oasis®MAX SPE	PFPA	GC–NCI–MS/MS	347 → 163 347 → 119	8.4
Álvarez *et al.*, 2009[60]	100	5 ml 0.1% polysorbate 80 (2×), 5 ml deionized water	Cut	4 ml deionized water, 4 ml hexane (11 min microwave-assisted extraction at 110 °C)	—	BSTFA	GC–EI–MS	261; 160; 405	300.0

Shi *et al.*, 2010[61]	20	Deionized water, dichloromethane, methanol	Cut	1 ml deionized water (1 h sonication, overnight incubation)	—	BSTFA	LVI-GC–EI–MS/MS	261 → 143 261 → 73	10.0
Agius *et al.*, 2010[42]	10–50	2 ml deionized water, 2 ml acetone	Pulverized	2 ml deionized water (2 h sonication at 40 °C)	Cleanscreen EtG SPE	HFBA	HS-SPME-GC–NCI–MS/MS	596 → 427 596 → 288	2.8
Mönch *et al.*, 2013[62] Mönch *et al.*, 2014[54]	50	1 ml dichloromethane, 1 ml methanol	Pulverized	450 µl deionized water (48 h incubation) 450 µl deionized water (30 min ball-milling, 10 min centrifugation)	—	PFPA	GC–NCI–MS	496; 347	2.4
Cappelle *et al.*, 2015[55]	30	Deionized water, acetone	Pulverized	2 ml deionized water (1.5 h sonication)	Oasis®MAX SPE	HFBA	GC–NCI–MS/MS	596 → 213 397 → 213	0.2
Vignali *et al.*, 2018[63]	30	1 ml dichloromethane, 1 ml methanol	Pulverized	1 ml deionized water (2 h sonication)	Strata®X-A SPE	PFPA	GC–NCI–MS/MS	347 → 163 347 → 119	3.0
Mattia *et al.*, 2022[64]	50	10 m in washes with 5 ml dichloromethane +10-min washes with 5 ml methanol OR 1-min wash in 5 ml dichloromethane	Cut	Extraction overnight at 60 °C with 2 ml of deionized water OR Extraction for 2 h under sonication at 50 °C with 2 ml of deionized water	Strata®X-A-33	MSTFA	GC–EI–MS/MS	405 → 359 410 → 359 405 → 331 405 → 287	6.0

[a] MSTFA: *N*-trimethylsilyl-*N*-methyl trifluoroacetamide; BSTFA: *N*,*O*-bis(trimethylsilyl)trifluoroacetamide; PFPA: pentafluoropropionic anhydride; HFBA: heptafluorobutyric anhydride.

Table 7.2 Overview of the available LC–ESI–MS/MS methods for determination of hETG.

Authors, year [ref.]	Amount of hair (mg)	Decontamination	Homogenization	Extraction	Clean-up	Chromatographic column	Monitored ions/transitions (m/z)	LLoQ ($pg\,mg^{-1}$)
Janda *et al.*, 2002[50]	100	Methanol, acetone	Cut	1.5 ml deionized water, 3.5 ml acetonitrile, 0.1 ml 3 M HCl (3 h sonication at 50 °C)	Isolute NH_2 SPE	Synergy polar-RP	221 → 75	102
Morini *et al.*, 2006[9]	100	1 ml dichloromethane, 1 ml methanol	Cut	700 µl deionized water (overnight incubation, 2 h sonication)	—	Inertsil ODS-3	221 → 85 221 → 75	3
Bendroth *et al.*, 2007[65]	100	Dichloromethane, methanol	Cut	700 µl deionized water (overnight incubation, 2 h sonication)	—	Acquity UPLC HSS T3	221 → 85 221 → 75	2.5
Kintz *et al.*, 2008[66]	50	5 ml dichloromethane (2×)	Cut	2 ml deionized water (2 h sonication)	Oasis®MAX SPE	Acquity UPLC BEH HILIC	221 → 85 221 → 75	10
Concheiro *et al.*, 2009[67]	100	5 ml dichloromethane, 5 ml methanol	Pulverized	2 ml deionized water (2 h sonication, overnight incubation)	—	Hypercarb PGC	221 → 85 221 → 75	50
Lamoureux *et al.*, 2009[68]	30	10 ml deionized water (2×), 10 ml dichloromethane (2×)	Cut	2 ml deionized water (2 h sonication)	Cleanscreen EtG SPE	Uptisphere-3SI	221 → 85 221 → 75	10
Tarcomnicu *et al.*, 2010[69]	25	5 ml deionized water, 5 ml acetone	Cut	1 ml deionized water, 1 ml acetonitrile (overnight incubation, 1.5 h sonication)	—	Luna HILIC	221 → 85 221 → 75	20
Albermann *et al.*, 2010[70]	100	1 ml dichloromethane, 1 ml methanol	Cut	700 µl deionized water (overnight incubation, 2 h sonication)	—	Synergi Polar-RP	221 → 113 221 → 85 221 → 75	4

Pragst *et al.*, 2010[71]	30–50	1 ml dichloromethane, 1 ml methanol	Cut	1 ml deionized water (30 min sonication, 48 h incubation)	Oasis®MAX SPE	Hypercarb PGC	221 → 85 221 → 75	6
Pirro *et al.*, 2011[52]	50	3 ml dichloromethane, 3 ml methanol	Cut	700 µl deionized water (overnight incubation, 1.5 h sonication)	—	Purospher STAR RP-18	221 → 85 221 → 75	10
Albermann *et al.*, 2012[72]	75	4 ml dichloromethane, 4 ml methanol	Pulverized	500 µl deionized water (overnight incubation, 2 h sonication)	—	Hypercarb PGC	221 → 113 221 → 85 221 → 75	2.3
Suesse *et al.*, 2012[41]	50	3 ml acetone, 3 ml deionized water	Cut	2 ml deionized water, 3.5 ml acetonitrile, 0.3 ml 1 N HCl (4 h sonication at 50 °C)	Isolute NH_2 SPE	Zorbax Eclipse XDB-C8	221 → 75 221 → 57 221 → 55	2.6
Kronstrand *et al.*, 2012[73]	30	1 ml dichloromethane, 1 ml methanol	Cut	1.5 ml deionized water (overnight incubation at 37 °C)	Cleanscreen EtG SPE	Acquity UPLC HSS T3	221 → 85 221 → 75	2
Hastedt *et al.*, 2012[29]	20–50	1 ml dichloromethane, 1 ml methanol	Cut	500 µl deionized water (24 h incubation)	—	Hypercarb PGC	221 → 85 221 → 75	5
Pirro *et al.*, 2013[10]	50	3 ml dichloromethane, 3 ml methanol	Cut	500 µl deionized water/ methanol (35 : 1 v/v) (overnight incubation, 1.5 h sonication)	—	Acquity UPLC BEH C18	221 → 85 221 → 75	1
Yaldiz *et al.*, 2013[53]	200	5 ml deionized water (2×)	Cut	2 ml acetonitrile/deionized water (80 : 20 v/v) (15 min sonication)	—	Zorbax RRHD HILIC Plus	221 → 85 221 → 75	0.18
Cabarcos *et al.*, 2013[74]	100	Deionized water (2×), dichloromethane	Cut	2 ml deionized water (15 min sonication)	Oasis®MAX SPE	Inertsil ODS-3	221 → 203 221 → 85 221 → 75	20
Imbert *et al.*, 2014[75]	30	10 ml deionized water (2×), 10 ml dichloromethane (2×)	Cut	2 ml deionized water (2 h sonication in ice)	Cleanscreen EtG SPE	Uptisphere-3SI	221 → 113 221 → 85 221 → 75	3

Table 7.2 (*Continued*)

Authors, year [ref.]	Amount of hair (mg)	Decontamination	Homogenization	Extraction	Clean-up	Chromatographic column	Monitored ions/transitions (m/z)	LLoQ ($pg\,mg^{-1}$)
Binz *et al.*, 2014[76]	20	15 ml deionized water, 15 ml acetone	Pulverized	1.5 ml deionized water (15 min ball-milling, 25 min centrifugation)	Oasis®MAX SPE	Hypercarb PGC	221 → 85 221 → 75 221 → 55	2.5
Kummer *et al.*, 2015[77]	50	Dichloromethane, methanol	Pulverized	1.5 ml deionized water (2 h sonication at 40 °C)	BondElut® SAX SPE	Acquity UPLC HSS T3	221 → 85 221 → 75	10
Joya *et al.*, 2016[78]	25	5 ml methanol, 2.5 ml ether	Cut	500 µl M3 reagent (1 h incubation at 100 °C)	—	Acquity UPLC HSS C18	221 → 85 221 → 75	7
Oppolzer *et al.*, 2017[79]	30	Deionized water, methanol, acetone, dichloromethane	Cut	2 ml deionized water (2 h sonication)	Oasis®MAX SPE	Synergi Fusion-RP	221 → 85 221 → 75	3
Meier *et al.*, 2017[80]	20	3 ml deionized water, 3 ml 2-propanol, 3 ml acetone	Pulverized	1.5 ml deionized water (1–1.5 h sonication)	Oasis®MAX SPE	Hypercarb PGC	221 → 113 → 85 (LC–MS^3) 221 → 75 221 → 55	4
Palumbo *et al.*, 2018[81]	50	20 ml methanol (3×)	Cut	350 µL deionized water (overnight incubation, 2 h sonication at 50 °C)	—	Poroshell 120 HILIC-Z	221 → 113 221 → 85 221 → 75	4.5

Bastiani *et al.*, 2020[82]	50	Dichloro-methane for 30 min + 1 min	Pulverized	350–1500 mL water with 0–10% of methanol for 6–24 h	OASIS®MAX SPE	Acquity I+ Xevo TQS-Micro triple quadrupole mass spectrometer	221.0 → 74.9 221.0 → 84.9	4
Iglesias *et al.*, 2020[83]	5–40	Water and acetone (5 + 5 min)	Cut	1 mL of water, agitation for 30 min at 30 Hz in a Retsch 400MMball mill	Evolute® Express AX SPE	Chromolith1 RP-C18e	220.9 → 112.8 220.9 → 84.9 220.9 → 74.8	4
Pascali *et al.*, 2020[84]	50	20 mL methanol for 5 min	Cut	350 µL deionized water (overnight incubation, 2 h sonication at 50 °C)	—	Poroshell 120 HILIC-Z	221 → 75 221 → 85 221 → 113	4.5
Müller *et al.*, 2020[85]	50–100	5 mL portions of ethanol, water, acetone and dichloromethane	Cut	4 mL water (5 min centrifugation)	—	Waters Acquity® UPLC with a C18 Acquity® HSS T3	221 → 75 221 → 85 221 → 57 221 → 113	4.7
Pragst *et al.*, 2020[86]	25	Dichlorometane and methanol	Cut + pulverization	500 µL water in 5 µL acetonitrile/water (9 : 1 v/v)	—	Hypercarb	221.2 → 75.1 226.2 → 85.1	2.3

methods (see Table 7.2) failed to improve analytical sensitivity significantly (despite some utilizing ultrapressure liquid chromatography (UPLC) technology or re-introducing the additional step of SPE clean-up employed by Janda *et al.*[50]), until in 2013 Pirro *et al.*[52] and Yaldiz *et al.*[53] devised two different LC–MS/MS protocols – respectively employing UHPLC and hydrophilic interaction liquid chromatography (HILIC) technology – that reached LLOQ ≤ 1.0 pg mg^{-1}, thus providing the opportunity to investigate basal levels of hEtG in abstinent populations and reassess the cut-off value for alcohol abstinence. GC analytical methods (see Table 7.1) are more costly and time-consuming than LC methods, and are thus rarely used nowadays in routine hEtG analysis; differences in sample preparation and analytical methods exist, as for LC–MS/MS methods, however, all GC protocols developed since 2010 guarantee very high analytical sensitivity (LLOQ ≤ 3 pg mg^{-1}) while requiring only 30–50 mg of hair. In 2014 Mönch *et al.*[54] devised a relatively simple GC–NCI–MS method (simultaneous pulverization and extraction by ball milling, without SPE clean-up) with LOD = 0.68 pg mg^{-1} and LLOQ = 2.4 pg mg^{-1}, while the following year Cappelle *et al.*[55] published an extremely sensitive GC–NCI–MS/MS method reaching a LLOQ of 0.2 pg mg^{-1}; as part of the same study, extracts were also analyzed by GC–NCI–MS in order to reliably assess performance differences between single and tandem MS, achieving LLOQ = 1.5 pg mg^{-1} and therefore demonstrating that, contrary to previous understanding and analogously to what already observed with EI methods, the use of tandem MS in NCI methods can significantly improve analytical sensitivity (although not in such a way as to influence interpretation of hEtG results according to current cut-offs).

7.2.2 Factors Affecting Result Reproducibility

Several different protocols for hEtG analysis – employing GC–MS, GC–MS/MS or LC–MS/MS – are currently in use, and despite all reaching similar limits of detection and quantification, they vary greatly in terms of sample preparation procedures and analytical methods (see Tables 7.1 and 7.2). Although much research is still needed to elucidate the impact of such differences on hEtG results, in recent years a growing number of studies has focused on this matter, with the aim of identifying significant biases, increasing inter- and intra-laboratory reproducibility and possibly developing standardized and shared protocols for quantification of hEtG. Nowadays, several different proficiency tests, reference standards and certified materials are available in the market for all the laboratories that include the hEtG analysis in their routine workflow.

7.2.3 Differences in Sample Preparation

Measured hEtG values can vary significantly based on sample treatment prior to analysis, since the extraction yield can be greatly affected by factors such as analyte wash-out during decontamination, degree of sample homogenization and extraction conditions. The pulverization technique is the most investigated pre-analytical variable so far, and recommendations to favor automated pulverization have been included in the SoHT consensus since 2014.[87]

7.2.3.1 Decontamination

Washing the hair sample prior to analysis is necessary to minimize external contamination, which may cause false-positive results as well as increase analytical noise/background; as for most other substances, the analytical outcome of hEtG analysis can be strongly affected by the washing procedure used.[85,88] Decontamination protocols for EtG hair testing described by different laboratories vary significantly (see Tables 7.1 and 7.2), especially in terms of: polarity of the used solvents; volume of the solvents; number of washes. In 2014, Bossers *et al.*[88] investigated the impact on hEtG results of differences in washing procedure (selecting the most common solvents employed in hEtG testing) by concomitantly analyzing EtG concentrations in the ultrasonication extract and in the corresponding washing solvent, and found that: different decontamination procedures vary greatly in extraction capability, from dichloromethane and *n*-heptane ('high washing-out effect', *i.e.*, wash concentration of EtG $\geq$ 2 extract concentration of EtG) to methanol and water + *n*-heptane ('low washing-out effect', *i.e.*, extract concentration of EtG $>$ 3 wash concentration of EtG); most decontamination procedures have 'medium washing-out effect', removing roughly half of the EtG during the washing step, in contrast with previous evidence by Morini *et al.*[9] that EtG loss in washes is negligible – although such difference can be partly explained by the high alcohol consumption of the subjects in the study by Bossers *et al.*,[89] resulting in significant EtG hair incorporation through sweat, hence more superficially than with blood incorporation. Further investigations are necessary to assess the practical implications of such differences in EtG extraction yield, possibly identify an optimal decontamination protocol and evaluate the need for additional hEtG cut-offs based on the sample preparation protocol employed.

7.2.3.2 Pulverization

Hair pulverization is crucial to improve sample homogeneity and increase the sample surface in contact with the extraction solvent, hence optimizing analytical results. In 2012, Albermann *et al.*[72] suggested that increased particle size reduction may lead to higher hEtG values, and Kronstrand *et al.*,[73] investigating hEtG extraction kinetics, demonstrated that EtG extraction from automatically-pulverized hair reaches a plateau already after four hours of water incubation, compared to 18 h for intact hair. In 2013, Mönch *et al.*[90] reported, in seven hair samples from male volunteers, a significant increase in extraction yield regardless of the extent of automated pulverization (137% or higher compared to manual cutting) and of variations in other steps of the sample preparation, while in 2015 Kummer *et al.*[77] obtained somewhat different findings, demonstrating that weak pulverization (one cycle of 30 s at a speed of 6500 rpm) results in similar hEtG values as manual cutting, whereas extensive pulverization (three cycles of 60 s, at a speed of 6500 rpm, with a cooling time of 2 min in between) leads to a significantly higher amount of measured hEtG: for the two analyzed QC samples, analysis of the manually-cut and weakly-pulverized samples yielded concentrations below the target value, while the extensively-pulverized samples measured significantly above the target value – in line with the fact that certified target values reflect the mean calculated concentration of results obtained by different laboratories, working with either cut hair samples or pulverized hair samples; for the two volunteers' hair samples, the results of manually-cut and weakly-pulverized sample analysis measured below LLOQ (*i.e.*, $<$10 $pg\,mg^{-1}$), while the extensively-pulverized samples yielded significantly different values (14 and 40 $pg\,mg^{-1}$). In 2015, Salomone *et al.*[91] expanded on these preliminary studies and demonstrated, in a much larger population, that the percentage of head hair samples with concentration of EtG $>$ 30 $pg\,mg^{-1}$ shows an average 45% increase upon application of the automated milling procedure with respect to manual cutting; notably, when applying a cut-off of 40 $pg\,mg^{-1}$ to automatically-pulverized hair samples, the percentage of positive hEtG results aligns with that obtained in manually cut samples, leading the authors to suggest inclusion of the new cut-off for automatically-pulverized samples in the SoHT recommendations. In 2017, Mueller *et al.*[92] confirmed that the pulverization method is among the most significant factors that should be homogenized in order to maximize reproducibility of hEtG measurements, however, they emphasized what had already been found by Kronstrand *et al.*,[73] *i.e.*, that the impact on measured hEtG of different pulverization techniques varies greatly based on other experimental factors (see section 7.2.3.3). In 2018, Vignali *et al.*[63] also demonstrated a significant increase (mean $\Delta = +20\%$) in hEtG extraction yield for milled hair compared to manual cutting.

It might be appropriate to adopt, in the future, different cut-offs for automatically-pulverized and manually-cut hair samples, to standardize the homogenization method requiring forensic laboratories to perform automated pulverization, and to provide QC hair samples specifying the method to be used to obtain the certified reference value (in order to avoid bias upon accuracy assessment).

7.2.3.3 *Extraction*

After washing, drying, and cutting/grinding, the resulting hair material is spiked with deuterated internal standard (EtG-D5) and subjected to the extraction procedure, which involves addition of a suitable extraction solvent (typically deionized water) plus incubation and/or ultrasonication (microwave-assisted extraction has also been described); most protocols also include SPE clean-up of the extract. Unsurprisingly, analytical methods proposed by different laboratories vary in terms of solvent(s) chosen for extraction, amount of solvent used, duration of ultrasonication (from 15 minutes to several hours), presence and temperature of incubation (from ice to room temperature to 100 °C), as well as choice of SPE sorbents (see Tables 7.1 and 7.2). In 2014, Bossers *et al.*[89] evaluated different extraction solvents and found that the best performance was achieved with water (second-best with solvents containing methanol); the potential benefit of sonication was also assessed, with extraction yield after 4 h ultrasonication not demonstrating significant improvement over 30 min ultrasonication. In 2017, Mueller *et al.*[92] investigated the influence of six experimental factors – particle size ("high" = powder; "low" = snippets), extraction solvent ("high" = water; "low" = methanol), sample/solvent ratio ("high" = 25 mg ml^{-1}; "low" = 12.5 mg ml^{-1}, ultrasonication ("high" = 1 h; "low" = 0 h), incubation temperature ("high" = 60 °C; "low" = 20 °C) and time ("high" = 24 h; "low" = 2 h) – on quantitative hEtG results, and found that considerable differences (including concentrations below and above cut-off values) can be obtained from the same hair materials using different extraction conditions, with the most influential factors appearing to be extraction solvent and incubation temperature. The protocol yielding the greatest extraction recoveries had "high" levels of all experimental factors except for particle size, although automated pulverization resulted in hEtG values superimposable to manual cutting under these extraction conditions; the lowest extraction yields were obtained with the protocol involving "low" levels of all six factors and the protocol involving manual hair cutting, 1 h ultrasonication, 24 h incubation in methanol at 20 °C, and high sample/solvent ratio. The protocol involving no sonication, 24 h incubation in water at 20 °C and low sample/solvent ratio achieved medium extraction recovery and showed an appreciable impact of pulverization technique on extraction recovery, with better performance obtained with automated grinding. Another factor that may affect result reproducibility is the presence and type of SPE clean-up of the extract, however, little literature exists about this variable; the most commonly used sorbents are Oasis®MAX, Isolute NH^2 and Cleenscreen EtG. Overall, extraction conditions (especially extraction solvent and incubation temperature) can significantly influence hEtG results, and homogenization of this pre-analytical step would greatly increase inter-laboratory measurement reproducibility; future studies should also focus on identifying the optimal extraction conditions to minimize intra-laboratory variability.

7.2.3.4 *Derivatization*

For analytical methods involving gas chromatography, derivatization of the extract is required due to EtG's non-volatile nature; the most common derivatizing agents used in hEtG analysis include PFPA, HFBA, BSTFA, and MSTFA (see Table 7.1). In 2004, Jurado *et al.*[51] evaluated the performance of three derivatizing agents, concluding that, despite their efficacy being similar, PFPA should be preferred due to HFBA's high LOD and LOQ and to BSTFA's short period of derivative stability. More research is needed to understand the impact of different derivatizing agents on measured hEtG results.

7.2.4 Differences in Instrumental Method

Liquid chromatography (see Table 7.2) is typically performed in reverse phase, with ESI in negative mode, and it is followed by tandem MS to minimize sensitivity problems, however, great differences exist in the chromatographic columns employed (including HILIC and PCG columns to optimize ionization conditions); gas chromatography (see Table 7.1) is performed in either EI or NCI modes and followed by either single or tandem MS, and it can be preceded by large volume injection or by HS-SPME. Scarce research has been carried out concerning the impact of different analytical methods on measured hEtG values. As observed by Cappelle *et al.*,[55] the use of negative chemical ionization (NCI) improves analytical sensitivity of GC methods compared to electron impact (EI), and for both NCI and EI methods, use of tandem MS instead of single MS further improves sensitivity. In 2018, Vignali *et al.*[63] conducted a preliminary study on the differences in measured hEtG between two validated analytical methods (LC–ESI–MS/MS and GC–NCI–MS/MS) by dividing, after extraction, each of 46 hair samples into two aliquots and analyzing each aliquot by one of the two methods: results from most (85%) of the samples were within the acceptable threshold of variability (20%), as compared to less than 2/3 of the samples in a different group aliquoted before extraction and undergoing different extraction procedures (*i.e.*, additional SPE clean-up for samples destined for GC analysis).

7.2.5 Establishment of hEtG Cut-offs

In 2009, the Society of Hair Testing (SoHT) published its first Consensus on Hair Testing for Chronic Excessive Alcohol Consumption, with the aim of regulating and standardizing the use of hEtG determination in the forensic context. Based on findings from available studies, the cut-off value to discriminate between moderate (<60 g per day) and heavy drinking was set at 30 $pg\,mg^{-1}$ (first proposed by Bendroth *et al.*[65]), since such a value offered the best diagnostic performance to minimize the rate of false positive results – *i.e.*, the highest diagnostic specificity (0.97) combined with an acceptable sensitivity (0.85). Optimal hair sample length for EtG analysis was initially indicated as the proximal 3 cm of scalp hair, however the 2011 revision extended the accepted length to the proximal 6 cm, warning that for analysis of samples <3 cm caution is required; later editions of the SoHT consensus also advise caution for samples >6 cm.

As for discrimination between complete abstinence and social/moderate drinking, a cut-off of 7 pg mg^{-1} was introduced in 2012 by the SoHT,[93] with the caveat that all analytical methods of hEtG determination to be used for abstinence assessment should be validated with a LOQ $\leq$ 3 pg mg^{-1}; such a prudent value was first proposed in 2008 by Pragst *et al.*[57] in light of the paucity of studies concerning basal hEtG levels in teetotalers and possible factors creating analytical interference, with the aim of minimizing false positive results. The 7 pg mg^{-1} cut-off was subsequently adopted by the GTFCh (German Society of Toxicological and Forensic Chemistry) and by several studies investigating hEtG performance in abstinence monitoring, however, its diagnostic sensitivity for abstinence assessment was immediately questioned, with various authors proposing lower cut-off values as more precise analytical methods became available; for example, in 2012 Kharbouche *et al.*[94] proposed that only hEtG values < LOQ can verify complete abstinence (in their study, LOQ = 4 pg mg^{-1}), while the following year Pirro *et al.*[10] (using a method validated with LOQ = 1 pg mg^{-1}) suggested 2 pg mg^{-1} as the cut-off for abstinence in forensic settings (1 pg mg^{-1} in clinical settings). In 2012 Kronstrand *et al.*[73] found hEtG levels <7 pg mg^{-1} in volunteers with a controlled daily consumption of 16 g of ethanol (LOQ = 2 pg mg^{-1}), and in 2016 Crunelle *et al.*[12] conducted two studies also revealing hEtG concentrations <7 pg mg^{-1} in moderate drinkers (LOQ = 2.1 pg mg^{-1} and LOQ = 0.2 pg mg^{-1}), proposing a new abstinence cut-off of 5 pg mg^{-1}. The 2016 consensus for the use of alcohol markers in hair for assessment of both abstinence and chronic excessive alcohol consumption confirmed the 7 pg mg^{-1} cut-off, and the expert group of SoHT members drafting the consensus revision explained this choice – in a commentary published the following year – by emphasizing the serious legal consequences of hEtG results and the need for larger studies and more sensitive analytical methods before any change can be adopted. Lastly, in 2017 Crunelle *et al.*[95] pointed out that, when discussing lowering the abstinence cut-off, attention should be paid to the effects – on measured EtG in the proximal 3 cm of hair – of differences in the amounts of alcohol consumed prior to the abstinence period, since the few millimetres of hair lying immediately below the scalp surface still contain EtG from pre-study alcohol consumption and are inevitably included in the subsequent hair sample; according to their findings, while an abstinence cut-off of 4.5 pg mg^{-1} could be suitable for individuals without alcohol dependence, the percentage change in hEtG levels may serve as a more reliable marker in detoxifying alcoholics. The current consensus of the SoHT on alcohol biomarkers, approved in 2019, confirmed the two different cut-offs for hEtG, by stating that "a concentration lower than or equal to 5 pg mg^{-1} EtG in the proximal head hair segment with a length of 3 cm up to 6 cm does not contradict self-reported abstinence" and that "a concentration greater than or equal to 30 pg mg^{-1} EtG in the proximal head hair segment with a length of 3 cm up to 6 cm strongly suggests chronic excessive alcohol consumption. If sample lengths less than 3 cm or greater than 6 cm are used, the results should be interpreted with caution. Segmentation may provide additional information''.[96]

7.2.6 Factors Affecting Result Interpretation

Given the potentially serious legal and social consequences of positive hEtG measurements, any factor that may contribute to false positives/negatives must be carefully considered when interpreting hEtG results, in particular body mass index (BMI), kidney disease, hair hygiene, and regular use of hair products and chemical treatments; in case of dubious findings, measurement of fatty acid ethyl esters in hair (hFAEEs), which it will be discussed later, and of traditional blood markers of alcohol abuse is advised to confirm hEtG results or prompt further investigations.

7.2.6.1 Factors Affecting Ethanol Metabolism and Circulating [EtG]

Ethanol is a small, water-soluble molecule, mainly administered orally (in the form of alcoholic beverages), absorbed through the GI tract (stomach and small intestine), readily distributed throughout the body (except into adipose tissue) and quickly eliminated by the liver and kidneys.

7.2.6.2 Body Mass Index

The volume of distribution of ethanol decreases proportionally to increasing BMI – in line with the known preferential distribution of this hydrophilic substance to non-fat tissues – possibly leading, for the same body weight and dosage of alcohol, to higher blood alcohol concentrations (BAC) and hEtG levels in individuals with elevated BMI. In fact, in 2017 Crunelle *et al.*[95] found, across a wide range of amounts of alcohol consumed, that the average concentration of hEtG was significantly higher in individuals with BMI ≥ 25 kg m^2 compared to those with BMI < 25 kg m^2.

7.2.6.3 Gender, Age, and Ethnicity

These factors are usually assumed to play no confounding role in interpretation of hEtG results.

Differences in the rate of EtOH metabolism strongly depend on gender and age: women metabolize alcohol faster than men, probably due to differences in sex hormones, however, the faster metabolism is at least partly offset by women's lower volume of distribution (*i.e.*, higher proportion of adipose tissue), resulting in shorter but higher BAC peaks; aging, on the other hand, gradually reduces speed and efficacy of alcohol metabolism, along with total body water and hence ethanol's volume of distribution.

Despite the known gender differences in EtOH metabolism and EtG incorporation, several studies have consistently reported no significant gender influence on hEtG test performance. In 2014, Gareri *et al.*[97] analyzed the discordance between hEtG and hFAEEs results in women and men, finding the former group to have higher discordance rates; concordance was, however, significantly improved when excluding colored hair.

In 2015, Salomone *et al.*[91] evaluated over 20 000 subjects and found higher average EtG concentrations, in both hair and urine, in subjects >40 years old with respect to younger groups. This finding, however, can be explained by multiple interpretations, including different drinking behaviours, tolerance development, and age-related metabolic changes.

Ethnic differences in hEtG concentration have been poorly investigated. It is now widely accepted that alcohol dehydrogenase (ADH) and aldehyde dehydrogenase ALDH variants (*e.g.*, ALDH2*2) favouring acetaldehyde accumulation in blood, particularly common in Asian populations, may have a protective role against heavy drinking and development of alcohol dependence, however, the influence of such genotypic differences on the production of non-oxidative EtOH metabolites (*e.g.*, EtG, EtS) is typically not regarded as significant.

Renal excretion is the major elimination route for glucuronides, including EtG, hence patients with decreased kidney function can be expected to show higher blood concentrations, prolonged urinary excretion, and increased incorporation into hair of this EtOH metabolite. In fact, Høiseth *et al.*[98] demonstrated both longer EtG half-life in blood and prolonged EtG urinary detection window in patients with compromised renal function.

In 2016, Fosen *et al.*[99] carried out a study on non-dialyzed renal-disease patients reporting moderate alcohol consumption in the past three months, to demonstrate the possibility of obtaining false-positive hEtG results of heavy drinking in social drinkers with decreased kidney function. EDI was estimated based on self-reported alcohol consumption and, in the patient group, also on measurement of objective markers (*i.e.*, GGT: gamma-glutamyl transpeptidase, AST: aspartate aminotransferase, ALT: alanine aminotransferase). The inclusion criteria for the patient group were: serum creatinine >90/100 $\mu mol\, l^{-1}$ (respectively for women/men); estimated glomerular filtration rate (GFR) < 60 $ml\, min^{-1}/1.73\ m^2$; established a diagnosis of decreased kidney function. Information about hair treatments and use of hair products was collected, along with information about co-morbidities potentially influencing kidney function. In the age-matched control group, all subjects tested negative (or barely positive) for hair EtG – as expected from the estimated EDI – while in the patient group all subjects showed normal levels of ALT, AST, and GGT, but 20% (eight patients) resulted as heavy drinkers with respect to hEtG (≥30 $pg\, mg^{-1}$), with a significant dose–response relation between degree of kidney impairment (GFR) and measured hEtG (corrected for EDI); this correlation persisted when subjects with hair coloring were excluded. Altogether, the studies performed by the Norwegian group suggest that caution is needed when interpreting hEtG measurements obtained in renal patients, due to the possibility of false-positive results.

In contrast to most of the traditional biomarkers for alcohol abuse (with the accepted exception of serum CDT in non-cirrhotic patients), hEtG concentration doesn't appear to be influenced by hepatic diseases, including alcoholic cirrhosis, and can therefore be used reliably also when monitoring patients on liver transplant lists.[100]

Gilbert's syndrome is an inherited liver condition, usually asymptomatic, that occurs in about 5% of the population and results in decreased levels (up to 80%) of UGT isotype 1A1; given that UDP-glucuronosyltransferases are responsible for ethanol conversion to EtG, lower levels of EtG with respect to ethanol intake may be expected in such patients. In 2015, however, Huppertz *et al.*[101] analyzed urine samples from 30 patients suffering from Gilbert syndrome after consuming one drink (9 g of ethanol) and found no evidence of impaired EtG formation.

7.2.7 Factors Affecting EtG Incorporation and Stability into Hair

Hair is a proteinaceus fiber with a highly hierarchical composition of tightly-packed keratinized cells arranged into three concentric structures: cuticle (the outermost), cortex and medulla (the innermost). Notably, the cuticle is susceptible to damage from a number of sources, such as exposure to light and heat, mechanical stress (*e.g.*, long-term effects of weather, vigorous hair drying, use of aggressive hair products) and chemical treatments (*e.g.*, bleaching, perming, coloring), hence over time it may become – especially in the distal ends – compromised enough to expose the inner layers and more prone to release previously incorporated substances (especially hydrophilic molecules as EtG). Incorporation of EtG into hair occurs mainly by passive diffusion from: blood capillaries into the maturing keratinocytes (to a lesser degree than basic or lipophilic substances) – which appears to result in a mostly-central EtG deposition (*i.e.*, into the hair's forming medulla); sweat (and sebum) into the completed hair shaft – which results in a more peripheral (*i.e.*, mainly in the cuticle), hence more unstable, EtG deposition. It is known for several classes of drugs that – assuming a constant dosage over time and normal hair keeping (*i.e.*, absence of aggressive cosmetic treatments) – the mean hair concentration slowly decreases with increasing distance from the hair root, resulting in roughly 4% of the drug being detectable after one year in the corresponding hair segment (*i.e.*, 12–15 cm); stability of substances incorporated in hair appears to depend mainly on the chemical properties of the substance (in particular, it increases with polarity) and on the hair quality and hair-care practices of the individual (most notably chemical treatments such as bleaching and perming, but also excessive exposure to sunlight or water, can significantly decrease detection window).[102]

7.2.7.1 *Natural Hair Color*

Melanin content is known to influence the deposition in hair of many drugs, in particular basic substances, however pigmentation has been shown, both in animal and human studies, not to affect EtG concentration in hair.[40,103]

7.2.7.2 *Chemical Treatments and Hair Products*

Alcohol-free hair products (*e.g.*, hairspray, gel, wax, oil, grease) do not seem to affect hEtG concentration. On the other hand, the concentration of hEtG may be lowered by chemical hair treatments such as bleaching, perming, and dying (which damage the cell-membrane complex that glues cortex and cuticle cells to each other), leading to false-negative results;[104] notably, also temporary and henna colorations may have significant effects on EtG stability in hair. It is therefore essential to document presence and types of chemical treatments along with the hair sampling, and, in case of suspected false negatives, analysis of hFAEEs (in particular ethyl palmitate) can be used to confirm the negative result or prompt further investigations.

Occupational exposure to alcohol-containing products (*e.g.*, hand sanitizers) may falsely increase urine [EtG], but doesn't appear to affect hEtG concentration. In 2012, Martins Ferreira *et al.*,[105] expanding on previous findings by Gareri *et al.*,[106] evaluated and rejected the possibility of obtaining false positive hEtG results after prolonged use of ethanol-containing hair products, however a study published in 2018 by Sempio *et al.*[107] demonstrated the possibility of false positives if applying, on scalp hair, alcohol-based perfumes with high EtOH content (80% *vs.* 44% in the hair tonic analyzed by Martins Ferreira *et al.*[105]); in such cases, hFAEEs analysis is unhelpful, while quantification of EtG in chest hair may help assess the reliability of hEtG results.

In 2012, Sporkert *et al.*[108] reported a case of cosmetic hair product directly contaminated by EtG – leading to falsely increased concentration of hEtG – and subsequent targeted analysis of market products for the presence of EtG has shown that this EtOH metabolite is frequently present in commercially available herbal hair tonics; in dubious cases, analysis of the involved product, as well as EtG testing on non-head hair, may be indicated.

7.2.7.3 *Hair Hygiene*

According to SoHT recommendations, EtG can be reliably detected in the 0–6 cm proximal hair segment, while if samples <3 cm or >6 cm are used, results should be interpreted with caution, due to the unclear influence of, respectively, EtG incorporation from sweat or sebum, and wash-out from routine activities such as hair washing or swimming.

The stability of EtG in hair over several months (where each month is roughly represented by 1 cm of hair) is assumed based on several studies; in particular, in 2012 Agius *et al.*,[21] building on a previous segmental study carried out in 2007 by Appenzeller *et al.*,[109] suggested that EtG can be reliably determined in hair samples up to 12 cm long and that segmentation can provide further information about alcohol consumption trends due to the (near) absence of wash-out effects.

In 2017, however, Luginbühl *et al.*[110] set out to investigate the wash-out effects caused by chlorinated water, expanding on previous findings from two separate studies carried out in 2009 which indicated that hair hygiene may influence the reliability of hEtG results, and in particular frequent hair washing can significantly reduce EtG concentration in distal hair segments; the results of this recent *in vitro* study indicate that wash-out effects of EtG in hair are proportional to the duration of water incubation (with a loss of up to 1/3 after 2 hours of incubation in chlorinated water and up to 2/3 after 10 hours, and slightly smaller losses with deionized water), regardless of the type of water (*i.e.*, chlorinated or Milli-Q water) or the application of hygiene products (*e.g.*, shampoos or conditioners), and that, consequently, even extensive hair washing or bathing may lead to false negative hEtG results. Compatible findings were obtained during the same year by Meier *et al.*,[111] who performed segmental analysis of seven 3 cm hair samples and demonstrated decreasing proximal-to-distal EtG concentrations, suggesting analyte wash-out as the probable explanation. In 2018, Fosen *et al.*[112] carried out a longitudinal study to evaluate hEtG changes over time, starting from the assumption that, given an average hair growth of 1 cm per month, measurement of the 1–4 cm segment of a freshly collected hair sample should yield roughly the same hEtG concentration (*i.e.*, should represent the same period of alcohol consumption) as the 0–3 cm segment of a first sample collected four weeks earlier at the beginning of a facility-assisted abstinence period; their results, however, indicate a statistically significant decrease in hEtG concentration in most (18 out of 27) patients (median $\Delta[\text{hEtG}] = -46\%$), regardless of the presence of chemical hair treatments, suggesting that interpretation of EtG values in hair samples other than the proximal 0–3 cm segment should always be performed with caution, and that further research on the mechanisms and extent of EtG wash-out is necessary to assess adequacy of current guidelines.

7.2.7.4 Solar Irradiation

In 2019, Miolo *et al.*[113] evaluated the photostability of hEtG by artificially irradiating 28 real hair samples for 48 hours (obtaining roughly the equivalent of three months of solar exposition) and then analyzing them by UPLC–MS/MS ($\text{LOQ} = 5\ \text{pg mg}^{-1}$). The concentrations of EtG in hair samples following irradiation displayed a mean increase of 37%, however, great variability was found among post-irradiation results (from −36% to +255%); the average hEtG increase could be explained as improved extraction yield following hair matrix denaturation, while the variability of the impact of irradiation may be ascribed to several factors (*e.g.*, degree of hair damage, hair thickness, hair color), of which one (hair color) was assessed during the study and shows a promising correlation with ΔhEtG: the largest increases were observed for gray and grizzled hair (which also tends to be thinner and more damaged, hence more prone to release EtG), while the seven samples showing concentration of hEtG decreases were all black or brown, suggesting that melanin content could boost production of reactive oxygen species (ROS) responsible for EtG photodegradation.

7.2.8 Other Potential Influencing Factors

Other factors such as tobacco smoking, drinking pattern, and type of consumed beverages may have confounding effects on measured hair EtG concentration, however, their influence has not yet been adequately explored.

In 2017, Meier *et al.*[114] found evidence of EtG concentration gradients (with a factor of three) across the head, underlining the need for consistency in sampling site (ideally the posterior vertex) to avoid false hEtG results.

Expanding on preliminary observations by Salomone *et al.*[91] that the rate of hEtG positivity increases significantly during winter with respect to warm months, in 2018 Własiuk *et al.*[115] analyzed seasonal variability in a seven year data series and postulated a concentration of hEtG cycle with troughs around February and peaks around August, and with an estimated deviation from the overall mean of $\pm 2.78\ \text{pg}\,\text{mg}^{-1}$.

7.2.9 hEtG in Driving Under the Influence Issues

An in-depth diagnostic study is appropriate, in particular, in alcohol-related pathologies/issues or when the positivity of indicators (*e.g.*, laboratory tests) leads to such conditions, in relation to the supposed aptitude of the subject to manifest and/or to repeat behaviours that compromise vigilance and driving performance, or in the presence of a driving-centered work activity (*e.g.*, road transport).

The laboratory analyses, in the context of health checks for driving fitness, have the purpose of making available the results of the research of some biological markers of alcoholic beverage intake, with the aim of helping to identify the conditions specified by the different national and international laws and rules, as incompatible with the fitness to drive: state of alcohol dependence or inability to dissociate driving from alcohol consumption.

The laboratory tests can be divided into two macro categories:

- Non-specific biochemical markers: liver enzymes (GGT, AST and ALT); mean corpuscular volume erythrocyte; serum protein electrophoresis (to identify conditions of hypoalbuminemia and hypergammaglobulinemia). Although these indicators cannot provide direct and quantitative data on the usual consumption of alcohol, they are nevertheless useful for identifying the existence of alcohol-related organic diseases.

- Specific indicators of chronic alcoholic abuse: the most reliable laboratory tests are currently the search in the hair of EtG, which represents a direct marker of alcohol consumption (not only by drinking but also through food), and the dosage of the serum concentration of carbohydrate-deficient transferrin (CDT).

Since EtG was widely discussed, it is useful to give a brief introduction to CDT, which is the most extensively used marker together with EtG.

7.2.10 Carbohydrate-deficient Transferrin

To date, the exact mechanism by which the excessive use of ethyl alcohol causes the increase of carbohydrate-deficient transferrin (CDT) glycoforms is not yet fully understood. According to the most recent literature, this is a complex process that involves both the intracellular transport of proteins and the enzymatic activity of glycosylation rather than the synthesis of the protein chain.[116] There is also evidence that the concentration of CDT increases, while remaining within the common limits of normal, even in the case of moderate consumption of alcoholic beverages. The serum concentration of CDT is reduced following abstinence from ethyl alcohol with an average half-life of about 14 days. The individual variability of the pathophysiological mechanisms underlying the increase due to alcohol consumption is not yet clear. From recent studies, there are no significant differences in the basal concentrations of CDT in different geographical areas, even including those in which a deficit of the alcohol oxidative systems is known. Both asialo- and disyalotransferrin are clearly correlated with chronic alcoholic consumption, although the two molecules show different sensitivity and specificity.[117] The IFCC group (International Federation of Clinical Chemistry and Laboratory Medicine) has identificd disialotransferrin as the target analyte for CDT to refer to for the standardization of the measurement of the molecule as a biomarker of alcohol abuse. The IFCC study group recommends calculating the CDT as a percentage ratio with respect to total transferrin (% CDT), to compensate for false positives or false negatives linked respectively to high or low total transferrin values. Cut-offs are method-dependent as well as population-dependent.[118–124]

The literature reports that the most accepted cut-off values are around 1.8–2.1% in separative methods and around 2.5% in immunometric methods.[116] Also, they differ when using the % CDT or the CDT index. The reasons for these differences must probably be sought in the unavoidable interferences determined by the main glycoforms (which represent over 90% of the transferrin) on the immunological determination. To date it is now proven that if appropriate methods are used, the possibility of having false negatives or false positives in the case of primary biliary cirrhosis, chronic viral hepatitis, hepatocellular carcinoma, hypertransferrinemia, pregnancy, use of estrogen, iron deficiency anemia, hypoferritinemia, transplantation, and use of antiepileptics is extremely low or even nil. CDT is certainly altered in cases of chronic alcohol abuse, genetic variants of transferrin and congenital disorder of glycosylation (CDG I and II).

7.3 Alcohol Biomarkers - Fatty Acid Ethyl Esters

FAEEs, direct ethanol metabolites, are formed from free fatty acids, triglycerides, lipoproteins, and phospholipids, by the action of different enzymes, such as cytosolic and microsomal FAEE synthases, cholesterol and carboxyl esterases, lipoprotein and carboxylester lipases, in the presence of ethanol. The mechanisms of formation, incorporation, and accumulation of FAEEs into the hair matrix have been extensively studied and different hypotheses are still considered valid. In particular, FAEEs can be directly incorporated into the hair root through the bloodstream. Moreover, since FAEEs synthase activity is observed in many human tissues, the production of these direct metabolites could also occur into hair root or, alternatively, in the sebaceous glands after ethanol diffusion. Auwarter and coauthors observed that a high concentration of FAEEs in the sebum favours their incorporation into keratin matrix. Moreover, the reason for the increase of the concentration of hFAEEs from proximal to distal segments can be justified by the hair aging process, that on the one hand prolongs the deposition, and on the other hand enhances the incorporation of the lipids into the hair matrix, due to its texture deterioration.

FAEEs were detected in hair samples for the first time by Pragst and coauthors in 2001.[125] In particular, the authors decided to monitor four different compounds, namely ethyl myristate, ethyl palmitate, ethyl oleate, and ethyl stearate. Since FAEEs undergo a hydrolyzation process in alkaline solutions, the authors developed a new head space solid phase microextraction (HS-SPME) procedure, followed by an injection in a GC–MS system, operating in single ion monitoring (SIM) mode. Since this first study, the four monitored FAEEs in hair (hFAEEs) were considered suitable direct metabolites for the evaluation of chronic heavy drinking behaviours. Hence, the consensus on alcohol biomarkers, edited by SoHT in 2009, focused on the analysis of either hEtG and hFAEEs.[126]

7.3.1 Advancements in hFAEE Analysis

The analysis of hFAEEs is generally performed through GC–MS methods (see Table 7.3). The aforementioned method developed by Pragst and coauthors[125] was adopted by the majority of the following studies involving the detection of these metabolites in keratin matrices. In particular, different organic solvents, mostly non-polar ones, such as *n*-heptane and *n*-hexane were chosen for decontamination procedures. In contrast to hEtG analysis, hair cutting was preferred to pulverization process. Hair was then generally extracted in mixtures constituted of *n*-hexane or *n*-heptane and dimethyl sulfoxide (DMSO). Many authors highlighted the importance of a long

Table 7.3 Overview of the available methods for determination of hFAEEs.

Authors, year [ref.]	Amount of hair (mg)	Decontamination	Homogenization	Extraction	Clean-up	Analytical technique	Monitored substances (LOQ (ng mg^{-1}))
Pragst *et al.*, 2001[125]	50	Solvent not specified	Cut	500 µL DMSO, 2000 µL *n*-hexane – 14 h incubation	HS-SPME	GC–MS–SIM	Ethyl myristate (0.050) Ethyl palmitate (0.070) Ethyl oleate (0.120) Ethyl stearate (0.040)
Hartwig *et al.* 2003[128]	20 to 40	1 mL *n*-heptane	Cut	500 µL DMSO, 2000 µL *n*-heptane – 15 h incubation	HS-SPME	GC–MS–SIM	Ethyl myristate (0.050) Ethyl palmitate (0.070) Ethyl oleate (0.120) Ethyl stearate (0.040)
De Giovanni *et al.* 2007[129]	30	*n*-Hetane	Cut	500 µL DMSO, 2000 µL *n*-heptane – 20 h incubation	HS-SPME	GC–MS–SIM	Ethyl myristate (0.050) Ethyl palmitate (0.050) Ethyl oleate (not validated) Ethyl stearate (0.20)
Politi *et al.* 2011[130]	100	Methanol	Cut	500 µL DMSO, 4000 µL *n*-hexane – overnight incubation	NH^2 SPE	GC–MS–SIM	Ethyl myristate (0.013) Ethyl palmitate (0.019) Ethyl oleate (0.031) Ethyl stearate (0.011)

Hastedt *et al.* 2012[29]	20	Heptane	Cut	500 µL DMSO, 2000 µL *n*-heptane – 15h incubation	HS-SPME	GC–MS–SIM	Ethyl myristate (0.050) Ethyl palmitate (0.050) Ethyl oleate (0.050) Ethyl stearate (0.050)
Bertol *et al.* 2014[131]	50	Methanol	Cut	500 µL DMSO, 4000 µL *n*-hexane – overnight incubation	NH^2 SPE	GC–MS–SIM	Ethyl myristate (0.010) Ethyl palmitate (0.010) Ethyl oleate (0.010) Ethyl stearate (0.010)
Oppolzer *et al.* 2017[79]	30	Water, methanol, acetone, dichlormethane	Cut	500 µL heptane – 15 h incubation	NH^2 SPE	GC–MS/MS (MRM mode)	Ethyl myristate (0.030) Ethyl palmitate (0.030) Ethyl oleate (0.030) Ethyl stearate (0.030)
Tsanaclis *et al.* 2020[132]	10	Heptane	Pulverized	1 mL heptane – 60 minutes in an ultrasonic bath	None	LC–MS/MS (MRM mode)	Ethyl myristate (0.27) Ethyl palmitate (0.207) Ethyl oleate (0.207) Ethyl stearate (0.207)

incubation (up to 20 h), rather than a sonication assisted extraction. Indeed, the sonication in solvents, such as methanol, could lead to transesterification of FAEEs. Before injection in the GC–MS system, the majority of the authors optimized different solid phase microextraction online procedures, that allow good sensitivity and clean chromatograms to be reached. Only a few authors developed a solid phase extraction procedure, generally carried out through aminopropryl-NH^2 cartridges.

Except for the study of Caprara *et al.*,[127] where six different FAEEs (ethyl laureate (E12:0), ethyl myristate (E14:0), ethyl palmitoleate (E16:1), ethyl palmitate (E16:0), ethyl oleate (E18:1), and ethyl stearate (E18:0)) were monitored, the other authors evaluated the diagnostic sensitivity and specificity on the sum of the four most specific compounds found by Pragst *et al.*[125]

Recently, Tsanaclis *et al.* developed a LC–MS/MS method, accredited to ISO/IEC 17025 standards, for the detection and quantification of ethyl myristate, ethyl oleate, ethyl palmitate, and ethyl stearate in hair. The sample preparation consisted of a washing procedure, followed by pulverization and a sonicated assisted extraction in 1 mL of heptane. The dried sample was reconstituted in mobile phase and injected in the LC–MS/MS system, operating in reverse phase/Selected Reaction Monitoring (SRM) mode. A LOD and a LOQ of 89 and 207 pg mg^{-1} were measured for each analyte, respectively.

7.3.2 Establishment of hFAEEs to be Detected and Their Cut-offs

In the first consensus on alcohol biomarkers, edited by the SoHT in 2009, a concentration of 0.5 ng mg^{-1} in the 0–3 cm proximal hair segment was chosen as the hFAEE cut-off for chronic excessive alcohol consumption diagnosis. The selected cut-off was referred to the sum of four different hFAEEs, namely ethyl myristate, ethyl oleate, ethyl palmitate and ethyl stearate.[126] In the revision edited in 2012, a further cut-off of 1.0 ng mg^{-1} was added to the 0.5 ng mg^{-1}.[93] The new cut-off was suggested for the interpretation of the results obtained from the analysis of the 0–6 cm proximal hair segment. Starting from 2012, the group of experts of the SoHT highlighted the scarce reliability of the results obtained from hair segments with a length shorter than 3 cm. In 2014 consensus revision the authors stated that the cut-off could be extended also to hair samples collected from other body areas and not only from head, but with the exception of axillary and pubic sites.[87]

The last SoHT consensus revision, edited in 2019, focused on the detection of only ethyl palmitate (EtPa), among the four selected FAEEs. Moreover, similarly to hEtG, the group of experts decided to suggest two different cut-offs. The current consensus on hEtPa states that "an EtPa concentration lower than or equal to 120 pg mg^{-1} for a 0–3 cm proximal head hair segment, or lower than or equal to 150 pg mg^{-1} for a 0–6 proximal head hair segment, does not contradict self report abstinence", and that "an

EtPa concentration greater than 120 pg mg^{-1} for a 0–3 cm proximal scalp hair segment, or 150 pg mg^{-1} for a 0–6 cm proximal head hair segment, strongly suggest repeated alcohol consumption". The same consensus states that, for chronic excessive alcohol consumption diagnosis, "an EtPa concentration greater than or equal to 350 pg mg^{-1} in head hair strongly suggests chronic excessive alcohol consumption when measured in the 0–3 cm segment. An EtPa concentration greater than or equal to 450 pg mg^{-1} in head hair strongly suggests chronic excessive alcohol consumption when measured in the 0–6 cm segment".[96]

7.3.3 Factors Affecting Result Interpretation

Nowadays, only a few protocols, all employing one single instrumental technique (GC–MS), are currently used for the detection and quantification of hFAEEs or for the determination of hEtPa. In contrast to hEtG analysis, there is still a lack of studies carried out on large population, thus, inevitably affecting the biomarker reliability. Also, proficiency tests (PTs) and reference standards for hFAEEs analyses are not common and only few laboratories perform the determination of these metabolites routinely.

7.3.4 Factors Affecting [hFAEEs]

7.3.4.1 Body Mass Index

Though a correlation between body mass index (body mass/squared body height) was not measured on a large population, Suesse *et al.*[133] observed, on a group of 644 monitored subjects, that weight did not influence the levels of the sum of the concentrations of the four monitored hFAEEs.

7.3.4.2 Gender, Age, and Ethnicity

To date, the potential influence of gender, age, and ethnicity on the levels of hFAEEs is poorly investigated. However, since sebum represents an important route of accumulation of FAEEs in hair, the measurement of sebum content in comparison to the concentration of hFAEEs could be worthy of further investigation. Indeed, different disorders and pathologies, such as seborrheic dermatitis, can significantly affect the production of sebum in the head area. Moreover, some authors observed differences in sebum content depending on gender, age, and ethnicity. Zhao *et al.*[134] stated that, based on their studies, sebum content seems to be higher in males than in females and in younger subjects with respect to adults. Yet, they observed that differences in lifestyle and dietary patterns can also increase or decrease the sebum content. These differences could impact on the concentration of hFAEEs and the reliability of this biomarker for alcohol repeated and chronic excessive consumption diagnosis.

7.3.4.3 Hygiene and Cosmetic Treatments

The influence of cosmetic treatments and hair products on hFAEEs is well known. Since the first SoHT consensus on alcohol biomarkers, the group of experts warned the international community about the potential false positive results due to the increase of the levels of hFAEEs after alcohol-based products, containing lipids esters themselves.

Salomone *et al.*[30] evaluated the potential effect of the prolonged and repeated use of alcohol-based hand sanitizers on subjects exposed for occupational reasons. The study did not reveal any false positive results on hair analysis, thus confirming that the concentrations of hEtG and hFAEEs are not influenced by the use of hand sanitizers.

Luginbuhl *et al.*[135] evaluated the potential effect of detox shampoos in hFAEEs measurement. The authors detected amounts of FAEEs in two different commercial products and confirmed that the use of these shampoos may result in a false positive result, by increasing the concentration of FAEEs in hair.

7.3.4.4 Use of Contraceptives in Women

Bertol *et al.*[131] observed a significantly high rate of false positive results for hFAEEs in women under estroprogestin replacement therapy or under treatment with different steroidal contraceptives. The authors suggested that the increase of concentration of hFAEEs could be related to the decreased ethanol elimination rate and the reduced ethanol metabolism. Hence, results from subjects under hormonal therapies should be interpreted with caution.

7.3.5 Stability of FAEEs in Hair

A recent study performed by Tsanaclis *et al.*[132] highlighted the scarce stability of FAEEs in hair after sample collections. The authors measured the concentration of hFAEEs in 50 different samples at two different time points. The results suggested that, in order to avoid a loss of diagnostic sensitivity, either concentrations of hFAEEs or hEtPa should be measured within one week from sample collection. Indeed, the stability of these compounds in hair appeared to be affected by prolonged storage, independently of the conditions.

7.4 Alcohol Biomarkers – Cocaethylene

The consumption of drugs of abuse is frequently related to alcohol abuse. When alcohol is consumed together with cocaine, the common metabolic pattern changes. The cocaine undergoes a transesterification process, thus producing a direct ethanol/cocaine metabolite. Cocaethylene (CE) is formed by the hepatic carboxyltransferase, an enzyme family also involved in the conversion of cocaine to benzoylecgonine. It has been estimated that approximately 17% of the intravenous cocaine is metabolized to CE.[136] The pharmacodynamic profile of CE does not significantly differ from cocaine, though it appears less potent and with a longer half-life.[137]

CE was identified in hair by Cone *et al.* in 1991.[138] The authors developed a GC–MS method for the detection of cocaine and its main metabolites in hair, and applied the analytical procedure to 10 cocaine users. CE in hair (hCE) was detected in 6 out of 10 samples collected from the 10 monitored subjects, and the measured concentration ranged from 0.4 to 2.6 ng mg^{-1}.

In a letter to the editor, by DiGregorio *et al.*,[139] CE was detected in the majority of the tested cocaine users (88%), declaring an alcohol past history. The concentration range of hCE was in the interval from 2.5 and 30.3 ng mg^{-1}. Gruszecki *et al.*[140] observed the presence of CE in two hair samples in a study carried out on 53 postmortem samples.

Bourland *et al.*[141] observed that, when CE is present in hair samples, it represents the cocaine metabolite with the highest relative concentration. Indeed, the mean values of the metabolite-to-cocaine ratios were: CE (15.4%), benzoylecgonine (BE – 12.8%), norcocaine (NCOC – 2.5%) and ecgonine methylester (EME – 1.8%). Since hCE and hBE are metabolized by the same enzyme family, the authors suggested that the formation of BE, in the presence of ethanol, may be reduced, providing a competitive transesterification process. The hCE/cocaine ratios were evaluated in several other studies and it was generally included in the range from 0.4% to 60%, rarely exceeding 100%.[36,40,142,143]

hCE was measured together with hEtG in a study developed by Politi *et al.*[36] in 2007, where 68 hair samples collected from cocaine users were tested. The authors did not observe any correlation between hEtG and hCE concentrations. The reason for the lack of correlation was attributed to different factors: the physicochemical characteristics of the two metabolites, the alcohol drinking behaviours and individual metabolism, the variables due to the period of time and amount of alcohol and cocaine intake. Moreover, it is also known that CE incorporation into the hair matrix is influenced by the melanin content,[144] in contrast to EtG that, as already reported, appears to be accumulated into hair independently of the natural hair color.[103]

Another study of Politi *et al.*[36] evaluated a potential correlation between the concentration of hCE and the concentration of hFAEEs. The authors measured hCE and hFAEEs in hair samples collected from 80 cocaine users. The quantitative results of hFAEEs ranged from 0.02 to 17.67 ng mg^{-1}, expressed as the sum of four different substances and the presence of esters was detected in about 69% of the tested samples. The authors stated that, based on the results, hCE appeared as a more sensitive biomarker than hFAEEs of alcohol use in a population of cocaine use, either by measurement of the absolute concentration of CE or by calculating the percentage of CE over other cocaine metabolites.

In a recent study carried out by Pragst *et al.*[86] hEtG, hEtPa and hCE were tested in a group of 126 children with drug/alcohol abusing parents or suspected consumers. Surprisingly, 15.5% of the monitored subjects provided a positive result for hEtG ($n = 25$). hEtG concentrations ranged from 3.0 to 42.6 pg mg^{-1} with two samples higher than the 30 pg mg^{-1} cut-off. The authors observed a significantly higher prevalence of positive cases among boys with short hair. This outcome can be indicative of an external contamination rather than a variable related to gender. Though it is not possible to exclude that EtG in hair of the tested children was due to an unattended consumption of alcohol-based beverages, there are also several potential sources of external contamination such as: parents' urine, through children's head hair petting with unclean hands in low hygiene conditions, accidental contact with red wine and other beverages[145] or lotions[108] and cosmetic products containing EtG.[107] This study highlighted the importance of checking different biomarkers in order to increase both sensitivity and specificity in the diagnosis of alcohol consumption.

References

1. A. Peacock, J. Leung, S. Larney, S. Colledge, M. Hickman, J. Rehm, G. A. Giovino, R. West, W. Hall, P. Griffiths, R. Ali, L. Gowing, J. Marsden, A. J. Ferrari, J. Grebely, M. Farrell and L. Degenhardt, *Addiction*, 2018, **113**, 1905–1926.
2. World Health Organization, Global status report on alcohol and health 2014, 2014, 1505, https://apps.who.int/iris/bitstream/handle/10665/112736/9789240692763_eng.pdf?sequence=1&isAllowed=y.
3. World Health Organization, Global status report on alcohol and health 2018, 2018, 1504, https://www.who.int/publications/i/item/9789241565639.
4. T. Takahashi, G. Lapham, L. J. Chavez, A. K. Lee, E. C. Williams, J. E. Richards, D. Greenberg, A. Rubinsky, D. Berger, E. J. Hawkins, J. O. Merrill and K. A. Bradley, *Addict. Sci. Clin. Pract.*, 2017, **12**, 17.
5. G. Skopp, G. Schmitt, L. Pötsch, P. Drönner, R. Aderjan and R. Mattern, *Alcohol Alcohol.*, 2000, **35**, 283–285.
6. A. Alt, I. Janda, S. Seidl and F. M. Wurst, *Alcohol Alcohol.*, 2000, **35**, 313–314.
7. F. M. Wurst, S. Seidl, A. Alt and J. Metzger, *Psychiatr. Prax.*, 2000, **27**, 367–371.
8. M. Yegles, A. Labarthe, V. Auwärter, S. Hartwig, H. Vater, R. Wennig and F. Pragst, *Forensic Sci. Int.*, 2004, **145**, 167–173.
9. L. Morini, L. Politi, A. Groppi, C. Stramesi and A. Polettini, *J. Mass Spectrom.*, 2006, **41**, 34–42.
10. V. Pirro, D. Di Corcia, F. Seganti, A. Salomone and M. Vincenti, *Forensic Sci. Int.*, 2013, **232**, 229–236.
11. C. L. Crunelle, M. Yegles, A. Nuijs, A. Covaci, M. De Doncker, K. E. Maudens, B. Sabbe, G. Dom, W. E. Lambert, P. Michielsen and H. Neels, *Drug Alcohol Depend.*, 2014, **134**, 1–11.

12. C. L. Crunelle, J. Verbeek, G. Dom, A. Covaci, M. Yegles, P. Michielsen, M. De Doncker, F. Nevens, D. Cappelle, A. L. van Nuijs and H. Neels, *Clin. Biochem.*, 2016, **49**, 554–559.
13. L. Berger, M. Fendrich, J. Jones, D. Fuhrmann, C. Plate and D. Lewis, *Addiction*, 2014, **109**, 425–431.
14. R. Paul, R. Kingston, L. Tsanaclis, A. Berry and A. Guwy, *Forensic Sci. Int.*, 2008, **176**, 82–86.
15. L. Morini, E. Marchei, L. Tarani, M. Trivelli, G. Rapisardi, M. R. Elicio, J. Ramis, O. Garcia-Algar, L. Memo, R. Pacifici, A. Groppi, P. Danesino and S. Pichini, *Ther. Drug Monit.*, 2013, **35**, 402–407.
16. X. Joya, B. Friguls, S. Ortigosa, E. Papaseit, S. E. Martínez, A. Manich, O. Garcia-Algar, R. Pacifici, O. Vall and S. Pichini, *J. Pharm. Biomed. Anal.*, 2012, **69**, 209–222.
17. X. Joya, E. Mazarico, J. Ramis, R. Pacifici, J. Salat-Batlle, C. Mortali, O. García-Algar and S. Pichini, *Drug Alcohol Depend.*, 2016, **158**, 45–51.
18. I. Jańczewska, J. Wierzba, M. Cichoń-Kotek and A. Jańczewska, *Dev. Period Med.*, 2019, **23**, 60–66.
19. F. M. Wurst, E. Kelso, W. Weinmann, F. Pragst, M. Yegles and I. Sundström Poromaa, *Am. J. Obstet. Gynecol.*, 2008, **198**, 407.e401–405.
20. F. Freni, M. Moretti, S. Scardo, C. Carelli, C. Vignali, M. C. Monti and L. Morini, *Drug Test. Anal.*, 2022, **14**, 1417–1428.
21. R. Agius, T. Nadulski, H. G. Kahl and B. Dufaux, *Forensic Sci. Int.*, 2012, **218**, 10–14.
22. B. Liniger, A. Nguyen, A. Friedrich-Koch and M. Yegles, *Traffic Inj. Prev.*, 2010, **11**, 123–126.
23. H. Andresen-Streichert, G. von Rothkirch, E. Vettorazzi, A. Mueller, A. W. Lohse, D. Frederking, B. Seegers, B. Nashan and M. Sterneck, *Ther. Drug Monit.*, 2015, **37**, 539–545.
24. L. Politi, L. Morini, F. Mari, A. Groppi and E. Bertol, *Int. J. Legal Med.*, 2008, **122**, 507–509.
25. H. Andresen-Streichert, A. Müller, A. Glahn, G. Skopp and M. Sterneck, *Dtsch. Arztebl. Int.*, 2018, **115**, 309–315.
26. A. Cinquetti, C. Terranova, A. Aprile and D. Favretto, *Drug Test. Anal.*, 2023, **15**, 931–932.
27. E. Lendoiro, A. de Castro, C. Jiménez-Morigosa, X. A. Gomez-Fraguela, M. López-Rivadulla and A. Cruz, *Forensic Sci. Int.*, 2018, **286**, 239–244.
28. A. Schröck, M. Pfäffli, S. König and W. Weinmann, *Int. J. Legal Med.*, 2016, **130**, 1527–1533.
29. M. Hastedt, S. Herre, F. Pragst, M. Rothe and S. Hartwig, *Alcohol Alcohol.*, 2012, **47**, 127–132.
30. A. Salomone, A. Bozzo, D. Di Corcia, E. Gerace and M. Vincenti, *J. Anal. Toxicol.*, 2018, **42**, 157–162.
31. P. Cabarcos, I. Álvarez, M. J. Tabernero and A. M. Bermejo, *Anal. Bioanal. Chem.*, 2015, **407**, 4907–4925.
32. L. Politi, F. Leone, L. Morini and A. Polettini, *Anal. Biochem.*, 2007, **368**, 1–16.

33. K. D. Tawiah, S. B. Riley and M. M. Budelier, *Clin. Chem.*, 2022, **68**, 635–645.
34. H. Sachs, *Drogennachweis in Haaren*, Schmidt-Römhild, Lübeck, Germany, 1997.
35. L. Politi, L. Morini, F. Leone and A. Polettini, *Addiction*, 2006, **101**, 1408–1412.
36. L. Politi, A. Zucchella, L. Morini, C. Stramesi and A. Polettini, *Forensic Sci. Int.*, 2007, **172**, 23–27.
37. L. Morini, C. Varango, C. Filippi, C. Rusca, P. Danesino, F. Cheli, M. Fusini, G. Iannello and A. Groppi, *Ther. Drug Monit.*, 2011, **33**, 654–657.
38. G. Høiseth, L. Morini, A. Polettini, A. Christophersen and J. Mørland, *Alcohol.: Clin. Exp. Res.*, 2009, **33**, 812–816.
39. L. Morini, L. Politi and A. Polettini, *Addiction*, 2009, **104**, 915–920.
40. F. Pragst and M. A. Balikova, *Clin. Chim. Acta*, 2006, **370**, 17–49.
41. S. Suesse, F. Pragst, T. Mieczkowski, C. M. Selavka, A. Elian, H. Sachs, M. Hastedt, M. Rothe and J. Campbell, *Forensic Sci. Int.*, 2012, **218**, 82–91.
42. R. Agius, T. Nadulski, H. G. Kahl, J. Schräder, B. Dufaux, M. Yegles and F. Pragst, *Forensic Sci. Int.*, 2010, **196**, 3–9.
43. C. Abernethy, K. E. McCall, G. Cooper, D. Favretto, F. Vaiano, E. Bertol and H. Mactier, *Arch. Dis. Child. Fetal Neonatal Ed.*, 2018, **103**, F216–F220.
44. L. Morini, E. Marchei, F. Vagnarelli, O. Garcia Algar, A. Groppi, L. Mastrobattista and S. Pichini, *Forensic Sci. Int.*, 2010, **196**, 74–77.
45. S. Pichini, E. Marchei, F. Vagnarelli, L. Tarani, F. Raimondi, R. Maffucci, B. Sacher, M. Bisceglia, G. Rapisardi, M. R. Elicio, P. Biban, P. Zuccaro, R. Pacifici, A. Pierantozzi and L. Morini, *Alcohol.: Clin. Exp. Res.*, 2012, **36**, 417–424.
46. S. Pichini, L. Morini, E. Marchei, I. Palmi, M. C. Rotolo, F. Vagnarelli, O. Garcia-Algar, O. Vall and P. Zuccaro, *Can. J. Clin. Pharmacol.*, 2009, **16**, e370–e375.
47. M. K. Woźniak, L. Banaszkiewicz, J. Aszyk, M. Wiergowski, I. Jańczewska, J. Wierzba, A. Kot-Wasik and M. Biziuk, *Anal. Bioanal. Chem.*, 2021, **413**, 3093–3105.
48. E. M. A. Henderson, D. Tappin, D. Young, D. Favretto and H. Mactier, *Arch. Dis. Child.*, 2023, **108**, 36–41.
49. M. Luginbühl, F. Stöth, A. Schröck, S. Gaugler and W. Weinmann, *Nat. Protoc.*, 2021, **16**, 283–308.
50. I. Janda, W. Weinmann, T. Kuehnle, M. Lahode and A. Alt, *Forensic Sci. Int.*, 2002, **128**, 59–65.
51. C. Jurado, T. Soriano, M. P. Giménez and M. Menéndez, *Forensic Sci. Int.*, 2004, **145**, 161–166.
52. V. Pirro, D. Di Corcia, S. Pellegrino, M. Vincenti, B. Sciutteri and A. Salomone, *Forensic Sci. Int.*, 2011, **210**, 271–277.
53. F. Yaldiz, N. Daglioglu, A. Hilal, A. Keten and M. K. Gülmen, *J. Forensic Leg. Med.*, 2013, **20**, 799–802.

54. B. Mönch, R. Becker and I. Nehls, *Int. J. Legal Med.*, 2014, **128**, 69–72.
55. D. Cappelle, H. Neels, M. Yegles, J. Paulus, A. L. van Nuijs, A. Covaci and C. L. Crunelle, *Forensic Sci. Int.*, 2015, **249**, 20–24.
56. R. Paul, L. Tsanaclis, R. Kingston, A. Berry and A. Guwy, *Drug Test. Anal.*, 2011, **3**, 201–205.
57. F. Pragst and M. Yegles, *Ther. Drug Monit.*, 2008, **30**, 255–263.
58. I. Kerekes, M. Yegles, U. Grimm and R. Wennig, *Alcohol Alcohol.*, 2009, **44**, 62–66.
59. H. Kharbouche, F. Sporkert, S. Troxler, M. Augsburger, P. Mangin and C. Staub, *J. Chromatogr. B: Anal. Technol. Biomed. Life Sci.*, 2009, **877**, 2337–2343.
60. I. Álvarez, A. M. Bermejo, M. J. Tabernero, P. Fernández, P. Cabarcos and P. López, *Anal. Bioanal. Chem.*, 2009, **393**, 1345–1350.
61. Y. Shi, B. Shen, P. Xiang, H. Yan and M. Shen, *J. Chromatogr. B: Anal. Technol. Biomed. Life Sci.*, 2010, **878**, 3161–3166.
62. B. Mönch, R. Becker, C. Jung and I. Nehls, *Forensic Sci. Int.*, 2013, **226**, 202–207.
63. C. Vignali, S. Ortu, C. Stramesi, F. Freni, M. Moretti, L. Tajana, A. M. M. Osculati, A. Groppi and L. Morini, *J. Pharm. Biomed. Anal.*, 2018, **159**, 18–22.
64. A. Mattia, C. Moschella, M. C. David, M. Fiore, S. Gariglio, A. Salomone and M. Vincenti, *Front. Chem.*, 2022, **10**, 858205.
65. P. Bendroth, R. Kronstrand, A. Helander, J. Greby, N. Stephanson and P. Krantz, *Forensic Sci. Int.*, 2008, **176**, 76–81.
66. P. Kintz, M. Villain, E. Vallet, M. Etter, G. Salquebre and V. Cirimele, *Forensic Sci. Int.*, 2008, **176**, 87–90.
67. M. Concheiro, A. Cruz, M. Mon, A. de Castro, O. Quintela, A. Lorenzo and M. López-Rivadulla, *J. Anal. Toxicol.*, 2009, **33**, 155–161.
68. F. Lamoureux, J. M. Gaulier, F. L. Sauvage, M. Mercerolle, C. Vallejo and G. Lachâtre, *Anal. Bioanal. Chem.*, 2009, **394**, 1895–1901.
69. I. Tarcomnicu, A. L. van Nuijs, K. Aerts, M. De Doncker, A. Covaci and H. Neels, *Forensic Sci. Int.*, 2010, **196**, 121–127.
70. M. E. Albermann, F. Musshoff and B. Madea, *Anal. Bioanal. Chem.*, 2010, **396**, 2441–2447.
71. F. Pragst, M. Rothe, B. Moench, M. Hastedt, S. Herre and D. Simmert, *Forensic Sci. Int.*, 2010, **196**, 101–110.
72. M. E. Albermann, F. Musshoff, L. Aengenheister and B. Madea, *Anal. Bioanal. Chem.*, 2012, **403**, 769–776.
73. R. Kronstrand, L. Brinkhagen and F. H. Nyström, *Forensic Sci. Int.*, 2012, **215**, 51–55.
74. P. Cabarcos, H. M. Hassan, M. J. Tabernero and K. S. Scott, *J. Appl. Toxicol.*, 2013, **33**, 638–643.
75. L. Imbert, S. Dulaurent, M. Mercerolle, J. Morichon, G. Lachâtre and J. M. Gaulier, *Forensic Sci. Int.*, 2014, **234**, 132–138.
76. T. M. Binz, M. R. Baumgartner and T. Kraemer, *Forensic Sci. Int.*, 2014, **244**, 20–24.

77. N. Kummer, S. M. Wille, V. Di Fazio, M. Ramírez Fernández Mdel, M. Yegles, W. E. Lambert and N. Samyn, *J. Anal. Toxicol.*, 2015, **39**, 17–23.
78. X. Joya, E. Marchei, J. Salat-Batlle, O. García-Algar, V. Calvaresi, R. Pacifici and S. Pichini, *Clin. Chem. Lab. Med.*, 2016, **54**, 427–435.
79. D. Oppolzer, M. Barroso, L. Passarinha and E. Gallardo, *Biomed. Chromatogr.*, 2017, **31**, DOI: 10.1002/bmc.3858.
80. U. Meier, T. Briellmann, E. Scheurer and F. Dussy, *Drug Test. Anal.*, 2017, **9**, 1594–1603.
81. D. Palumbo, P. Fais, A. Calì, M. Lusardi, E. Bertol and J. P. Pascali, *J. Chromatogr. B: Anal. Technol. Biomed. Life Sci.*, 2018, **1100–1101**, 33–38.
82. M. F. Bastiani, L. L. F. Lizot, A. C. C. Da Silva, R. Z. Hahn, S. S. Dries, M. S. Perassolo, M. V. Antunes and R. Linden, *Forensic Sci. Int.*, 2020, **306**, 110071.
83. S. Baggio, B. Trächsel, V. Rousson, S. Rothen, J. Studer, S. Marmet, P. Heller, F. Sporkert, J. B. Daeppen, G. Gmel and K. Iglesias, *Addiction*, 2020, **115**, 426–436.
84. J. P. Pascali, D. Palumbo, F. Umani Ronchi, F. Mari, A. Fioravanti and E. Bertol, *J. Forensic Sci.*, 2020, **65**, 189–192.
85. A. Müller and S. Iwersen-Bergmann, *Drug Test. Anal.*, 2020, **12**, 948–956.
86. F. Pragst, F. Krumbiegel, D. Thurmann, L. Westendorf, M. Methling, A. Niebel and S. Hartwig, *Int. J. Legal Med.*, 2020, **134**, 523–532.
87. P. Kintz, *Forensic Sci. Int.*, 2015, **249**, A1–2.
88. L. C. Bossers, R. Paul, A. J. Berry, R. Kingston, C. Middendorp and A. J. Guwy, *J. Chromatogr. B: Anal. Technol. Biomed. Life Sci.*, 2014, **953–954**, 115–119.
89. D. Bossers and D. R. Paul, *Forensic Sci. Int.*, 2014, **242**, e56–e58.
90. B. Mönch, R. Becker and I. Nehls, *Alcohol Alcohol.*, 2013, **48**, 558–563.
91. A. Salomone, V. Pirro, T. Lombardo, D. Di Corcia, S. Pellegrino and M. Vincenti, *Drug Test. Anal.*, 2015, **7**, 407–413.
92. A. Mueller, H. Jungen, S. Iwersen-Bergmann, L. Raduenz, S. Lezius and H. Andresen-Streichert, *Forensic Sci. Int.*, 2017, **271**, 43–48.
93. P. Kintz, *Forensic Sci. Int.*, 2012, **218**, 2.
94. H. Kharbouche, M. Faouzi, N. Sanchez, J. B. Daeppen, M. Augsburger, P. Mangin, C. Staub and F. Sporkert, *Int. J. Legal Med.*, 2012, **126**, 243–250.
95. C. L. Crunelle, M. Yegles, M. De Doncker, D. Cappelle, A. Covaci, A. L. van Nuijs and H. Neels, *Forensic Sci. Int.*, 2017, **274**, 107–108.
96. Consensus on alcohol markers, Society of Hair testing, 2019, http://www.soht.org/images/pdf/Revision_2019_Alcoholmarkers.pdf.
97. J. Gareri, C. Rao and G. Koren, *Drug Test. Anal.*, 2014, **6**(Suppl 1), 30–36.
98. G. Høiseth, L. Morini, R. Ganss, K. Nordal and J. Mørland, *Alcohol.: Clin. Exp. Res.*, 2013, **37**(Suppl 1), E14–E16.
99. J. T. Fosen, L. Morini, C. Sempio, R. Ganss, J. Mørland and G. Høiseth, *Alcohol.: Clin. Exp. Res.*, 2016, **40**, 451–456.
100. A. Biondi, F. Freni, C. Carelli, M. Moretti and L. Morini, *Forensic Sci. Int.*, 2019, **300**, 106–119.

101. L. M. Huppertz, B. Moosmann and V. Auwärter, *Drug Test. Anal.*, 2018, **10**, 206–211.
102. J. A. Ruth and P. R. Stout, *Forensic Sci. Rev.*, 2004, **16**, 115–133.
103. B. M. Appenzeller, M. Schuman, M. Yegles and R. Wennig, *Alcohol Alcohol.*, 2007, **42**, 326–327.
104. L. Morini, A. Zucchella, A. Polettini, L. Politi and A. Groppi, *Forensic Sci. Int.*, 2010, **198**, 23–27.
105. L. Martins Ferreira, T. Binz and M. Yegles, *Forensic Sci. Int.*, 2012, **218**, 123–125.
106. J. Gareri, B. Appenzeller, P. Walasek and G. Koren, *Anal. Bioanal. Chem.*, 2011, **400**, 183–188.
107. L. Morini, C. Sempio and M. Moretti, *Curr. Pharm. Biotechnol.*, 2018, **19**, 175–179.
108. F. Sporkert, H. Kharbouche, M. P. Augsburger, C. Klemm and M. R. Baumgartner, *Forensic Sci. Int.*, 2012, **218**, 97–100.
109. B. M. Appenzeller, R. Agirman, P. Neuberg, M. Yegles and R. Wennig, *Forensic Sci. Int.*, 2007, **173**, 87–92.
110. M. Luginbühl, S. Nussbaumer and W. Weinmann, *Drug Test. Anal.*, 2018, **10**, 689–693.
111. U. Meier, T. Briellmann, E. Scheurer and F. Dussy, *Drug Test. Anal.*, 2018, **10**, 701–710.
112. J. T. Fosen, G. Høiseth, C. Sempio, N. Giarratana, A. Enger, J. Mørland and L. Morini, *Drug Test. Anal.*, 2019, **11**, 112–118.
113. G. Miolo, G. Stocchero, S. Vogliardi, L. Menilli, S. Scrivano, M. Montisci and D. Favretto, *J. Anal. Toxicol.*, 2020, **44**, 58–64.
114. U. Meier, F. Colledge, S. Imfeld, T. Briellmann, K. Mercer-Chalmers-Bender, E. Scheurer and F. Dussy, *Drug Test. Anal.*, 2019, **11**, 1522–1541.
115. P. Własiuk, E. Alladio, A. Salomone, M. Vincenti and G. Zadora, *Drug Test. Anal.*, 2019, **11**, 77–85.
116. F. Bortolotti, D. Sorio, A. Bertaso and F. Tagliaro, *J. Pharm. Biomed. Anal.*, 2018, **147**, 2–12.
117. N. M. Porpiglia, F. Tagliaro, R. Micciolo, L. Canal, G. Musile and F. Bortolotti, *Forensic Sci. Int.*, 2022, **340**, 111438.
118. F. Bortolotti, G. De Paoli, J. P. Pascali and F. Tagliaro, *Clin. Chim. Acta*, 2007, **380**, 4–7.
119. F. Bortolotti, R. Micciolo, L. Canal and F. Tagliaro, *Alcohol.: Clin. Exp. Res.*, 2015, **39**, 2108–2114.
120. F. Bortolotti, F. Tagliaro, F. Cittadini, R. Gottardo, M. Trettene and M. Marigo, *Forensic Sci. Int.*, 2002, **128**, 53–58.
121. J. P. Pascali, F. Bortolotti, D. Sorio, M. Ivanova, T. M. Palmbach and F. Tagliaro, *Med., Sci. Law*, 2011, **51**, 26–31.
122. N. M. Porpiglia, E. F. De Palo, S. A. Savchuk, S. A. Appolonova, F. Bortolotti and F. Tagliaro, *Clin. Chim. Acta*, 2018, **483**, 256–262.
123. F. Tagliaro, F. Bortolotti, R. M. Dorizzi and M. Marigo, *Clin. Chem.*, 2002, **48**, 208–209.

124. F. Tagliaro, F. Bortolotti, M. Zuliani, F. Crivellente, G. Manett, V. L. Pascali and M. Marigo, *J. Capillary Electrophor. Microchip Technol.*, 1999, **6**, 137–143.
125. F. Pragst, V. Auwaerter, F. Sporkert and K. Spiegel, *Forensic Sci. Int.*, 2001, **121**, 76–88.
126. P. Kintz, *Forensic Sci. Int.*, 2010, **196**, 2.
127. D. L. Caprara, J. Klein and G. Koren, *Ann. Ist. Super. Sanita*, 2006, **42**, 39–45.
128. S. Hartwig, V. Auwärter and F. Pragst, *Alcohol Alcohol.*, 2003, **38**, 163–167.
129. N. De Giovanni, G. Donadio and M. Chiarotti, *J. Anal. Toxicol.*, 2007, **31**, 93–97.
130. L. Politi, F. Mari, S. Furlanetto, E. Del Bravo and E. Bertol, *J. Pharm. Biomed. Anal.*, 2011, **54**, 1192–1195.
131. E. Bertol, E. Del Bravo, F. Vaiano, F. Mari and D. Favretto, *Drug Test. Anal.*, 2014, **6**, 930–935.
132. L. Tsanaclis, K. Bagley, S. Bevan and J. Wicks, *J. Anal. Toxicol.*, 2020, **44**, 829–833.
133. S. Suesse, M. Blueml and F. Pragst, *Forensic Sci. Int.*, 2015, **249**, 1–5.
134. C. Zhao, X. Wang, Y. Mao, Z. Xu, Y. Sun, X. Mei and W. Shi, *J. Cosmet., Dermatol. Sci. Appl.*, 2021, **20**, 249–255.
135. M. Luginbühl, B. Bekaert, S. Suesse and W. Weinmann, *Drug Test. Anal.*, 2019, **11**, 870–877.
136. D. S. Harris, E. T. Everhart, J. Mendelson and R. T. Jones, *Drug Alcohol Depend.*, 2003, **72**, 169–182.
137. M. Perez-Reyes, A. R. Jeffcoat, M. Myers, K. Sihler and C. E. Cook, *Psychopharmacology*, 1994, **116**, 428–432.
138. E. J. Cone, D. Yousefnejad, W. D. Darwin and T. Maguire, *J. Anal. Toxicol.*, 1991, **15**, 250–255.
139. G. J. DiGregorio, E. J. Barbieri, A. P. Ferko and E. K. Ruch, *J. Anal. Toxicol.*, 1993, **17**, 445–446.
140. A. C. Gruszecki, C. A. Robinson Jr, J. H. Embry and G. G. Davis, *Am. J. Forensic Med. Pathol.*, 2000, **21**, 166–171.
141. J. A. Bourland, E. F. Hayes, R. C. Kelly, S. A. Sweeney and M. M. Hatab, *J. Anal. Toxicol.*, 2000, **24**, 489–495.
142. N. C. Rubio, M. Hastedt, J. Gonzalez and F. Pragst, *Int. J. Legal Med.*, 2015, **129**, 69–84.
143. O. López-Guarnido, I. Álvarez, F. Gil, L. Rodrigo, H. C. Cataño, A. M. Bermejo, M. J. Tabernero, A. Pla and A. F. Hernández, *J. Appl. Toxicol.*, 2013, **33**, 838–844.
144. K. B. Scheidweiler, E. J. Cone, E. T. Moolchan and M. A. Huestis, *J. Pharmacol. Exp. Ther.*, 2005, **313**, 909–915.
145. A. Müller and S. Iwersen-Bergmann, *Alcohol Alcohol.*, 2018, **53**, 532–538.

CHAPTER 8

Hair Analysis to Understand the Diffusion of New Psychoactive Substances

A. SALOMONE[a,b]

[a] Department of Chemistry, University of Turin, Via P. Giuria 5, 10125 Torino, Italy; [b] Centro Regionale Antidoping, Regione Gonzole 10/1, 10043 Orbassano (TO), Italy
Email: alberto.salomone@unito.it

8.1 Introduction

Since the publication of the first articles describing the existence of an unprecedented class of "new psychoactive substances" (NPS) distributed under the deceptive name of "legal highs", "designer drugs", "herbal highs", "bath salts", or "research chemicals",[1,2] forensic and clinical laboratories worldwide have had to face the analytical challenge of the identification and quantification of these new molecules in various biological matrices. At least until few years ago, laboratories performing drug testing in conventional and non-conventional biological matrices did not include screening procedures for NPS in their routine protocols in the context of workplace drug testing, driving re-licensing, roadside control, and withdrawal programs.[3] As a matter of fact, most analytical methods commonly used in forensic toxicology are generally based on the targeted detection of a limited and well-defined list of compounds to monitor, usually chosen on the base of the national or international reports, or alerts from national warning system. Therefore, it became difficult for laboratories to stay up to date because of the continued entry into the market of newly synthesized products, and the

Issues in Toxicology No. 47
Perspectives and Challenges of Hair Analysis
Edited by Pascal Kintz, Alberto Salomone and Marco Vincenti

Published by the Royal Society of Chemistry, www.rsc.org

consequent low availability (and high cost) of reference analytical standards. Furthermore, many of the new compounds are very potent, thus low doses ingested will lead to trace concentrations in biological matrices, especially in hair.[4] Nevertheless, the forensic toxicologist community has reacted promptly to propose several procedures for NPS detection in the keratin matrix,[5–8] which are now applicable in either routine or more complex cases.

8.2 The Diffusion of New Psychoactive Molecules

Among the newly synthetized compounds, synthetic cannabinoids, synthetic cathinones, phenethylamines, piperazines, and tryptamines are the most popular. Yet, the understanding of the real prevalence of NPS use is hampered by numerous objective difficulties, primarily the huge number of possible substances and the continuous turnover of molecules. While most NPS disappear after a short time, others seem to establish a niche market and only a few are used by large populations. Furthermore, more recently, there has been an outbreak of new synthetic opioids (NSO)[9] and designer benzodiazepine derivatives.[10,11] Fentanyl and novel synthetic opioids have appeared with increasing frequency in the illicit drug supply as standalone products or as adulterants in illicit and pharmaceutical drugs to increase their potency or as a substitute ingredient at little cost. Not only heroin but also several common drugs are frequently replaced or cut with NSO,[12–14] raising a major health concern for unaware users.[15,16] In parallel, some authors have pointed out that users' preference for fentanyl has increased, leaving the doubt about whether fentanyl spread is more demand-led or supply-led.[17] North America, in particular, is facing an unprecedented opioid crisis, with high levels of drug seizures and mortality related to opioid overdose, with high-potency fentanyl playing a primary role.[18–22] Therefore, a greater understanding of the co-use (intentional or not) of illicit fentanyl, opioids, and other psychoactive drugs is deemed as necessary from several authors in order to adapt and improve existing interventions aimed to reduce overdose mortality,[23,24] together with broad integrated public health strategy based on overdose education and prevention[25] and to support the drugs debate.

Currently, most information about NPS/NSO prevalence comes from the analysis of law enforcement seizures, while data on biological material (from living or deceased subjects) are limited. In this scenario, a primary goal is to incorporate and complement the existing monitoring system with new testing methods. A promising approach is the identification of NPS/NSO in different biological matrices to investigate their diffusion in the population. In particular, hair has proved to be a unique (and easier to implement today than in the past) tool, since the keratin matrix allows the detection of past drug exposure and the investigation of the chronological profile of the exposure to one or multiple compounds. Whilst it is not possible to accurately assess the level of consumption, the method can provide a geographical and temporal overview of substances diffusion. If used in combination with other tools, *e.g.* surveys, it can also validate self-reported information.[26–28]

8.3 Interpreting NPS Results in Hair

The results of hair analysis should be always interpreted considering all relevant factors surrounding the case. The meaning of quantitative results in the keratin matrix is often debated and subjected to controversy and different interpretation, and this is probably the beauty of hair analysis. A preliminary attempt to review the typical alternative circumstances (passive exposure *vs.* active consumption, mindful *vs.* unaware intake, and sporadic *vs.* chronic use) in hair analysis was made, trying to shed light on possible criteria needed before a definitive interpretation of either positive or negative results can be safely given.[4] From this and other studies published up to now, a few preliminary conclusions can be tentatively outlined. In particular, as usually happens with hair analysis, the detection of metabolites to exclude external contamination, and the comparison of the detected levels in single cases with larger populations to make inferences about the frequency of use, appear to be fundamental steps when testing for NPS and NSO. Provided that large panels of NPS/NSO (taking into account in a timely way the fluctuation of the market) are put in place, hair testing proved effective to monitor individuals undergoing periodic control and to check how frequently these compounds are taken, either knowingly or unknowingly.[29–35] When large populations are subjected to NPS/NSO hair analysis, then the relevant gathered results can provide epidemiological data on the NPS trends and extent of use in the community.[36] Certainly, the longer detection window of hair is a big advantage for prevalence studies. Even sporadic exposures occurring several weeks before the hair sampling can be identified, while this would not be possible with urine and blood. Therefore, exposure, voluntarily or not, to a certain NPS/NSO occurring in the weeks prior to the sampling, can be identified by means of hair analysis.

Hair analysis has already been used to speculate about NPS/NSO diffusion in selected populations. Reliable and sensitive methods dedicated to the identification and quantification of NPS/NSO in hair have been proposed, and several have been applied to real samples. In this chapter, the possible information and results interpretation obtainable from hair analysis in terms of trends and pattern of NPS/NSO use will be explored. In particular, (i) the diffusion of NPS/NSO, and (ii) the co-use of traditional and new drugs, will be discussed. Only articles published after 2016 have been considered.

8.4 Hair Analysis to Investigate the Prevalence of NPS/NSO

Although NPS/NSO have been the focus of forensic toxicological research for several years, information about their real role in the drug scene compared to more traditional compounds is still scarce. Among the synthetic cannabinoids, a certain interest has recently raised in the diffusion of MDMB-4en-PINACA[37,38] and the increase in reports of cannabis adulterated with synthetic cannabinoids. It is not known what might be driving this development, but it could possibly reflect both shortages of cannabis linked to the pandemic or, possibly in some countries, criminal groups exploiting the availability of low-THC cannabis products, which may be difficult to distinguish from cannabis sold on the drug market. In spite of the fact that the warning came initially from Europe, this phenomenon seems to be spread worldwide: from the analysis of 59 actual positive hair samples collected between 2020 and 2021 in China, the most frequently detected synthetic cannabinoid was indeed MDMB-4en-PINACA. Any scenario where people consume synthetic cannabinoids is worrying, given the toxicity of most substances.[39] In Italy, extensive research was performed on hair samples ($n = 82$) collected and analyzed in the frame of compulsory toxicological tests performed to verify the suitability of subjects to hold a driving license, according to the provisions for the Italian road traffic regulations.[40] All samples tested negative for the target analytes. The authors speculated that, due to low recovery of new synthetic cannabinoids, the method might have missed identifying 'positive' hair samples, or, alternatively, that the sensitivity of the investigated method did not meet the required sensitivity to detect new synthetic cannabinoids at the lowest concentrations (a few picograms per milligram) which reasonably would be required for identifying occasional use.

Different evidence was obtained from the analysis of authentic human hair samples over the period of three years (2016–2018) in South Korea, for the simultaneous detection of 18 synthetic cannabinoid parents and 41 of their metabolites.[41] AB-CHMINACA (32 cases), JWH-210 (12 cases), XLR-11 (2 cases), 5F-AKB-48 (2 cases), AKB-48 (1 case), PB-22 (1 case), AB-PINACA (1 case), AB-FUBINACA (1 case), and their metabolites were detected. The ages of the suspects ranged from 23 to 44 (mean = 33.1); 39 of them were male. AB-CHMINACA and JWH-210 were the most consistently detected synthetic cannabinoids in this study.

The situation with NPS belonging to classes other than synthetic cannabinoids is even more complex, and further demonstration of the fact that the drug scenario is continuously evolving was recently produced by another Italian study.[42] Despite the limited screened population, attention was focused on the multiple positive results observed in most samples after the screening for 32 synthetic cathinones and two metabolites. The most prevalent molecules were MDPHP, α-PHP, 4-MEC, and 3-MMC. Other findings included 3,4-DMMC, butylone, methylone and 4-MMC. The most frequent combination was composed of α-PHP and 3-MMC.

Another extensive investigation was performed on hair samples from subjects ($n = 100$) previously tested for driving re-licensing from March to June 2022 in Italy.[43] Only two samples were positive for ketamine and two for methylone and mephedrone, albeit the possible co-elution with 3-MMC was not resolved.

One study is also available exploring the diffusion of 35 synthetic cathinones and selected piperazines in Germany.[44] The method was applied to authentic hair samples from different forensic cases, including driving license renewals, post-mortem toxicological analyses, or the family court. Samples were obtained over a period from February 2015 to February 2019 and were selected according to case histories, which meant a known poly-toxicological drug use or an unclear cause of death. 92.5% of the subjects were male (37 cases) and only three cases were female (7.5%). Of the 40 analyzed cases, 28 were negative for synthetic cathinones and piperazines. Most of the positive cases (8 of 12) had two or more synthetic cathinones, and one synthetic cathinone was detected in four hair samples. Mephedrone was the most prevalent compound in the screened population.

Ketamine is often included in the category of NPS. In a recent study,[45] the aim was the identification of arylcyclohexylamines and their metabolites present in the analyzed hair samples by UHPLC–MS/MS, and contextually the study of polyconsumption prevalence, with special focus on ketamine and NPS. From the analysis of routine cases and from ten real cases related to criminal actions involving ketamine consumption, the authors concluded that (i) ketamine is the most common compound among the class of arylcyclohexylamines, in accordance with the evidence obtained from the analysis of non-biological samples, and (ii) ketamine users are also generally exposed to other psychoactive compounds, such as synthetic cathinones and hallucinogens.

A few researchers have focused their efforts on NSO in hair. Preliminary results obtained from a large dataset of real hair samples positive for fentanyl or its analogs were preliminarily presented and discussed.[46] In this study, among the 293 samples, two-thirds (67.6%; $n = 198$) tested positive for at least one "fentalog". Remarkably, all 198 samples tested positive for fentanyl, either alone or together with other molecules. The most common parent compounds detected were fentanyl, acetyl fentanyl, and furanylfentanyl. Among prescription opioids, oxycodone, hydrocodone and tramadol were also monitored. In particular, oxycodone was detected in 168 samples (57.3% of cases). Whenever a sample was positive for any NSO compound included in the panel, fentanyl was also always present, suggesting that fentanyl is the most prevalent molecule while the less common analogs are not consumed in isolation. The sporadic occurrence of the other fentalogs has different possible explanations: (i) the low prevalence at the time of the sample collection, (ii) poor incorporation or low stability in the keratin matrix and/or (iii) insufficient sensitivity of the analytical method in relation to the low effective dosage.

In a similar investigation on a selected population, 17 authentic samples were obtained from workplace drug testing in the United States.[34] The samples came from multiple U.S. states, and they had previously tested positive for heroin, buprenorphine, methadone, and fentanyl. The analysis targeted 16 synthetic opioids in segmented hair. Fentanyl ($n = 16$), norfentanyl ($n = 15$), acetylfentanyl ($n = 5$), valerylfentanyl ($n = 1$), furanylfentanyl

($n=5$), and tetrahydrofuranylfentanyl ($n=1$) were detected. After the same samples were also analyzed for other drugs of abuse, a poly-drug use in the fentanyl-analogue user population was observed. The mean value of detected compound in the seventeen samples was five, with the following distribution: amphetamine and/or methamphetamine ($n=10$), buprenorphine ($n=5$), cocaine ($n=8$), methadone ($n=8$), 6-MAM ($n=17$), meperidine ($n=1$), oxycodone ($n=11$), tramadol ($n=3$).

Less information about the diffusion of NSO in Europe is available. Preliminary results based on a population made of 97 hair samples collected from drug addicts under withdrawal treatment in the North of Italy suggested that fentalogs are still rare in the investigated territory.[32] Only two samples gave positive results: one for carfentanil, ocfentanil, tetrahydrofuranylfentanyl, and butyryl fentanyl, while the second one was positive for the metabolite of 3-methylfentanyl. Despite the few positive results, ocfentanil and butyryl fentanyl were detected in hair for the first time in Italy, suggesting that fentalogs are diffused among drug users. Therefore, it is recommended that screening methods for NSO are currently developed, updated with eventually emerging synthetic opioids and finally applied to a large population, to monitor their spread and the related risk for health.

The scarce diffusion of fentalogs in Europe seems confirmed by research on 64 samples collected in European countries (Switzerland and Spain). Only four samples were positive for fentanyl and one for furanyl fentanyl.[30] Sporadic positive findings were also found in France.[47] From the analysis of 134 samples screened during 2019 and three hair samples previously screened positive for fentanyl, six were positive for fentanyl alone ($n=4$) or in combination with five fentanyl analogs (3-fluorofentanyl, furanylfentanyl, methoxyacetylfentanyl, acetylfentanyl, and carfentanil) ($n=2$).

At risk populations are often selected to verify the applicability of newly developed methods for NPS/NSO and/or to assess their prevalence in those situations where psychoactive substances are common. To qualitatively and quantitatively measure drug use among men who have sex with men, hair analysis was used with 69 volunteers. Besides traditional drugs of abuse, at least one NPS was detected in at least one hair segment in 39% of cases.[48]

When exploring the prevalence of NPS in postmortem samples, the number of positive results increases. In an in-depth exploration performed in Germany,[49] 1203 hair samples were reanalyzed to screen for several NPS/NSO. Samples were selected based on age (samples from deceased persons in the range 14–60 years old were included) and case history, namely when the past use of common drugs of abuse (*e.g.*, amphetamines, cocaine, opiates/opioids, cannabinoids, and benzodiazepines) could be suspected or proven. The limits of detection (LODs) of the analytical method were used as the minimum criterion to identify positive samples. Overall, NPSs were detected in 381 cases (31.6%). Most positive findings were from male individuals, while the highest number of positive cases ($n=73$, 19%) was observed in the 31–35 age group. Among all NPS positive cases, 107 hair samples were positive for two or more (up to 10) NPS, and 48 different target analytes were identified in positive cases over the 13 year monitored period. The most frequently detected NPS was *N*-ethylamphetamine, which was found in 81 hair samples. A clear increase in frequency was observed from 2008 to 2020. While 4-FA and mCPP were the most frequent NPS up to 2011, *N*-ethylamphetamine took top spot in 2018 and 2020. Overall, synthetic cathinones were the most prevalent NPS, but other analytes of interest (*e.g.*, synthetic cannabinoids) were not included due to the lack of reference standards.

As previously mentioned, most studies aiming to gain information about the prevalence of NPS/NSO have focused on a selected population considered at risk or possibly exposed to NPS (*e.g.* postmortem samples, partygoers, heroin or MDMA users), generally composed of adults. An original approach focused on families where the use of conventional illegal drugs was known or suspected.[50] The extracts of 1537 hair samples from 318 children (age 1–14 years), 44 adolescents and 611 adults, which were collected and tested for conventional drugs between June 2016 and April 2021, were reanalyzed by a validated LC–MS/MS method for 33 cathinones, 10 phenylethylamines, 5 piperazines, 2 tryptamines, 9 designer benzodiazepines, 4 synthetic opioids and 4 ketamine-like substances including phencyclidine. Between one and up to five from 42 of these substances were detected in 227 samples (14.8%). The most frequently detected substances were benzedrone (62), α-pyrrolidinovalerophenone (41), *N*-ethylamphetamine (29), dimethyltryptamine (13) and pyrovalerone (11). From the regularly tested drugs, 6-AM, cocaine, amphetamine, methamphetamine, and MDMA occurred with a significantly higher percentage in the NPS-positive samples whereas there was no significant difference for methadone, THC and benzodiazepines. Concerning the additionally tested substances, ketamine occurred in a significantly higher frequency in the NPS-positive samples. There was no significant difference for buprenorphine, tilidine, tramadol, and zopiclone, whereas the total numbers of positive results for fentanyl, oxycodone and zolpidem were too small for statistical evaluation. Cocaine was the most frequent conventional drug with 56.6% in combination with cathinones, 86.8% with phenylethylamines and 100% with DMT. As the second most frequent conventional drug in this population, THC was found together with cathinones in 30.6%, with phenylethylamines in 55.2% and with DMT in 46.2% of the respective positive samples. The opioids U-47700, *p*-fluorofentanyl and MPPP were not detected together with methadone or the heroin metabolite 6-AM. In the same way, the nine positive samples for designer benzodiazepines did not contain a conventional benzodiazepine.

8.5 Hair Analysis to Understand the Co-use of Traditional Drugs of Abuse and New Substances

To circumvent the limitations of urine, blood, and saliva testing, the detection of NPS/NSO in hair samples is a practical means to provide preliminary information on black-market penetration. Thanks to the features of the keratin matrix, combined with the analytical performances of the last generation mass spectrometry instrument, even a single intake of any targeted NPS/NSO can be identified over an extended period of time.[51,52] Furthermore, the inclusion of both old and new drugs of abuse in the analytical methods allows different consumption patterns to be distinguished, including co-use, occasional intake, and frequent or heavy use. One typical example is related to MDMA and synthetic cathinones. The pioneer studies focused on a specific population of electronic dance music (EDM) nightclub and festival attendees in New York City (NYC), considered as a high-risk population particularly exposed to both intentional and unintentional NPS/NSO intake.[53–55] Preliminary results suggested that many ecstasy-using nightclub/festival attendees were unintentionally exposed to "bath salts" or other NPS, mainly methylone and butylone.[53] The findings were corroborated by subsequent research. Methylone, butylone and MXE were detected exclusively or more often concurrently in 25 out of 26 cases which tested positive for at least one NPS.[54] Instead, synthetic cannabinoids were not detected in any of the tested hair samples, suggesting two possible explanations. One is related to the targeted analytes: when exploring the NPS/NSO diffusion in selected populations, the panel of target molecules covered by the applied methods should fully reproduce the drug scenario at the time of sample collection. The second possible explanation is related to the targeted population. In fact, it should also be noted that different groups of NPS/NSO are expected to be used by different social groups. Nightclub attendees and partygoers are likely to use synthetic cathinones, not synthetic cannabinoids, whereas other social groups might resort to synthetic cannabinoids as a cheap but more potent substitute of THC to temporarily escape from personal or social problems, such as unemployment, homelessness, and/or incarceration.

It is indeed likely that users of traditional amphetamines or methamphetamines may also tend to experiment with the use of new stimulant, psychedelic substituted phenethylamines, and dissociative drugs. After the analysis of (i) 23 real samples taken from proven MDMA and ketamine abusers, and (ii) 54 real hair samples which had been previously tested negative during regular drug screening in driver's license recovery, six samples tested positive for at least one target NPS. Methoxetamine (MXE) was found in three cases (range of concentration 7.7–27 $pg\,mg^{-1}$); mephedrone (4-MMC) was found in two cases (50–59 $pg\,mg^{-1}$) while one sample tested positive for methylone at 28 $pg\,mg^{-1}$. Other positive findings included 4-methylethcathinone (4-MEC), alpha-pyrrolidinovalerophenone (α-PVP), 4-fluoroamphetamine (4-FA), 3,4-methylenedioxypyrovalerone (MDPV), and diphenidine.[56] Subsequent research performed on different populations confirmed the trend of old and new stimulants co-use, mostly with an apparently sporadic tendency.[6,7,44]

Probably the most comprehensive investigation of this kind was performed in France.[57] The aim of this study was to evaluate the prevalence of

NPS in high-risk populations from Paris and its suburbs over a six-year period using hair testing, and to compare it with that of conventional drugs of abuse within the same populations. 480 patients (280 male, 200 female) were included and screened for the most common NPS/NSO at that time. 141 patients tested positive for NPS/NSO, corresponding to 99 males and 42 females. When adjusting positive results according to gender, one in four female (42/200) and one third of male patients (99/280) tested positive for NPS/NSO. The mean age of NPS users was 33 years (range: 17–70 years; median: 33 years) and the highest number of positive cases ($n = 50$, 35%) was observed in the 20–30 age group. In comparison, the median ages of cocaine, amphetamine and opioid users within the same populations were 36, 38, and 40 years old, respectively. Among the 141 samples which tested positive for at least one NPS, 54% had been exposed to more than one NPS/NSO in the same period (at least two detected NPS in the same segment), mostly in combination with cocaine (66%), amphetamines (65%), or opioids (64%). NPS alone were found in 32% of patients, suggesting that these consumers would not be identified if only routine drug analysis, which does not include NPS/NSO screening, was performed. Synthetic cathinones were the most prevalent compounds and mephedrone, 4-MEC and methylone were the most detected molecules. The temporal trend observed is also noteworthy. The number of detected NPS increased from 2012 (9% of total cases) to 2017 (28%) with a peak of detection in 2015 where more than half of analyzed cases (51%) were positive for designer drugs. JWH-122 was the only synthetic cannabinoid detected during this study, while ketamine was detected in 104 cases (mainly in association with cocaine), of which only 10 were explained by hospitalization treatment. Interestingly, a warning about ketamine diffusion as an emerging abused drug in several countries has recently been issued.[58] Again, the low reporting of synthetic cannabinoids could be explained in part by the high potency and thus their use at very low doses, which reduces their detectability in hair. The low reporting could also be explained by the rapid and extensive metabolism of synthetic cannabinoids [59] leading to lower integration of the parent drug into hair.

Several further methods to detect NSO in hair have been developed. The analytical methods currently available include the well-known fentanyl and many clandestinely synthesized derivatives. This group of compounds is sometimes referred to as "fentalogs."

Preliminary research including common and new opioids/opiates in the analytical screening was put in place to distinguish different consumption patterns, including co-use, occasional NSO intake, and frequent use.[35] Among the 34 analyzed samples, 17 (50%) tested positive for at least one opioid compound. The molecules detected were oxycodone (nine samples, range of concentration: 13–780 $pg\,mg^{-1}$), tramadol (eight samples, in the range 2.0–3700 $pg\,mg^{-1}$), hydrocodone (four samples, in the range 13–71 $pg\,mg^{-1}$), fentanyl (two samples, respectively, 3 and 6 $pg\,mg^{-1}$), furanylfentanyl (one sample, 44 $pg\,mg^{-1}$), and 4-ANPP (two samples, respectively, 2 and 1 $pg\,mg^{-1}$). Only the two samples positive for fentanyl also showed high concentrations of heroin metabolites, but it was not possible to discriminate whether fentanyl was taken alone or as a heroin cutting agent.

More recent research focused on 293 real samples collected in the U.S.A. from subjects who had reported heroin use in the past year or had already tested positive in hair testing for common opiates.[46] A subset of 53 samples positive for fentanyl was also tested for other common drugs of abuse. The direct metabolite of heroin, namely 6-MAM, was detected in 49 samples (92.4% of cases). Benzoylecgonine, the main metabolite of cocaine, was detected in 45 samples (84.9% of cases). Methamphetamine and its metabolite amphetamine were detected in 20 samples (37.7% of cases). With respect to the prescription opioids included in the panel, at least one between oxycodone or hydrocodone was detected in 35 samples (66.0% of positive cases to fentanyl). Many of the tested subjects were positive for fentanyl and cocaine simultaneously. Again, it is not possible to deduce whether fentanyl was used separately from cocaine, or whether cocaine was cut with fentanyl. However, from a quantitative perspective, relatively low correlation coefficients may suggest that subjects who were more exposed to fentanyl were less exposed to cocaine, and *vice versa*, probably due to different sources of the substances. The highest correlation coefficient was instead obtained for the pair fentanyl/6-MAM ($r = 0.55$).

8.6 Conclusions

Hair testing offers a unique perspective in the investigation of drug consumption, provided that a large panel of target analytes is considered. Since it is highly likely that the prevalence of use of specific NPS—particularly synthetic cathinones—is constantly shifting, timely monitoring based on hair analysis should be constantly implemented. Updated targeted methods applied on different populations might become effective approaches for NPS/NSO screening to focus on patterns of drug exposure and produce valuable information for targeted harm-reduction and treatment approaches. Large screening panels using hair testing and taking into account the timely fluctuation of the market situation can also be applied to other individuals under periodic control (*e.g.*, in workplace control or driving relicensing) to check how frequently NPS/NSO are used, possibly in substitution of traditional drugs to escape a positive outcome.

Epidemiology studies based on hair analysis can increase the knowledge about NPS/NSO use among high-risk populations. New or especially problematic trends in drug use or adverse outcomes associated with use can be detected in a timely way, thus allowing quick reports and prompt alert dissemination to the public, health professionals, and national institutions. Results will inform prevention among this high-risk population, and possibly inform the general population about the presence of new psychoactive substances.

However, the prerequisite of targeted multi-analyte protocols used to investigate the diffusion of NPS/NSO in a selected population is that all the molecules of interest expected to be possibly present in the screened samples are included in the method. Unfortunately, the toxicological analyses dealing with the detection of NPS/NSO cannot be counted within such an ideal scenario. Instead, broad-spectrum HRMS screening methods[60,61] can become of particular interest owing to the challenges presented by NPS/NSO, and in particular for the continuous modification of drug scenarios in the black market. Furthermore, retrospective investigation represents added value for investigation of hair samples, especially when the small amount allows only one analysis, and especially in forensic labs where there is a greater need to maximize the range of detectable compounds.

References

1. M. Bossong, J. Van Dijk and R. Niesink, *Addict. Biol.*, 2005, **10**, 321–323.
2. V. Auwärter, S. Dresen, W. Weinmann, M. Müller, M. Pütz and N. Ferreirós, *J. Mass Spectrom.*, 2009, **44**, 832–837.
3. A. Salomone, J. J. Palamar and M. Vincenti, *Drug Test. Anal.*, 2020, **12**(2), 191–194.
4. A. Salomone, M. Vincenti and E. Gerace, *Toxicol. Anal. Clin.*, 2017, **29**, 4–10.
5. F. K. Nzekoue, M. Agostini, M. Verboni, C. Renzoni, L. Alfieri, S. Barocci, M. Ricciutelli, G. Caprioli and S. Lucarini, *J. Pharm. Biomed. Anal.*, 2021, **205**, 114310.
6. V. A. Boumba, M. Di Rago, M. Peka, O. H. Drummer and D. Gerostamoulos, *Forensic Sci. Int.*, 2017, **279**, 192–202.
7. G. Mannocchi, A. Di Trana, A. Tini, S. Zaami, M. Gottardi, S. Pichini and F. P. Busardò, *Anal. Bioanal. Chem.*, 2020, **412**, 5125–5145.
8. J. Musiał, J. Powierska-Czarny, J. Czarny, M. Raczkowski, N. Galant, B. Buszewski and R. Gadzała-Kopciuch, *Arch. Toxicol.*, 2022, **96**, 2927–2933.
9. M. M. Vandeputte, A. J. Krotulski, D. M. Papsun, B. K. Logan and C. P. Stove, *J. Anal. Toxicol.*, 2022, **46**, 115–121.
10. L. Orsolini, J. M. Corkery, S. Chiappini, A. Guirguis, A. Vento, D. De Berardis, D. Papanti and F. Schifano, *Curr. Neuropharmacol.*, 2020, **18**, 809–837.
11. M. K. Laing, L. Ti, A. Marmel, S. Tobias, A. M. Shapiro, R. Laing, M. Lysyshyn and M. E. Socías, *Int. J. Drug Policy*, 2021, **93**, 103169.
12. K. W. Tupper, K. McCrae, I. Garber, M. Lysyshyn and E. Wood, *Drug Alcohol Depend.*, 2018, **190**, 242–245.
13. S. Tobias, A. M. Shapiro, C. J. Grant, P. Patel, M. Lysyshyn and L. Ti, *Drug Alcohol Depend.*, 2021, **218**, 108300.
14. C. May, C. Downey, J. D. Power and P. V. Kavanagh, *Drug Test. Anal.*, 2020, **12**, 1387–1392.
15. T. J. Cicero, M. S. Ellis and Z. A. Kasper, *Drug Alcohol Depend.*, 2017, **177**, 101–103.
16. S. Moallef, E. Nosova, M. J. Milloy, K. DeBeck, N. Fairbairn, E. Wood, T. Kerr and K. Hayashi, *Public Health Rep.*, 2019, **134**, 423–431.
17. S. G. Mars, D. Rosenblum and D. Ciccarone, *Addiction*, 2019, **114**, 774–780.

18. J. J. Palamar, A. Le, T. H. Carr and L. B. Cottler, *Drug Alcohol Depend.*, 2021, **221**, 108580.
19. E. Gerace, A. Salomone and M. Vincenti, *Curr. Pharm. Biotechnol.*, 2018, **19**(2), 113–123.
20. M. H. Baumann, G. Tocco, D. M. Papsun, A. L. Mohr, M. F. Fogarty and A. J. Krotulski, *Brain Sci.*, 2020, **10**, 1–30.
21. J. B. Zawilska, K. Kuczyńska, W. Kosmal, K. Markiewicz and P. Adamowicz, *Forensic Sci. Int.*, 2021, **320**, 110715.
22. D. Ciccarone, *Curr. Opin. Psychiatry*, 2021, **34**, 344–350.
23. S. Martinez, J. D. Jones, L. Brandt, A. N. C. Campbell, R. Abbott and S. D. Comer, *Am. J. Addict.*, 2021, **30**, 65–71.
24. J. J. Palamar, A. Salomone and M. J. Barratt, *Curr. Opin. Psychiatry*, 2020, **33**, 301–305.
25. H. Bach, V. Jenkins, A. Aledhaim, S. Moayedi, S. M. Schenkel and H. K. Kim, *Clin. Toxicol.*, 2020, **58**, 460–465.
26. J. J. Palamar, A. Salomone, R. Bigiarini, M. Vincenti, P. Acosta and B. Tofighi, *Am. J. Drug Alcohol Abuse*, 2019, **45**, 90–96.
27. J. J. Palamar, A. Salomone, C. M. Cleland and S. Sherman, *Subst. Abuse*, 2019, **40**, 116–123.
28. J. J. Palamar, A. Le, H. Guarino and P. Mateu-Gelabert, *Drug Alcohol Depend.*, 2019, **200**, 161–167.
29. J. J. Palamar, A. Salomone, R. Bigiarini, M. Vincenti, P. Acosta and B. Tofighi, *Am. J. Drug Alcohol Abuse*, 2019, **45**(1), 90–96.
30. N. A. Platosz, T. M. Binz, M. R. Baumgartner, E. Lendoiro, A. De Castro and M. Concheiro, *J. Anal. Toxicol.*, 2021, **45**, 943–949.
31. G. Musile, M. Mazzola, K. Shestakova, S. Savchuk, S. Appolonova and F. Tagliaro, *J. Chromatogr. B: Anal. Technol. Biomed. Life Sci.*, 2020, **1152**, 122263.
32. F. Freni, M. Moretti, D. Radaelli, C. Carelli, A. M. M. Osculati, L. Tronconi, C. Vignali and L. Morini, *J. Pharm. Biomed. Anal.*, 2020, **189**, 113476.
33. N. Qin, M. Shen, P. Xiang, D. Wen, B. Shen, H. Deng, H. Qiang, F. Song and Y. Shi, *Sci. Rep.*, 2020, **10**, 1–13.
34. M. d M. Ramírez Fernández, S. M. R. Wille, D. Jankowski, V. Hill and N. Samyn, *Forensic Sci. Int.*, 2020, **307**, 110137.
35. A. Salomone, J. J. Palamar, R. Bigiarini, E. Gerace, D. Di Corcia and M. Vincenti, *J. Anal. Toxicol.*, 2019, **43**(4), 259–265.
36. D. Florou and V. A. Boumba, *Toxicol. Rep.*, 2021, **8**, 1699–1720.
37. Y. Wang, Y. Pan, H. Yang, J. Liu, A. Wurita and K. Hasegawa, *Forensic Toxicol.*, 2022, **40**, 340–348.
38. E. Gerace, F. Seganti, D. Di Corcia, M. Vincenti and A. Salomone, *Curr. Pharm. Des.*, 2022, **28**, 2618–2621.
39. https://www.emcdda.europa.eu/publications/risk-assessments/mdmb-4en-pinaca_en
40. G. Musile, C. Palacio, M. Murari, S. Appolonova and F. Tagliaro, *J. Anal. Toxicol.*, 2023, **47**(1), 72–80.

41. B. Cho, H. S. Cho, J. Kim, J. Sim, I. Seol, S. K. Baeck, S. In, D. H. Shin and E. Kim, *Forensic Sci. Int.*, 2020, **306**, 110058.
42. N. La Maida, G. Mannocchi, S. Pichini, G. Basile, A. Di Giorgi, F. P. Busardò and E. Marchei, *Eur. Rev. Med. Pharmacol. Sci.*, 2022, **26**, 5033–5042.
43. J. Rubicondo, L. Scuffi, L. Pietrosemoli, M. Mineo, F. Terranova, M. Bartucca, C. Trignano, E. Bertol and F. Vaiano, *J. Anal. Toxicol.*, 2023, **46**, e262–e273.
44. A. Niebel, F. Krumbiegel, S. Hartwig, M. K. Parr and M. Tsokos, *Forensic Sci. Med. Pathol.*, 2020, **16**, 32–42.
45. J. M. Matey, G. Montalvo, C. García-Ruiz, F. Zapata, A. López-Fernández and M. A. Martínez, *Forensic Sci. Int.*, 2021, **329**, 111080.
46. A. Salomone, R. Bigiarini, J. J. Palamar, C. McKnight, L. Vinsick, E. Amante, D. Di Corcia and M. Vincenti, *J. Anal. Toxicol.*, 2020, **44**, 362–369.
47. I. A. Larabi, M. Martin, I. Etting, G. Pfau, Y. Edel and J. C. Alvarez, *Drug Test. Anal.*, 2020, **12**, 1298–1308.
48. J. Chas, R. Bauer, I. A. Larabi, G. Peytavin, P. Roux, E. Cua, L. Cotte, A. Pasquet, C. Capitant, L. Meyer, F. Raffi, B. Spire, G. Pialoux, J.-M. Molina and J.-C. Alvarez, *JAIDS J. Acquir. Immune Defic. Syndr.*, 2021, **86**, 552–561.
49. A. Niebel, L. Westendorf, F. Krumbiegel, S. Hartwig, M. K. Parr and M. Tsokos, *Drug Test. Anal.*, 2022, **14**, 110–121.
50. A. Niebel, F. Pragst, F. Krumbiegel and S. Hartwig, *Forensic Sci. Int.*, 2022, **331**, 111148.
51. M. Vincenti, A. Salomone, E. Gerace and V. Pirro, *Mass Spectrom. Rev.*, 2013, **32**(4), 312–332.
52. M. Vincenti, A. Salomone, E. Gerace and V. Pirro, *Bioanalysis*, 2013, **5**(15), 1919–1938.
53. J. J. Palamar, A. Salomone, M. Vincenti and C. M. Cleland, *Drug Alcohol Depend.*, 2016, **161**, 200–205.
54. A. Salomone, J. J. Palamar, E. Gerace, D. Di Corcia and M. Vincenti, *J. Anal. Toxicol.*, 2017, **41**, 376–381.
55. J. J. Palamar, A. Salomone, C. Rutherford and K. M. Keyes, *J. Gen. Intern. Med.*, 2021, **36**(1), 235–237.
56. A. Salomone, G. Gazzilli, D. Di Corcia, E. Gerace and M. Vincenti, *Anal. Bioanal. Chem.*, 2016, **408**(8), 2035–2042.
57. I. A. Larabi, N. Fabresse, I. Etting, L. Nadour, G. Pfau, J. H. Raphalen, P. Philippe, Y. Edel and J. C. Alvarez, *Drug Alcohol Depend.*, 2019, **204**, 107508.
58. M. R. Varì, G. Ricci, M. Cavallo, S. Pichini, A. Sirignano and S. Graziano, *Curr. Pharm. Des.*, 2022, **28**, 1213–1220.
59. Z. Wang, E. Y. Q. Leow, H. Y. Moy and E. C. Y. Chan, *Adv. Clin. Chem.*, 2023, **115**, 1–32.
60. A. Salomone, D. Di Corcia, P. Negri, M. Kolia, E. Amante, E. Gerace and M. Vincenti, *Anal. Bioanal. Chem.*, 2021, **413**(1), 225–233.
61. N. Fabresse, I. A. Larabi, T. Stratton, R. Mistrik, G. Pfau, G. Lorin de la Grandmaison, I. Etting, S. Grassin Delyle and J. C. Alvarez, *Drug Test. Anal.*, 2019, **11**, 697–708.

CHAPTER 9

Hair Analysis and Doping Control

LAURIE GHEDDAR[a] AND PASCAL KINTZ*[a,b]

[a] Institute of Legal Medicine, rue Humann, Strasbourg, France; [b] X-Pertise Consulting, rue Principale, Mittelhausbergen, France
*Email: pascal.kintz@wanadoo.fr

9.1 Definition of Doping

In professional and amateur sports, some substances are banned by the World Anti-Doping Agency (WADA) because of "(1) the potential to enhance sport performance, (2) an actual or potential health risk to the athlete and (3) the lack of ethic".[1] At least two of these above criteria must be satisfied for a substance to be banned by the WADA.

WADA was created in 1998 to answer an urgent issue in sport following a scandal in the cycling sport with the Festina case.

Because of potential cheating, anti-doping controls are regularly carried out to avoid the consumption of prohibited substances. These molecules are classified by family and by period of prohibition. Some of them are banned in- and out-of-competition, including non-approved substances S0, anabolic agents S1, peptide hormones, growth factors and mimetics S2, β2-agonists S3, hormonal and metabolic modulators S4, and diuretics and masking agents S5. Other substances are prohibited only in competition, including stimulants S5, narcotics S7, cannabinoids S8, and glucocorticoids S9. Finally, β-blockers, classified P1, are banned both in- and out-of-competition or only in competition depending on the practiced sport. The term "in-competition" includes "the period commencing just before midnight (at 11:59 p.m.) on the day before a competition in which the Athlete is

Issues in Toxicology No. 47
Perspectives and Challenges of Hair Analysis
Edited by Pascal Kintz, Alberto Salomone and Marco Vincenti

Published by the Royal Society of Chemistry, www.rsc.org

schedule to participate until the end of the competition and the sample collection process'' related to that competition. These rules are described in a document drafted by WADA called the Code.[1]

In addition, each year, WADA publishes a report that summarizes the analyses conducted and the results obtained, including the number of adverse analytical findings (AAF) per class. Anabolic agents represent the most identified class in cases of AAF over the past 10 years (see Figure 9.1).

Moreover, the list of prohibited substances is reviewed and updated every year by the WADA.[2] For example, tramadol will be included as a prohibited substance on the 2024 prohibited list, whereas it is still under surveillance today.[3]

9.2 The Place of Hair Analysis in Doping Control

Anti-doping laboratories use the traditional biological matrices (blood and urine) for routine analysis. Hair is not part of the routine analysis; the reasons will be discussed in this chapter.

Indeed, the analysis of hair in the fight against doping is quite recent and is not part of the routine of anti-doping laboratories.

In 2000, hair analysis and urinalysis results were compared for the first time in a doping context. This study showed that hair analysis was more effective to detect the use of anabolic agents.[4]

According to the WADA, other alternative matrices such as hair can be used for additional analysis at the request of the athlete to give evidence of innocence following an AAF in urine. Indeed, it is incumbent upon the athlete to prove their innocence, to demonstrate that they have consumed a prohibited substance without their knowledge. However, the Code specifies that a negative test result in an alternative matrix cannot contradict an AAF in urine.[1]

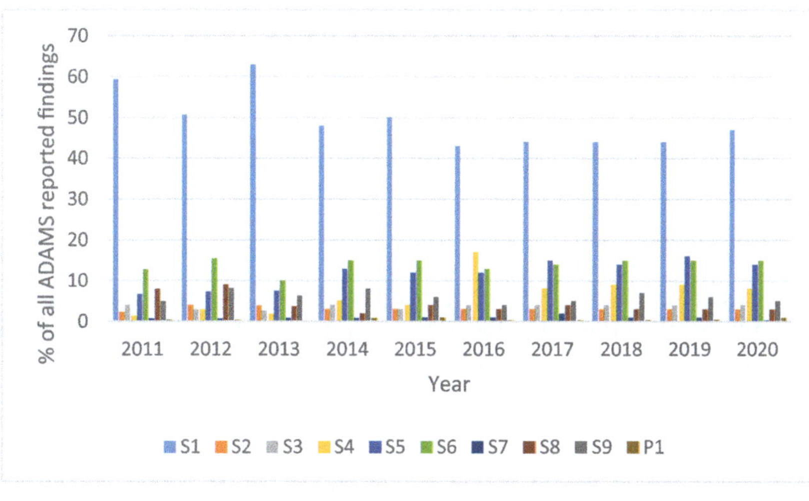

Figure 9.1 Graphical representation of substances identified as AAFs per year.

In 1999, the Society of Hair Testing (SoHT) published good practices on hair analysis in the doping context. This consensus stipulates that: "(1) hair analysis can essentially contribute to doping analysis in special cases, in addition to urine, (2) hair specimens are not suitable for general routine control and (3) in case of positive urine results, the negative hair result cannot exclude the administration of the detected drug and cannot overrule the positive urine result".[5]

Today, hair analysis can be accepted in some doping cases since the Richard Gasquet award, in which the Court of Arbitration for Sport (CAS) accepted, for the first time, hair testing in a contamination context with the theory of kissing defense.[6]

Despite the advantages that hair analysis offers such as retrospective analysis, simple and non-invasive sampling, or easy storage,[7] doping issues require other specificities. Some substances are only banned by the WADA in competition, such as narcotics or stimulants (*e.g.*, opioids, amphetamines), while others are within a standard therapeutic use exemption (TUE), such as some glucocorticoids. Segmental analysis could help with interpretation in some cases. Nevertheless, there are still many limitations that may explain why hair is not accepted by sports authorities.

The prohibited substances are varied in terms of pharmacology, chemical, and physical properties. As such, not all molecules can be detected in hair.

9.2.1 Issues in Doping

Despite the advantages of hair analysis, there are still issues. When hair is analyzed, the expert or analyst may be confronted with three major situations:

- First situation: the hair test is positive; the hair testing is consistent with the urinary analysis and in favour of a doping practice.
- Second situation: the quantities found in the hair are low and can correspond to a contamination. This situation is close to forensic situations where further investigation is necessary.
- Third situation: the hair is negative for the implicated molecule in the urine. This situation is highly controverted. Indeed, it can be interpreted in two ways. The athlete could use it to defend themselves, but the sport institutions will argue that the lack of sensitivity of the method or its lack of validation, such as the absence of a known minimum detectable amount, or external factors may have an impact on the concentrations found.

These limitations are rightly recognized and have been listed by the scientific community in recent years,[8–10] including the question on the minimal detectable amount in hair, or the unfairness dealing with the different incorporation of molecules in dark and light hair.

To be able to overcome the limitations of hair analysis, especially for substances for which there is little or no information in the literature, and to use a hair result, it is encouraged to answer the following questions:[10]

- is a single pharmacological dose detectable in hair?
- what is the minimal detectable dose?
- what are the expected hair concentrations of patients using the same drug for therapeutic purposes?
- what (if any) is the influence of hair colour?
- what (if any) is the influence of hair cosmetic treatment?

Nevertheless, for some classes of substances, it is complicated to address the above questions, which are necessary for the interpretation of the results. The next part of this chapter will highlight the limitations and the advantages related to hair analysis in the field of doping.

9.3 Limitations and Advantages

9.3.1 Limitations

9.3.1.1 Incorporation of Analytes Into Hair

There are many limitations to consider when hair is analyzed. First, what is detectable in hair? Not all molecules prohibited by the WADA are incorporated or are not detectable in hair. This is the case for molecules classified S0, which are not, by definition, screened. It is the same issue for class S2 molecules, which are hormones and do not pass through the blood capillaries and cannot cross the blood capillaries due to space overcrowding. To date, the largest molecule that has been detected in hair is cyclosporin A, an immunosuppressive agent, with a mass of 1202.6 Da.[11]

It was also stated that hair incorporation is limited for the analysis of acidic molecules like diuretics or carboxy THC. Indeed, the acidic molecules are poorly incorporated into hair due to the absence of an electrical bond between the melanin that is negatively charged and the acid molecules that are not positively charged at the acid pH of hair (pH 5.0). In addition, acidic molecules have difficulty passing through the membrane of the blood capillary, based on the pH gradient between the blood and the hair follicle.[7]

However, it has been published that some diuretics can be incorporated into hair. This is the case for furosemide, an acidic substance that is classified as S5 class and banned at all times by the WADA. This study demonstrated that furosemide incorporates into hair and the interpretation criteria have been established.[12]

Hair colour is another limiting factor, particularly for the incorporation of basic drug. This is an important issue for sport federations and sport authorities because of the individual inequality it creates in relation to doping control.[13] Indeed, molecules with a nitrogen atom, mostly basic, are incorporated into hair by binding to melanin that is not the case for neutral (*e.g.*, anabolic steroids) and acid (*e.g.*, diuretics) molecules. For example, phenobarbital, which is an acidic molecule, is incorporated at almost the same concentration in light hair and dark hair, in contrast to codeine, a basic molecule, which is incorporated better in dark hair *versus* light hair.[14] It was demonstrated that after administration of the same dose, codeine concentrations are 44 times higher in a subject with dark hair than in a subject with light hair.[14] Another study concerning the incorporation of basic metabolites (amphetamine and methamphetamine) of a stimulant, selegiline, showed a strong correlation with the melanin content.[15]

Among the anabolic agents, only stanozolol incorporates well into the hair thanks to the presence of a nitrogen atom. This particularity of the chemical structure can lead to a possible difference in incorporation between dark and light hair.

9.3.1.2 Minimal Detectable Amount

The lack of data on the minimum detectable dose and the dose-concentration relationships have been pointed by some authors. For many substances banned by the WADA, there is a lack of information due to the hazard of carrying out controlled studies such as for anabolic agents. This limited knowledge makes it very difficult to interpret a single hair concentration. Nevertheless, there are some data in the literature for a few substances.

Letrozole, an aromatase inhibitor, classified S4 and prohibited in- and out-of-competition, is one of them. Indeed, the criteria of interpretation were established, and the study demonstrated that a single intake of 0.62 mg, 1.25 mg, and 2.5 mg of letrozole was detectable in hair between 16 and 60 $pg\,mg^{-1}$. Moreover, it is possible to distinguish a single intake (in the range 16–60 $pg\,mg^{-1}$) and a repetitive consumption (higher than 160 $pg\,mg^{-1}$) thanks to the hair concentration ranges.[16]

Another study about hydrochlorothiazide, a diuretic, classified S5 by the WADA, enabled the difference between a single intake and repetitive use to be made; however, no dose-concentration relationship was established.[17]

The question of the minimum detectable quantity has been addressed several times and some authors have demonstrated that a single dose (at a therapeutic dosage) was not detectable in the hair. This is the case for hydrochlorothiazide,[17] betamethasone,[18] S23,[19] testosterone esters,[20] and nandrolone undecanoate. In contrast, after a single administration of nandrolone undecanoate, the metabolites norandrosterone and noretiocholanone are detectable in the urine for at least eight months at the threshold of 2 $ng\,ml^{-1}$.[21]

The minimal detectable amount has been also demonstrated for cocaine. It was published that a 4 mg dose is not detectable in hair.[22] For cocaine, the minimal detectable dose in hair was estimated at 25 mg.[23]

Few studies reported the presence of the analyte in hair after a single intake, including selegiline,[15] cocaine,[22] 25i-NBome,[24] and adrafinil.[25]

9.3.1.3 External Factors

Beyond the intrinsic limitations of hair, environmental and cosmetic issues are important factors to consider.

Indeed, natural factors (ultraviolet radiation, pollution, water...) and cosmetic treatments can alter the concentrations in hair.

The use of chemical and external substances like shampoo, permanent waving products, and coloration can damage the hair cuticle because of their basic properties that can extract molecules from hair. A study showed that there was a 50–80% decrease in the initial concentrations of the substances following cosmetic treatment.[26] Considering these issues, this could have a real impact on the concentrations found in the hair. For some molecules such as acidic and neutral molecules that do not incorporate well in hair, external factors could lead to a negative result in hair.

However, it is possible to use other body hair that is less exposed. The detection window is different for body hair and nails, and varies between 4 and 12 months.

Analysis of body hair and nails has already been demonstrated for ostarine in nails[27] but also for stanozolol in body hair (pubic hair, arm, leg, and axillary hair)[28] (see Figure 9.2). Moreover, it was demonstrated that hair

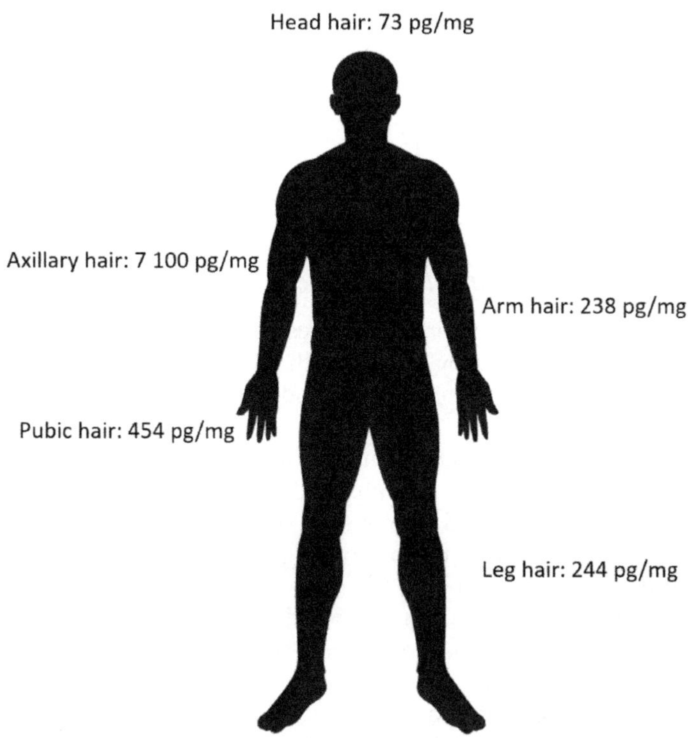

Figure 9.2 Stanozolol concentrations in body hair following repetitive use.

concentrations for anabolic agents were higher in body hair than in head hair thanks to a larger window of detection.[29]

Finally, environmental contamination is another issue. This is particularly the case with cocaine or THC, which are smoked or snorted.[30,31] It is for this reason that hair preparation requires washing and decontamination steps. Different types of washes have been described: the use of dichloromethane, sodium dodecyl sulfate, water, and methanol. However, it has been shown that even after several washes, some contaminants may remain.[32] To circumvent this problem, a segmental analysis can be performed to obtain the concentration profile. This was described for cocaine, for example.[30]

9.3.1.4 Routes of Administration

The differentiation of routes of administration is a challenge for analysts. This differentiation would be necessary for the glucocorticoids. Glucocorticoids are administered *per os*, per injection, per inhalation, per intranasal spray, with ophthalmological drop, or topically. Not all routes of administration are prohibited. The route by injection, *per os* and rectal are prohibited at all times whereas other routes like topical administration are permitted.[33] As with urinalysis, it is impossible to differentiate the route of administration with hair analysis. Nonetheless, there is a washout to respect before a competition. It is therefore possible to carry out a segmental analysis to obtain the consumption profile. For example, a result involving three consecutive positive segments is very indicative of repeated consumption and therefore of doping practice.

9.3.2 Advantages

9.3.2.1 Testing for Parent Molecules

Hair analysis has other major advantages in addition to retrospective analysis. It has been accepted by scientists that the main analyte found in hair is the parent molecule and not the metabolites. This has been verified over 30 years.

As a consequence, hair analysis makes it possible to evidence the parent compound, a situation that can help finding the source of contamination. This specificity has been used to go back to the source of exposure. Kintz and his team analyzed the parent compound (nandrolone, norandrostenediol, and norandrostenedione) in hair after an AAF involving their common metabolite: 19-norandrosterone. This allowed documentation of the source of exposure; however, there was no consequence on the doping offense since the three substances are permanently prohibited by WADA.[34]

The detection of testosterone esters is an irrefutable proof of exogenous testosterone intake. Their detection in blood is recent because the use of dried blood spots[35] avoids the rapid hydrolysis by esterase present in the blood. Hair analysis allows the identification of testosterone esters to prove exogenous intake.[36]

Along the same lines, hair analysis enabled toxicologists to distinguish the use of zeranol and zearalenone. These two substances share a common metabolite: β-zeranol, which is a target compound for anti-doping laboratories as it is a non-steroidal synthetic estrogen, permanently banned by WADA and classified as an S1.2 other anabolic agent. Zearalenone is a non-WADA prohibited mycotoxin. Thanks to hair analysis, it has been possible to document the origin of the β-zeranol found in the urine of an athlete and characterize a contamination, with the presence of zearalenone in hair. In another case, a doping practice was noticed.[37,38]

9.3.2.2 Contamination by Food and Supplements

Contamination by food and supplements is an important issue in doping control. These issues are recognized by the WADA. Athletes are advised that there is a possibility of contamination through food and supplements. Some breeders use growth promoters such as anabolic agents to improve the meat quality of their animals. Moreover, according to several studies, many substances prohibited by WADA, have been identified in supplements when they were not declared on the label.[39] The more frequent products identified in dietary supplement are stimulants, anabolic agents, and diuretics.[40]

In a WADA document, it is indicated that clenbuterol, a beta-agonist with anabolic properties, can be present in meat originating in Mexico, Guatemala, and China.[41,42] Hair analysis can, in some contexts, help to show contamination by food or dietary supplements. For instance, simultaneous cases of AAF for clenbuterol in several athletes in the same team were reported. The athletes disputed the findings and additional hair analyses were consistent with a contamination by food.[43] In 2014, Krumbholz published that it was possible to discriminate contamination from regular use of clenbuterol.[44] Following repeated therapeutic administration (20 μg) of clenbuterol in healthy subjects, hair was analyzed. Hair specimens of several volunteer Mexican football players ("exposed to clenbuterol") were also analyzed. Concentrations ranged up to 4.76 $pg\,mg^{-1}$.[44] These results enabled a difference to be made in a real doping case.

Another case of AAF was attributed to a contamination by food with the presence of trenbolone, an anabolic androgenic steroid. Jarrion Lawson, a long jumper, was contaminated by trenbolone in a restaurant and hair analysis proved that he was not abusing the drug on a long-term basis.[45]

The same strategy was carried out by the team of Segura in the context of contamination by ecstasy (MDMA).[46]

Finally, in a context of contamination by food or dietary supplements, it is very important, in addition to hair analysis supporting it, to identify the source, something that can be very complicated due to the delay between the offence and the notification.

9.3.2.3 *Hair Testing: Examples of Contamination*

The interest of hair analysis in doping cases has been established for several years and most of the challenged cases involve contamination (environmental, kissing theory, food, dietary supplement).

In these cases of possible unintentional doping, the approach is as follows. First, only a small amount of the prohibited substance must be present in the urine. A segmental hair analysis must be negative or consistent with very low concentrations when compared to what is expected after a therapeutic dose. The source of contamination must be identified and finally, all findings, including the delay between exposure and urine collection must be in accordance with the available scientific literature.[47]

For instance, this was the case for canoeist Laurence Vincent Lapointe, who tested positive in urine at 4 pg mg^{-1} for ligandrol or LGD-4033, a selective androgen receptor modulator (SARM), permanently prohibited and classified as an S1.2 other anabolic agent by the WADA. The analysis of her strand hair was negative. It turned out that it was her life partner who was a hidden regular user of ligandrol (supported by a positive hair test for ligandrol) and who contaminated her. The CAS accepted the defence of contamination by sex in this specific case.[48] Other examples of contamination cases[6,30,48–52] are presented in Table 9.1.

9.4 Hair Analysis – Analytical Aspects

Hair analysis for drugs has been conducted for more than 50 years. Many analytical methods to test for drugs of abuse, medicines, and doping substances in hair have been published. In this section, methods and analytical systems for each class prohibited by the WADA will be reviewed.

9.4.1 Substances Prohibited in Competition

Substances prohibited only in competition include stimulants S6 (*e.g.*, adrafinil, amphetamine, cocaine), narcotics S7 (*e.g.*, morphine, fentanyl), cannabinoids S8 (*e.g.*, THC, synthetic cannabinoids but not cannabidiol) and glucocorticoids S9 (*e.g.*, betamethasone, prednisolone, triamcinolone acetonide). These substances accounted for 15, 0.3, 3, and 5% of AFF, respectively, in 2020.

9.4.1.1 *Stimulants, Narcotics, and Cannabinoids*

Hair testing for narcotics, stimulants, and cannabinoids has been well described over time. Their detection and quantification limits are established in the literature. Moreover, the SoHT has published a consensus for the analysis of common drugs in which the positive cut-offs for identification are established (50–500 pg mg^{-1} depending on the drug).[53] Many papers have described the procedure for the identification of stimulants, narcotics, and

Table 9.1 Examples of hair testing after AAF in urine.

Sport	Substance	Urine concentration	Hair result	Vector	Conclusion	Ref.
Tennis	BZE	$151\ ng\,ml^{-1}$	Negative	Kissing defense	No fault	6
Long jumper	Trenbolone	$0.65\ ng\,ml^{-1}$	Negative	Meat	No fault	45
Canoe	LGD-4033	$4\ pg\ ml^{-1}$	Negative	Sex defense	No fault	48
Pole vault	BZE	$96\ ng\,ml^{-1}$	Negative	Sex defense	No fault	49
Tennis	BZE	$76\ ng\,ml^{-1}$	External contamination profile	Environment	No fault	30
Softball	LGD-4033	Not informed	Negative	Sex defense	No fault	50
Boxing	Letrozole + GW1516	Not informed	Negative	Sex defense	No fault	51
Hammer thrower	β-Zeranol	Not informed	Positive for zearalenone	Food	No fault	52

cannabinoids in hair using liquid chromatography coupled with tandem mass spectrometry (LC–MS/MS) and gas chromatography coupled with tandem mass spectrometry (GC–MS/MS) systems.[54,55] It remains one class for which little data is available, namely the new psychoactive substances (NPS). Some data are available, such as, *e.g.*, for 25I-NBome[25] or synthetic opioids.[56]

9.4.1.2 *Glucocorticoids*

Glucocorticoids, classified as S9, are prohibited in-competition with TUE for some routes of administration. Methods to test for glucocorticoids in human hair were developed using LC–MS and LC–MS/MS. Screening of ten glucocorticoids (triamcinolone, prednisolone, prednisone, cortisol, cortisone, methylprednisolone, beta, dexamethasone, flumethasone, and beclomethasone) was achieved using 50 mg of hair. The limits of detection (LOD) were in the range 0.03–0.17 ng ml^{-1}.[57] In two other papers, the identification of several glucocorticoids was described. Using 20–50 mg of hair LODs were 0.1–1 ng mg^{-1}.[58,59]

Betamethasone, cortisol, cortisone, triamcinolone acetonide, methylprednisolone, and hydrocortisone acetate were identified in authentic hair at 4.7 pg mg^{-1} and 1.31 ng mg^{-1}, from 5 to 91 pg mg^{-1}, from 12 to 163 pg mg^{-1}, at 0.28 ng mg^{-1}, 1.35 ng mg^{-1}, and 0.43 ng mg^{-1} respectively.[18,57,58] Budesonide was identified in authentic hair specimens with concentrations in the range 6.1–8.5 pg mg^{-1}.[59]

9.4.2 Substances Prohibited at All Times

Concerning the other prohibited classes, and more particularly the all-times prohibited classes, there are data in the literature for all classes except for the class S2 (*e.g.*, hormones) due to the non-incorporation into hair matrices.

9.4.2.1 *Anabolic Agents S1*

Anabolic agents include exogenous (trenbolone, metandienone, stanozolol, nandrolone, steroid esters...) and endogenous (testosterone, DHEA, DHT...) anabolic androgenic steroids (AAS). Analytical strategies have been described for the detection of doping agents and especially for anabolic agents, which are the most common class, detected after a doping control. Anabolic agents account for the largest percentage of AAF in urine. In 2020, they represented 47% of total AAF (see Figure 9.1).

For anabolic agents including exogenous and endogenous AAS, several screening and specific methods have been described. The first tests were performed by GC–MS and GC–MS/MS.[60] With the advance of technology, several methods have been developed on LC–MS/MS in order to avoid the derivatization step. However, some substances do not ionize well in electrospray; therefore, some laboratories simultaneously use multiple systems to detect all AAS. Regarding the derivatization agent, the most frequently used by laboratories is MSTFA.[61]

The detection limits must be much lower than those for narcotics and represent an analytical challenge in laboratories. Indeed, these substances are present in hair in small quantities, of the order of pg mg^{-1}, due to their weak incorporation rate. The amount of hair needed for analysis varies according to the extraction methods, 20–100 mg with LODs at 0.1 pg mg^{-1} for some substances. Anabolic agents were analyzed with GC–MS,[34,62] GC–MS/MS,[29,62–65] LC–MS/MS,[28,29,66] and liquid chromatography coupled with high-resolution mass spectrometry[67–69] (LC–HRMS) systems. Many references in the literature report hair concentrations, however, there is still a lack of data from controlled studies. To date, the more frequently identified molecules after doping control are stanozolol, dehydrochlormethyltestosterone, drostanolone, metandienone, oxandrolone, trenbolone, and boldenone (with % AAF higher than 5% in 2020 compared to total AAF in class S1). The identification of the latter was carried out in authentic hair following doping practices involving AAF in the athlete's urine but also cases of death or aggression in forensic medicine. The concentrations reported in the literature for repeated use include:

- nandrolone: 1–260 pg mg^{-1} [62–64,67,70]
- trenbolone: 2–230 pg mg^{-1} [66,67,70,71]
- metandienone: 4–2800 pg mg^{-1} [65,66,70,72]
- stanozolol: 1–1992 pg mg^{-1} [28,62–64,70,72]
- drostanolone: 2–17 pg mg^{-1} [70,73]

Finally, anabolic agents have the reputation of being used only by bodybuilders in large quantities, whereas they have been identified in boxers[65] and cyclists,[4] for instance.

9.4.2.2 Other Anabolic Agents

Substances included in this section are clenbuterol, ractopamine, osilodrostat, zeranol, zilpaterol, and SARMs. SARMs have not passed clinical trial tests and are extensively sold *via* the internet. A screening to test for nine SARMs (AC262536, andarine, ACP-105, ligandrol, MK-0773, MK677, ostarine, RAD-140, and S23) using 20 mg of hair has been described with LOD in the range 0.1–20 pg mg^{-1}.[74] Nevertheless, very little information and few case reports exist to better interpret the results. Ostarine, andarine, ligandrol, S23, and GSK2881078 were identified in authentic human hair with concentrations at 1–168 pg mg^{-1}, 0.1–1644 pg mg^{-1}, 14–42 pg mg^{-1}, 0.6–16 pg mg^{-1}, and at 1.7 pg mg^{-1}, respectively.[74–77]

Typical chromatograms for doping results in hair, after an AFF for andarine in urine, are presented in Figure 9.3.

Concerning the other anabolic agents, there is analytical information for clenbuterol and zeranol. Clenbuterol[62,78] and zeranol[38] were already

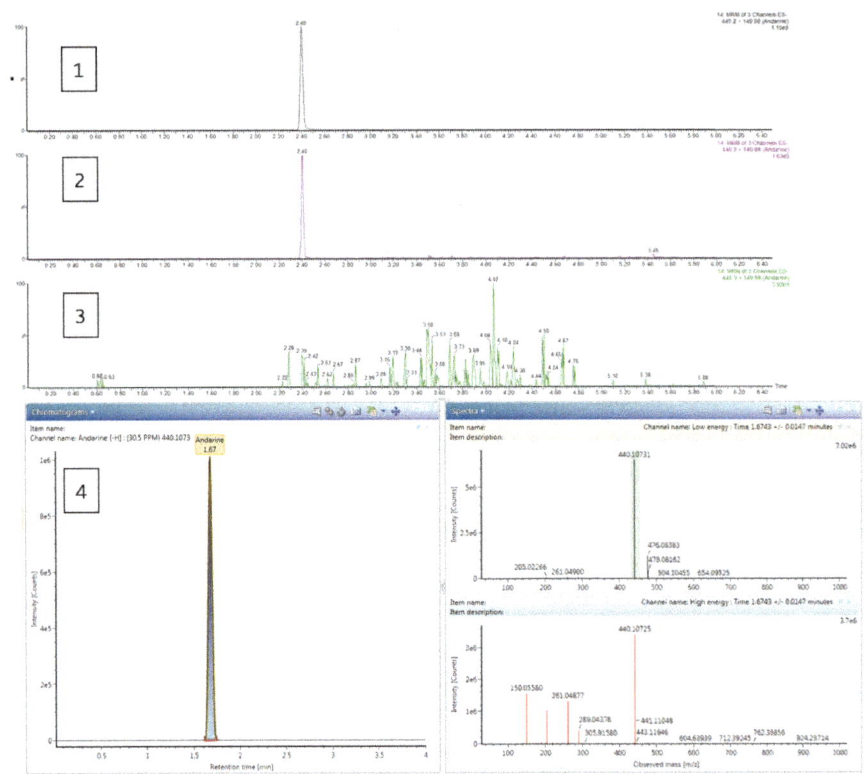

Figure 9.3 Typical chromatograms for doping results in hair. Testing for andarine. (1) Authentic hair specimens with andarine concentration 582 pg mg^{-1}. (2) Quality control at 1 pg mg^{-1}. (3) Blank hair. (4) Chromatogram for authentic hair with LC–HRMS.

identified in authentic hair at the following concentrations: 15–122 pg mg^{-1} for clenbuterol and 11–34 pg mg^{-1} for zeranol.

Thanks to hair analysis and literature data, it is now possible, in most circumstances, to differentiate between a doping practice and a contamination with clenbuterol.[43,44]

Zilpaterol, osilodrostat, and ractopamine were never reported in human hair.

9.4.2.3 *β2-Agonists*

With regard to the S3 class (β2-agonists such as salbutamol, tulobuterol), several substances have already been identified in authentic hair cases. Three papers have described the analysis of these molecules in human hair. The amount of hair required was 20–100 mg of hair. Drugs were analyzed by GC–MS with LODs in the range 2–10 pg mg^{-1}.[79] and by LC–MS/MS with LODs in the range 0.8–80 pg mg^{-1}.[59,80] Salbutamol was identified in asthmatic patients and in a swimmer with concentrations in the range

6.4–210 pg mg^{-1}.[59,79] These concentrations were reported as long-term exposure. Formoterol, salmeterol, and vilanterol have been reported in asthmatic patients with concentrations in the ranges 0.8–500 pg mg^{-1}, 6–20 pg mg^{-1}, and approximately at 2 pg mg^{-1}, respectively.[80]

9.4.2.4 Hormones and Metabolic Modulators

Hormones and metabolic modulators S4 include aromatase inhibitors (*e.g.*, letrozole), anti-estrogens and selective estrogen receptor modulators (SERMs) (*e.g.*, clomifen, tamoxifen), agents preventing activin receptor IIB activation, metabolic modulators (*e.g.*, SR 9009, GW1516), insulins and insulin-mimetics, meldonium and trimetazidine.

Letrozole has been identified in authentic hair in doping cases and in hair specimens from patients under therapeutic treatment. The quantities required were in the range 10–20 mg of hair and the authors analyzed them by LC–MS/MS and LC–HRMS.[16,81]

Tamoxifen is the only SERMs that was identified in authentic hair, at 174 pg mg^{-1}.[67] The published methods used 10–30 mg of hair. Testing for tamoxifen was realized with LC–HRMS/MS[67] and LC–MS/MS.[82]

A method to test for SR-9009 and GW 1516 was also described using 20 mg of hair with LC–MS/MS. The LODs were 10 and 3 pg mg^{-1}.[74]

Finally, insulin was never identified in human hair.

9.4.2.5 Diuretics and Masking Agents

Regarding the S5 class, there is only limited data on the analysis of diuretics in hair with the first publication, dating to 2019, reporting the identification of hydrochlorothiazide in human hair during the investigation of a doping case.[17]

The analysis of some diuretics like furosemide, hydrochlorothiazide, acetazolamide, brinzolamide, and dorzolamide has been described in hair with the necessary amount of hair in the range 20–30 mg. Analyses were performed with LC–MS/MS systems. LODs varied from 3 pg mg^{-1} to 0.15 ng mg^{-1} according to these papers.[12,17,83,84]

Furosemide, hydrochlorothiazide, acetazolamide, dorzolamide, and brinzolamide were identified in authentic hair specimens with concentrations from 5 to 1110 pg mg^{-1},[12] from 12 to 1845 pg mg^{-1},[17] from 24 pg mg^{-1} to 2.3 ng mg^{-1},[83,84] at 5.94 ng mg^{-1} [83] and at 3.26 ng mg^{-1},[83] respectively.

9.4.2.6 Beta-blockers

Beta-blockers are prohibited in-competition or at all times depending on the sport. Testing for beta-blockers in hair was described with LC–MS/MS and GC–MS systems. To test for several beta-blockers with LODs of 2–10 pg mg^{-1} the amount of hair required was 25–100 mg. Metoprolol and sotalol were identified in an authentic doping case with hair concentration at 8.41 ng mg^{-1} and 261 pg mg^{-1}, respectively.[79,85] Atenolol and bisoprolol were identified in patients under therapy with concentrations of 0.62–3.99 ng mg^{-1} and 1.30–26.7 ng mg^{-1},[79,85] respectively.

9.4.3 Advances in Technologies

Technological advances in recent years have allowed working on a single hair, which may address the lack of hair quantity that analysts may face on a routine basis. Single hair analysis was performed using standard technologies such as LC–MS/MS[86] and GC–MS.[87] Moreover, a recent paper described the usefulness of single hair analysis in the context of contamination by amphetamine-based dietary supplements.[88] The use of matrix-assisted laser desorption/ionization (MALDI) associated with imaging has grown in recent years. This approach could be used to monitor drugs in doping control and to document possible external contamination. Some authors have proposed assays to monitor drug use by direct MALDI–MS in human hair including cocaine,[89] synthetic cannabinoids,[90] and methamphetamine.[91]

9.5 Hair Testing in Animal Sports

Doping control applies to all sports, including those involving animals. However, the Code is less specific regarding anti-doping rules for animals; it is up to the federation of the sport concerned to establish and enforce anti-doping rules for animals.

Hair analysis has become more popular in doping laboratories and especially in horse sports where many applications have been described. Some analytical methods have been described targeting up to 121 prohibited products with detection limits in the range 0.1–10 pg mg^{-1} with a hair quantity of 100 mg.[92] There are many publications in equine sports for the identification in mane or tail hair of anabolic agents,[93] SARMs,[94] opioids,[95] or zilpaterol.[96]

The analysis of keratinized matrices was also reported to test for anabolic steroids in dog hairs,[97] glucocorticoids in racing camel hairs,[98] and in pigeon feathers[99] or to test for salbutamol in cattle hair.[100]

9.6 Perspectives

Hair is an interesting matrix to assess most of the substances banned by WADA. At least one substance per prohibited class has already been identified in human hair and published, as described in Section 9.4. In the case of permanently banned substances, this matrix is of real interest to provide additional findings in cases of doping challenge.

Hair analysis should, therefore, be considered as a complement and not as an alternative as stated in the Code. The detection window of a drug can be increased, depending on the length of hair available. The possibility of segmentation is an advantage to document the consumption profile.

However, this matrix cannot be used as a routine screening matrix because of the costs involved. In addition, it is not possible to take repeated hair samples from the same subject because of the obvious aesthetic problem associated with specimen collection.

On the other hand, hair analysis has been the key element in various cases of contamination as it is the only discriminating method known to date.

Interest in hair analysis is growing. The legal authorities of the sporting federations and the CAS have recognized the validity of hair analysis since the case of Richard Gasquet.[6]

Abbreviations

AAF	Adverse analytical finding
AAS	Anabolic androgenic steroid
CAS	Court of Arbitration for Sport
GC–MS/MS	Gas chromatography coupled with tandem mass spectrometry
HRMS	High-resolution mass spectrometry
LC–MS/MS	Liquid chromatography coupled with tandem mass spectrometry
LOD	Limit of detection
MALDI	Matrix-assisted laser desorption/ionization
NPS	New psychoactive substance
SERM	Selective estrogen receptor modulators
SoHT	Society of Hair Testing
TUE	Therapeutic use exemption
WADA	World Anti-Doping Agency

References

1. World Anti-Doping CODE 2021. https://www.wada-ama.org/sites/default/files/resources/files/2021_wada_code.pdf (accessed December 2022).
2. World Anti-Doping CODE. International standard. Prohibited list 2022. https://www.wada-ama.org/sites/default/files/2022-01/2022list_final_en_0.pdf (accessed December 2022).
3. WADA. WADA publishes 2023 Prohibited List. https://www.wada-ama.org/en/news/wada-publishes-2023-prohibited-list (accessed December 2022).
4. Y. Gaillard, F. Vayssette and G. Pépin, Compared interest between hair analysis and urinalysis in doping controls. Results for amphetamines, corticosteroids and anabolic steroids in racing cyclists, *Forensic Sci. Int.*, 2000, **107**, 361.

5. SoHT. Consensus of the Society of Hair Testing on hair testing for doping agents. https://www.soht.org/images/pdf/Consensus_Doping_Agents.pdf (accessed December 2022).
6. Court of Arbitration for Sport. https://jurisprudence.tas-cas.org/Shared%20Documents/1926,%201930.pdf (accessed December 2022).
7. A. Gail and A. Cooper, Hair Analysis in clinical and forensic toxicology, in *Anatomy and Physiology of hair, and principles for its collection*, ed. P. Kintz, A. Salomone and M. Vincenti, Elsevier, Amsterdam, 1st edn, 2015, ch. 1, pp. 1–17.
8. D. Thieme, Potential and limitations of alternative specimens in doping control, *Bioanalysis*, 2012, **4**, 1613.
9. M. Thevis, H. Geyer, L. Tretzel and W. Schänzer, Sports drug testing using complimentary matrices: Advantages and limitations, *J. Pharm. Biomed. Anal.*, 2016, **130**, 220.
10. P. Kintz, L. Gheddar, A. Ameline, N. Arbouche and J. S. Raul, Hair testing for doping agents. What is known and what remains to do, *Drug Test. Anal.*, 2020, **12**, 316.
11. A. Müller, H. Jungen, S. Iwersen-Bergmann, M. Sterneck and H. Andresen-Streichert, Analysis of cyclosporin a in hair samples from liver transplanted patients, *Ther. Drug Monit.*, 2013, **35**, 450.
12. L. Gheddar, M. O. Batt, J. S. Raul and P. Kintz, Identification of furosemide in hair in a post-mortem case by UHPLC–MS/MS with guidance on interpretation, *J. Forensic Sci.*, 2021, **66**, 272.
13. L. Rivier, Is there a place for hair analysis in doping controls?, *Forensic Sci. Int.*, 2000, **107**, 309.
14. S. P. Gygi, D. G. Wilkins and D. E. Rollins, A comparison of phenobarbital and codeine incorporation into pigmented and nonpigmented rat hair, *J. Pharm. Sci.*, 1997, **86**, 209.
15. R. Kronstrand, M. C. Andersson, J. Ahlner and G. Larson, Incorporation of Selegiline Metabolites into Hair after Oral Selegiline Intake, *J. Anal. Toxicol.*, 2001, **25**, 594.
16. D. Favretto, R. Snenghi, R. Pertile, R. El Mazloum, M. Tucci, S. Visentin and S. Vogliardi, Hair analysis to discriminate voluntary doping vs inadvertent ingestion of the aromatase inhibitor letrozole, *Drug Test. Anal.*, 2019, **11**, 762.
17. L. Gheddar, J. S. Raul and P. Kintz, First identification of a diuretic, hydrochlorothiazide, in hair: Application to a doping case and interpretation of the results, *Drug Test. Anal.*, 2019, **11**, 157.
18. J. S. Raul, V. Cirimele, B. Ludes and P. Kintz, A single therapeutic treatment with betamethasone is detectable in hair, *J. Anal. Toxicol.*, 2002, **26**, 582.
19. A. Ameline, L. Gheddar, J. S. Raul and P. Kintz, In vitro characterization of S-23 metabolites produced by human liver microsomes, and subsequent application to urine after a controlled oral administration, *J. Pharm. Biomed. Anal.*, 2022, **212**, 114660.

20. J. Segura, S. Picchini, S. H. Peng and X. de la Torre, Hair analysis and detectability of single dose administration of androgenic steroid esters, *Forensic Sci. Int.*, 2000, **107**, 347.
21. P. Kintz, V. Cirimele and B. Ludes, Pharmacological criteria that can affect the detection of doping agents in hair, *Forensic Sci. Int.*, 2000, **107**, 325.
22. E. Feisthauer, A. Ameline, L. Gheddar, N. Arbouche, J. S. Raul and P. Kintz, Analysis of Cocaine and Its Metabolites in Urine after Consumption of Coca Tea by Five Subjects and Subsequent Hair Testing, *J. Anal. Toxicol.*, 2022, **46**, 108.
23. G. L. Henderson, M. R. Harkey, C. Zhou, R. T. Jones and P. Jacob III, Incorporation of isotopically labelled cocaine and metabolites into human hair. 1. Dose–response relationships, *J. Anal. Toxicol.*, 1996, **20**, 1.
24. A. Ameline, A. Farrugia, J. S. Raul and P. Kintz, Retrospective Demonstration of 25I-NBOMe Acute Poisoning Using Hair Analysis, *Curr. Pharm. Biotechnol.*, 2017, **18**, 786.
25. A. Ameline, L. Gheddar, J. S. Raul and P. Kintz, Identification of adrafinil and its main metabolite modafinil in human hair. Self-administration study and interpretation of an authentic case, *Forensic Sci. Res.*, 2020, **5**, 322.
26. P. Kintz, Hair Analysis in Forensic Toxicology: An Updated Review with a Special Focus on Pitfalls, *Curr. Pharm. Des.*, 2017, **23**, 5480.
27. P. Kintz, L. Gheddar, A. Ameline and J. S. Raul, Identification of S22 (ostarine) in human nails and hair using LC–HRMS. Application to two authentic cases, *Drug Test. Anal.*, 2020, **12**, 1508.
28. L. Gheddar, J. S. Raul and P. Kintz, Testing for stanozolol, using UPLC–MS-MS and confirmation by UPLC-q-TOF-MS, in hair specimens collected from five different anatomical regions, *J. Anal. Toxicol.*, 2020, **44**, 834.
29. P. Kintz, L. Gheddar and J. S. Raul, Simultaneous testing for anabolic steroids in human hair specimens collected from various anatomic locations has several advantages when compared with the standard head hair analysis, *Drug Test. Anal.*, 2021, **13**, 1445.
30. P. Kintz, A. Ameline, L. Gheddar, P. Escudero, L. Ferrari and J. S. Raul, Cocaine External Contamination Can Be Documented by a Hair Test, *J. Anal. Toxicol.*, 2020, **44**, e4–e5.
31. L. Tsanaclis, J. Nutt, K. Bagley, S. Bevan and J. Wicks, Differentiation between consumption and external contamination when testing for cocaine and cannabis in hair samples, *Drug Test. Anal.*, 2014, **6**(Suppl. 1), 37.
32. D. Mantinieks, D. Gerostamoulos, P. Wright and O. Drummer, The effectiveness of decontamination procedures used in forensic hair analysis, *Forensic Sci., Med. Pathol.*, 2018, **14**, 349.
33. World Anti-Doping Agency. Glucocorticoids and therapeutic use exemptions. https://www.wada-ama.org/sites/default/files/resources/files/glucocorticoids_and_therapeutic_use_exemptions_final_20oct21.pdf (accessed December 2022).

34. P. Kintz, V. Cirimele and B. Ludes, Discrimination of the nature of doping with 19-norsteroids through hair analysis, *Clin. Chem.*, 2000, **46**, 2020.
35. S. A. Solheim, M. C. S. Levernaes, J. Mørkeberg, A. Juul, E. N. Upners, N. B. Nordsborg and Y. Dehnes, Stability and detectability of testosterone esters in dried blood spots after intramuscular injections, *Drug Test. Anal.*, 2021, **14**, 1926.
36. L. Gheddar, A. Blanchot, A. Ameline, N. Arbouche, J. S. Raul and P. Kintz, The Power of Keratinous Matrices (Head Hair, Body Hair and Nail Clippings) Analysis in a Case of Death Involving Anabolic Agents, *J. Anal. Toxicol.*, 2023, **46**, e307.
37. Y. Shi, presented in part at 54th annual meeting of the international association of forensic toxicologists (TIAFT) Segmental hair analysis of zearalenone by Q-Exactive high resolution mass spectrometry (HRMS): a doping offence case or not? Brisbane August–September, 2016.
38. P. Kintz, A. Ameline and J. S. Raul, Discrimination between zeranol and zearalenone exposure using hair analysis. Application to an adverse analytical finding case, *Drug Test. Anal.*, 2018, **10**, 906.
39. H. Geyer, M. K. Parr, K. Koehler, U. Mareck, W. Schänzer and M. Thevis, Nutritional supplements cross-contaminated and faked with doping substances, *J. Mass Spectrom.*, 2008, **43**, 892.
40. U. Mareck, H. Geyer, T. Schertel, S. Petring, O. Krug and M. Thevis, Detection of undeclared doping substances in nutritional supplements in the context of follow-up investigations concerning adverse analytical findings, *Drug Test. Anal.*, 2021, **13**, 1911.
41. USADA. Clenbuterol and meat contamination, 2021. https://www.usada.org/spirit-of-sport/clenbuterol-and-meat-contamination/ (accessed December 2022).
42. World Anti-Doping Agency. https://www.wada-ama.org/en/news/wada-publishes-stakeholder-notices-regarding-potential-contamination-cases-related-meat-and (accessed December 2022).
43. M. Thevis, L. Geyer, H. Geyer, S. Guddat, J. Dvorak, A. Butch, S. S. Sterk and W. Schänzer, Adverse analytical findings with clenbuterol among U-17 soccer players attributed to food contamination issues, *Drug Test. Anal.*, 2013, **5**, 372.
44. A. Krumbholz, P. Anielski, L. Gfrerer, M. Graw, H. Geyer, W. Schänzer, J. Dvorak and D. Thieme, Statistical significance of hair analysis of clenbuterol to discriminate therapeutic use from contamination, *Drug Test. Anal.*, 2014, **6**, 1108.
45. Court of Arbitration for Sport. https://www.doping.nl/media/kb/6463/CAS%202019_A_6313%20Jarrion%20Lawson%20vs%20IAAF%20%28OS%29.pdf (accessed December 2022).
46. J. Segura, Is antidoping analysis so far from clinical, legal or forensic targets?: the added value of close relationships between related disciplines, *Drug Test. Anal.*, 2009, **1**, 479.
47. P. Kintz, What are the prerequisites to account for "no fault" in doping control after an adverse analytical finding possibly due to drug

contamination? Perspective from a hair testing analyst, *J. Anal. Toxicol.*, 2021, **45**, e3–e5.
48. International Canoe federation. https://www.canoeicf.com/news/icf-decision-laurence-vincent-lapointe (accessed Jun 6, 2020). (accessed December 2022).
49. Sport dispute resolution centre of Canada. http://crdsc-sdrcc.ca/resource_centre/pdf/English/814_SDRCC_DT_16-0249.pdf (accessed January 2023).
50. USADA. https://www.usada.org/sanction/madilyn-nickles-found-at-no-fault/ (accessed January 2023).
51. USADA. https://www.usada.org/sanction/virginia-fuchs-found-not-at-fault/ (accessed January 2023).
52. http://en.people.cn/n/2015/0506/c90779-8888225.htmlF (accessed January 2023).
53. 2021 SoHT consensus on drug of abuse (DOA) testing in hair. https://www.soht.org/images/pdf/Consensus_DoA_2021.pdf (accessed January 2023).
54. M. Wada, R. Ikeda, N. Kuruda and K. Nakashima, Analytical methods for abused drugs in hair and their applications, *Anal. Bioanal. Chem.*, 2010, **397**, 109.
55. I. Shah, B. Al-Dabbagh, A. E. Salem, S. A. A. Hamid, N. Muhammad and D. P. Naughton, A review of bioanalytical techniques for evaluation of cannabis (Marijuana, weed, Hashish) in human hair, *BMC Chem.*, 2019, **13**, 106.
56. A. Salomone, J. J. Palamar, R. Bigiarini, E. Gerace, D. Di Corcia and M. Vincenti, Detection of fentanyl analogs and synthetic opioids in real hair samples, *J. Anal. Toxicol.*, 2019, **43**, 259.
57. V. Cirimele, P. Kintz, V. Dumestre, J. P. Goulle and B. Ludes, Identification of ten corticosteoids in human hair by liquid chromatography-ionspray mass spectrometry, *Forensic Sci. Int.*, 2000, **107**, 381.
58. F. Bevalot, Y. Gaillard, M. A. Lhermitte and G. Pepin, Analysis of corticosteroids in hair by liquid chromatography-elecrospray ionization mass spectrometry, *J. Chromatogr. B: Anal. Technol. Biomed. Life Sci.*, 2000, **740**, 227.
59. E. Lamy, A. Mahjoubi, H. Salvator, F. Lambinet, P. Devillier and S. Grassin-Delyle, Liquid chromatography method for the quantification of corticosteroids, β2-adregenoreceptor agonists and anticholinergics in human hair, *J. Pharmaceutical Biomed. Anal.*, 2020, **190**, 113530.
60. Y. Gaillard, F. Vayssette, A. Balland and G. Pépin, Gas chromatographic-tandem mass spectrometric determination of anabolic steroids and their esters in hair. Application in doping control and meat quality control, *J. Chromatogr. B: Anal. Technol. Biomed. Life Sci.*, 1999, **735**, 189.
61. J. Segura, R. Ventura and C. Jurado, Biomedical Sciences Applications Derivatization procedures for gas chromatographic-mass spectrometric determination of xenobiotics in biological samples, with special

attention to drugs of abuse and doping agents, *J. Chromatogr. B: Anal. Technol. Biomed. Life Sci.*, 1998, **713**, 61.
62. V. Dumestre-Toulet, V. Cirimele, B. Ludes and P. Kintz, Hair analysis of seven bodybuilders for anabolic steroids, ephedrine, and clenbuterol, *J. Forensic Sci.*, 2002, **47**, 211.
63. N. Deshmukh, I. Hussain, J. Barker, A. Petroczi and D. P. Naughton, Analysis of anabolic steroids in human hair using LC–MS/MS, *Steroids*, 2010, **75**, 710.
64. P. Kintz, V. Cirimele, H. Sachs, T. Jeanneau and B. Ludes, Testing for anabolic steroids in hair from two bodybuilders, *Forensic Sci. Int.*, 1999, **101**, 209.
65. M. Bresson, V. Cirimele, M. Villain and P. Kintz, Doping control for metandienone using hair analyzed by gas chromatography-tandem mass spectrometry, *J. Chromatogr. B: Anal. Technol. Biomed. Life Sci.*, 2006, **836**, 124.
66. F. Aknouche, L. Gheddar, A. Kernalléguen, C. Maruejouls and P. Kintz, Anabolic steroids and extreme violence: a case of murder after chronic intake and under acute influence of metandienone and trenbolone, *Int. J. Legal Med.*, 2021, **135**, 1449.
67. N. Fabresse, S. Grassin-Delyle, I. Etting and J. C. Alvarez, Detection and quantification of 12 anabolic steroids and analogs in human whole blood and 20 in hair using LC–HRMS/MS: application to real cases, *Int. J. Legal Med.*, 2017, **131**, 989.
68. P. Kintz, L. Gheddar, A. Ameline and J. S. Raul, Identification of S22 (ostarine) in human nails and hair using LC–HRMS. Application to two authentic cases, *Drug Test. Anal.*, 2020, **12**, 1508.
69. S. Strano-Rossi, Screening for exogenous androgen anabolic steroids in human hair by liquid chromatography/orbitrap-high resolution mass spectrometry, *Anal. Chim. Acta*, 2013, **793**, 61.
70. D. Thieme and P. Anielski, Doping, applications of hair analysis, in *Hair Analysis in Clinical and Forensic Toxicology*, Academic Press, Elsevier, London UK, 2015, ch. 10, pp. 275–299.
71. L. Gheddar, A. L. Pélissier, J. Desfeux, F. Niort, J. S. Raul and P. Kintz, Testing for Trenbolone, an Anabolic Steroid, in Biological Fluids and Head Hair in a Post-Mortem Case, *J. Anal. Toxicol.*, 2021, **19**, bkab091.
72. P. Kintz, A new series of hair test results involving anabolic steroids, *Toxicol. Anal. Clin.*, 2017, **29**, 320.
73. P. Kintz and L. Gheddar, Evidence of use of drostanolone, an anabolic steroid, at the time the subject commited a murder: place of hair analysis, *Toxicol. Anal. Clin.*, 2021, **33**, 222–225.
74. L. Gheddar, J. S. Raul and P. Kintz, Development and validation of SARMs and metabolic modulators screening in hair using UHPLC–MS/MS: Application to a doping case and first identification of S23 in authentic human hair, *J. Chromatogr. B: Anal. Technol. Biomed. Life Sci.*, 2021, **1187**, 123048.

75. P. Kintz, L. Gheddar, A. Ameline and J. S. Raul, Perspective in evaluating selective androgen receptor modulators in human hair: a short communication, *Ther. Drug Monit.*, 2021, **43**, 298.
76. P. Kintz, A. Ameline, L. Gheddar and J. S. Raul, Testing for GW501516 (cardarine) in human hair using LC/MS-MS and confirmation by LC/HRMS, *Drug Test. Anal.*, 2020, **12**, 980.
77. A. Rading, P. Anielski, D. Thieme and A. M. Keiler, Detection of the selective androgen receptor modulator GSK2881078 and metabolites in urine and hair after single oral administration, *Drug Test. Anal.*, 2021, **13**, 217.
78. P. Kintz, L. Gheddar, A. Ameline, V. Dumestre-Toulet, M. Verschoore, J. Comte and J. S. Raul, Complete Post-mortem Investigations in a Death Involving Clenbuterol after Long-term Abuse, *J. Anal. Toxicol.*, 2019, **43**, 660.
79. P. Kintz, V. Dumestre-Toulet, C. Jamey, V. Cirimele and B. Ludes, Doping control for beta-adregenic compounds through hair analysis, *J. Forensic Sci.*, 2000, **45**, 170.
80. D. Hassall, N. Brealey, W. Wright, S. Hughes, A. West, R. Ravindranath, F. Warren and P. Daley-Yates, Hair analysis to monitor adherence to prescribed chronic inhaler drug therapy in patients with asthma or COPD, *Pulm. Pharmacol. Ther.*, 2018, **51**, 59.
81. A. Ameline, L. Gheddar, J. S. Raul and P. Kintz, Analytical Technology Biomedical Life Sciences Characterization of letrozole in human hair using LC–MS/MS and confirmation by LC–HRMS: Application to a doping case, *J. Chromatogr. B: Anal. Technol. Biomed. Life Sci.*, 2021, **1**, 1162.
82. J. C. Drooger, A. Jager, M. H. Lam, M. D. den Boer, S. Sleijfer, R. H. J. Mathijssen and P. de Bruijn, Development and validation of an UPLC–MS/MS method for the quantification of tamoxifen and its main metabolites in human scal hair, *J. Pharm. Biomed. Anal.*, 2015, **144**, 416.
83. A. F. Lo Faro, A. Tini, M. Gottardi, F. Pirani, A. Sirignano, R. Giorgetti and F. P. Busardò, Development and validation of a fast ultra-high-performance liquid chromatography tandem mass spectrometry method for determining carbonic anhydrase inhibitors and their metabolites in urine and hair, *Drug Test. Anal.*, 2021, **13**, 1552.
84. P. Kintz, L. Gheddar, A. Ameline and J. S. Raul, Hair testing for acetazolamide as an evidence of the use of a contaminated dietary supplement, *Drug Test. Anal.*, 2021, **13**, 1584.
85. F. Taus, R. Gottardo, M. Ballotari, C. Utzeri and F. Tagliaro, Hair analysis for beta-blockers and calcium-channel clockers by using liquid chromatography-tandem mass spectrometry as a tool for monitoring adherence to antihypertensive therapy, *Drug Test. Anal.*, 2022, **14**, 1773.
86. D. Thieme, C. Baumer, H. Sachs and J. Teske, Screening and long-term retrospection for psychoactive drugs in presumptive drug-facilitated crimes using segmented single hairs, *Drug Test. Anal.*, 2013, **5**, 736.

87. S. B. Wainhaus, N. Tzanani, S. Dagan, M. L. Miller and A. Amirav, Fast analysis of drugs in a single hair, *J. Am. Soc. Mass Spectrom.*, 1998, **9**, 1311.
88. C. Wiedfeld, G. Skopp, D. Thieme and F. Musshoff, Application of single hair analysis in a doping case involving amphetamine, *Drug Test. Anal.*, 2022, **14**, 781.
89. F. Musshoff, T. Arrey and K. Strupat, Determination of cocaine, cocaine metabolites and cannabinoids in single hairs by MALDI Fourier transform mass spectrometry – preliminary results, *Drug Test. Anal.*, 2013, **5**, 361.
90. A. Kernalléguen, C. Enjalbal, J. C. Alvarez, O. Belgacem, G. Léonetti, D. Laffite and A. L. Pélissier-Alicot, Synthetic cannabinoid isomers characterization by MALDIMS3 imaging. Application to single scalp hair, *Anal. Chim. Acta*, 2018, **1041**, 87.
91. H. Wang and Y. Wang, Matrix-assisted laser desorption/ionization mass spectrometric imaging for the rapid segmental analysis of methamphetamine in a single hair using umbelliferone as a matrix, *Anal. Chim. Acta*, 2017, **975**, 42.
92. J. K. J. Wong, T. L. S. Choi, K. Y. Kwok, L. Eny and T. S. M. Wan, Doping control analysis of 121 prohibited substances in equine hair by liquid chromatography-tandem mass spectrometry, *J. Pharm. Biomed. Anal.*, 2018, **158**, 189.
93. T. L. S. Choi, K. Y. Kwok, W. H. Kwok, Y. Y. K. Tsoi, J. K. Y. Wong and T. S. M. Wan, Detection of seventy-two anabolic and androgenic steroids and/or their esters in horse hair using ultra-high performance liquid chromatography-high resolution mass spectrometry in multiplexed targeted MS2 mode and gas chromatographytandem mass spectrometry, *J. Chromatogr. A*, 2018, **1566**, 51.
94. C. Cutler, M. Viljanto, P. Hincks, J. Habershon-Butcher, T. Muir and S. Biddle, Investigation of the metabolism of the selective androgen receptor modulator LGD-4033 in equine urine, plasma and hair following oral administration, *Drug Test. Anal.*, 2020, **12**, 247.
95. M. M. Madry, B. S. Spycher, J. Kupper, A. Fuerst, M. R. Baumgartner, T. Kraemer and H. Naegeli, Long-term monitoring of opioid, sedative and anti-inflammatory drugs in horse hair using a selective and sensitive LC–MS/MS procedure, *BMC Vet. Res.*, 2016, **12**, 84.
96. B. C. Moeller, A. Clifford, R. T. Emery, G. Alarcio, G. Favro and R. M. Arthur, Detection and confirmation of zilpaterol in equine hair using liquid chromatography-mass spectrometry, *Drug Test. Anal.*, 2022, **14**, 31.
97. B. Gray, L. Tuckley, C. Cutler, S. Biddle, S. Hudson, S. Gower and L. Vanhaecke, Investigations into the analysis of intact drug conjugates in animal sport doping control – development and assessment of a rapid and economical approach for screening greyhound urine, *Drug Test. Anal.*, 2020, **12**, 731.

98. I. Shah, J. D. Haddow, H. A. Ibrahim, M. V. A. Sheikh and F. S. A. Alhemeiri, A novel and innovative hair test to determine glucocorticoid levels in racing camels for use in assessment of doping, health, and disease, *Drug Test. Anal.*, 2018, **10**, 742.
99. F. X. Moreira, H. Carmo, A. Melo, R. Silva, Z. Azevedo, M. L. Bastos and P. G. de Pinho, The use of feathers from racing pigeons for doping control purposes, *J. Anal. Toxicol.*, 2019, **43**, 307.
100. K. Zhang, X. Liang, J. Zhang, Q. Zhao, S. Liu, C. Tang, C. Su and Q. Meng, Hair Analysis to Monitor the Illegal Use of Salbutamol in Beef Cattle, *J. Anal. Toxicol.*, 2017, **41**, 65.

CHAPTER 10

Hair Versus *Nails: Is There a Place for Other Keratin Matrices?*

M. COBO-GOLPE, E. LENDOIRO AND A. DE-CASTRO-RÍOS*

Toxicology Service, Institute of Forensic Sciences, Universidade de Santiago de Compostela, C/San Francisco, s/n, 15782 Santiago de Compostela, Spain
*Email: ana.decastro@usc.es

10.1 Nail General Aspects

Nails are skin appendages that have a common origin and common anatomical structures with hair. Both are predominantly epithelial structures derived from primitive epidermis and made up of keratinous fibrils embedded in a sulfur-rich matrix.[1] As keratinized matrices, nails, like hair, stably accumulate xenobiotics and store them for long periods of time. However, while the use of hair for forensic and clinical purposes is nowadays widespread in toxicological analysis in many laboratories, the use of nails is limited and has been less investigated.

Nails have been used since the 1960s to detect the exposure and intoxication to metals (arsenic, cadmium, copper, lead, zinc…), but in recent years different new applications for the detection of pharmaceuticals and drugs of abuse have been described.[2] For example, nails have been used for the monitoring of antimycotics[3] and other pharmaceuticals of toxicological interest, for drugs of abuse in drug-treatment programs, for identification of *in utero* drug exposure, for the detection of doping agents and in forensic toxicology as an alternative to hair.[4]

Issues in Toxicology No. 47
Perspectives and Challenges of Hair Analysis
Edited by Pascal Kintz, Alberto Salomone and Marco Vincenti

Published by the Royal Society of Chemistry, www.rsc.org

10.1.1 Structure and Formation of Nails

The nail is part of the distal phalanx and forms a functional unit with the joint, tendons, ligaments, bone, and nerves to achieve different functions, such as physical protection of the distal fingers and toes; as a sensory organ, enhancing tactile discrimination and fine manipulation; and as a tool for scratching. Also, nails contribute to the aesthetic appearance of the hand and foot.[1,5]

The parts of the nail are shown in Figure 10.1 and include:[1,6]

- Nail plate: fully keratinized structure which continues growing throughout life and covers the dorsal distal phalanges.
- Lateral nail folds: the cutaneous folded structures providing the lateral borders of the nail.
- Proximal nail fold: the cutaneous folded structure providing the visible proximal border of the nail.
- Eponychium: the upper portion of the ventral aspect of the proximal nail fold.
- Cuticle: the desquamating tissue that seals the nail cul-de-sac, formed by the stratum corneum of both the dorsal and the ventral side of the proximal nail fold.
- Nail matrix: the specialized epithelial structure that lies underneath the proximal nail fold (germinal matrix).
- Lunula: the convex margin of the matrix seen through the nail. It is commonly visible on the thumb and the great toe, and it is paler than the adjacent nail bed.
- Nail bed: the vascular bed upon which the nail rests extending from the lunula to the hyponychium.
- Hyponychium: the cutaneous margin underlying the free nail.

The nail anlage starts to develop in the fingers of the embryos at 9 weeks and the primary nail field externally appears at 10 weeks. At 13 weeks the

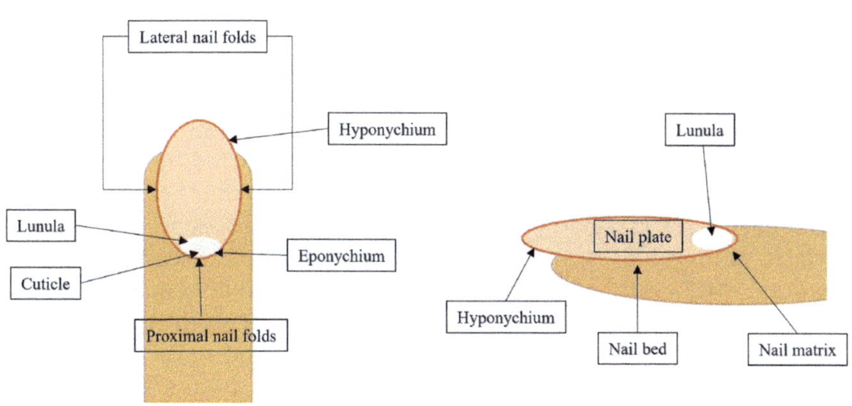

Figure 10.1 Scheme of the parts of the nail.

nail field is well defined in the finger and at 20 weeks all the matrix primordium cells have matured into matrix.[6,7] After that, nails grow continuously until birth.

The nail plate is continuously formed by the proliferation, maturation, and differentiation of epithelial cells produced by the nail matrix, a highly proliferative epidermal tissue. The nail matrix cells become larger and paler, and eventually the nucleus disintegrates, and the cells keratinize. The keratinization occurs both on the dorsal and ventral side of the forming nail plate. Most of the keratin fibers are perpendicularly oriented to the axis of growth, but others are oriented in the outermost segment, contributing to the fusing of the layers. The structure is so tightly knit that there is no cell exfoliation.[3] The nail plate gains thickness and density as it grows distally. Several authors have described that about 80% of the final nail mass and thickness of the nail plate is produced by the nail matrix and the other 20% by the nail bed.[8,9] Sweat ducts are present at the distal margin of the nail bed.[6]

According to several authors, nails can be compared in some respects to the hair follicle if the hair follicle was sectioned longitudinally and laid on its side.[6,10] However, some major differences between both keratinized matrices can be described, such as the presence or not of the cuticle layer and melanin.

The cuticle in hair is a protective outer layer which forms an outer barrier to avoid drug loss during washing and their penetration from external contamination.[1,10] This layer is missing in nails and for this reason, nails can be considered as a one compartment unit.[10]

Another major difference is the presence of melanin. Melanin is produced by melanocytes in the hair matrix, where they are usually very active, while in nails the melanocytes in the nail matrix lie dormant.[11] However, melanin can be produced in nails for several reasons and this phenomenon is called melanonychia. The main physiological cause of melanonychia is race. Darkly pigmented individuals, such as Black, Asian, Hispanic, and Middle Eastern people, frequently have longitudinal pigmented bands.[11] These pigmented bands are benign, and their number and width increase with age. Indeed, 77% of Black people will have melanonychia by the age of 20 and almost 100% at the age of 50, while 10–20% of Japanese people will have it developed in adulthood.[6,11] On the other hand, the appearance of melanonychia in Caucasians is more related to neoplasms including nevi, atypical melanocytic hyperplasia, and acral lentiginous melanoma.[12] Other causes of melanonychia are associated with dermatological diseases (onychomycosis, psoriasis, paronychia, *etc.*), systemic diseases (Addison's disease, nutritional disorders, AIDS, *etc.*) or iatrogenic causes (chemotherapeutic agents, phototherapy, X-ray exposure or administration of therapeutic drugs).[11]

The growth rate is also different between nails and hair. While hair is characterized by a growth cycle with different phases, nails grow continuously at a constant rate. However, the speed is not the same for fingernails and toenails, and it has been estimated that the growth rate of toenails is approximately 60% slower than fingernails.[13] On average, the fingernail growth rate is approximately 3 mm per month (0.1 mm per day) and for toenails is approximately 1.1 mm per month (0.03–0.04 mm per day).[3,4] The thickness also increases constantly at an average rate of 0.027 mm per mm of length.[3] Therefore, the time it takes for a nail to renew completely is 3–5 months for fingernails and 8–16 months for toenails.[4,9] Table 10.1 shows the growth rate calculated in different studies.

Variations in nail growth rate can be affected by several factors, although their effect is limited due to the slow growth rate on nails. Moreover, the number of studies evaluating these factors is scarce and in some of them the statistical significance is not evaluated, or their results were not confirmed in later studies. In general, some factors that may increase the growth rate are: sex (faster growth in males), age (faster in young people), pregnancy, climate with high temperatures, hyperthyroidism, onycholysis, onychophagia, psoriasis, minor trauma, regeneration after avulsion, and some drugs (biotin, terbinafine, itraconazole, fluconazole). In addition, other factors may be associated with a decrease in the growth rate, like malnutrition, hypothyroidism, decrease in blood supply, acute infection, onychomycosis, immobilization, and some drugs (methotrexate, azathioprine, cyclosporine).[4,6,13] On the other hand, there are no significant differences in growth rate between left and right fingernails or toenails,[2] but slight variations have been observed between the fingers of the same hand or foot, with higher growth rates for the longer fingers.[6]

10.1.2 Drug Incorporation Into Nails

Drugs and other xenobiotics present in the bloodstream can be incorporated into nail cells from the nail matrix and the nail bed *via* the capillary vessels and can be retained stably in the nail structure. Although the mechanism for drug uptake is not completely elucidated, similar to hair, three routes of incorporation can be proposed: from the capillary vessels during nail

Table 10.1 Average growth rate found in different studies for fingernails and toenails.

Average growth rate			
Fingernails	Toenails	Year of the study	Ref.
—	1.65 mm per month	1937	14
3.00 mm per month	—	1938	15
3.06 mm per month (men) 2.94 mm per month (women)	—	1955	16
3.51 mm per month	—	1958	17
3.28 mm per month	—	1977	18
3.47 mm per month	1.62 mm per month	2010	2
0.113 mm per day	—	2015	19
—	0.079 mm per day	2020	20

growth, from sweat and from environmental contamination (external exposure to drugs). These routes have been proved by the collection and analysis of nail clippings after single dose ingestion of zolpidem.[21,22] Hang *et al.*[21] administrated a single dose of zolpidem (10 mg) to seven individuals and fingernails were sampled after 1 week, 2 weeks and then every 2 weeks until the 20th week. Zolpidem was detected in nail clippings from the first week to at least the 18th week. The maximum peak concentration (0.40–1.74 $pg\,mg^{-1}$) was found in the first sample (1 week after consumption), then a decrease was detected during the next 6–12 weeks, and a new peak concentration was found between the 10th and 15th weeks (<0.37 $pg\,mg^{-1}$). Similarly, Madry *et al.*[22] administered a single dose of zolpidem (10 mg) to nine individuals and nail clippings were collected 24 hours after intake and then weekly until samples were negative. Zolpidem was detected in the 24 hours after drug intake, showing the maximum peak concentration (11.6–15.1 $pg\,mg^{-1}$), then the concentrations decreased, and a new concentration peak was found after 2–3 weeks in five of those individuals. Finally, 10–18 weeks after intake another concentration peak was found (0.15–0.9 $pg\,mg^{-1}$). Zolpidem was detected until 11–20 weeks after ingestion. These two studies indicate that incorporation of drugs into nails occurs immediately after intake by sweat (24 hours peak), second through the nail bed (which contributes to the increase in nail thickness, detectable after 2 weeks of ingestion), and third through the nail matrix (detectable after 10–18 weeks of ingestion).[21,22]

In cases of chronic intake, the concentration of nail clippings reflects a superposition of sweat incorporation, nail bed incorporation, nail matrix incorporation and, in some cases, incorporation through external contamination. Therefore, this superposed incorporation comprehends the total exposure to drugs from approximately the last 3–5 months for fingernails and 8–16 months for toenails.[4,9]

Several factors can influence the incorporation of drugs into nails. Firstly, the physicochemical properties of the drugs, because lipophilic drugs can cross membranes easily and they will diffuse from the blood into the cells of the nail matrix and nail bed. In general, relatively high concentrations of parent compounds are detected in comparison to the concentration of metabolites in nails. Secondly, it has been demonstrated that for some drugs a lower rate of incorporation occurs in those hair types with less melanin.[3] Consequently, the absence or low presence of melanin in nails can reduce the incorporation of these drugs into this matrix. Thirdly, nails are permeable to water, and they can retain water and increase their weight up to 21%,[6] so the incorporation by sweat gains importance in circumstances of increased sweat production (*e.g.*, in warm climates). Likewise, drugs may be eliminated by normal hygiene conditions or after prolonged immersion in water (wash out effect). Fourthly, the physical state and integrity of the nail may have an effect in the retention of the drugs incorporated. Finally, the use of some cosmetic treatments (use of nail polish, nail polish remover, permanent or semi-permanent polish, *etc.*) might alter drug incorporation or remove drugs from the nail.

10.2 Analytical Methods for the Determination of Drugs in Nails

As for hair, nail samples must be submitted to several steps for their analysis, including decontamination, drying, homogenization, incubation, and sample clean-up, before instrumental analysis.

In this section, the main characteristics of the aforementioned processes for different drugs of abuse and medicines are described, which are summarized in Table 10.2.

10.2.1 Alcohol

Detection of alcohol in nails was based on ethyl glucuronide (EtG) identification.[23–26] Morini *et al.*[26] also included ethyl sulphate (EtS) in a method based on LC–MS/MS, but this analyte had to be finally excluded due to the presence of interferences at the selected MRM transitions.

10.2.1.1 Pre-analytical Steps

The amount of sample employed varied from 10 to 50 mg of nail clippings. Decontamination was performed with two to three solvent steps, vortex-mixing and, in one case, also ultrasonicating. Solvents included hexane, followed by dichloromethane (DCM) and methanol,[23] DCM and methanol,[26] or water followed by acetone.[24] Cappelle *et al.*[24] evaluated the three procedures in nail specimens from alcohol users ($n = 5$) and did not find statistical differences in the decontamination efficiency, although EtG concentrations in the washes were lower with hexane–DCM–methanol in four samples, with no decrease in nail concentrations. The nail sample is then cut and/or pulverized. Cappelle *et al.*[24] compared EtG concentrations when the sample was cut or pulverized for 3–12 min. Higher EtG concentrations were observed in pulverized nails, although pulverization for more than 3 min did not lead to a significant increase in EtG concentrations. In most methods, the nail was incubated with water or water/acetonitrile (50 : 50, v/v), sonicating the samples for 2 h at room temperature. No EtG degradation was observed with these conditions.[26] The digested extract was cleaned-up with solid phase extraction (SPE) using anion exchange cartridges,[23,24] or directly injected into the analytical instrument,[25,26] with or without dilution.

10.2.1.2 Analytical Technique

Most methods employed LC–MS/MS with electrospray (ESI) in negative mode, using reversed-phase columns for polar analytes[26] or HILIC analytical columns.[23,25] As the mobile phase, acetonitrile, and water or 0.1% formic acid in water were employed, achieving a LOQ from 1.6 to 10 $pg\,mg^{-1}$.

Table 10.2 Summary of sample pretreatment steps, analytical instruments, and sensitivity for nail analytical methods for drugs of abuse and psychoactive medicines.[a]

Drug group	Analytes	Sample weight (mg)	Washing procedure	Homogenization	Digestion	Clean-up	Analytical technique	LOD ($pg\,mg^{-1}$)	LOQ ($pg\,mg^{-1}$)	Ref.
Alcohol	EtG	10–50	Hexane DCM MeOH	Pulverization	Water (ultrasound, 40 °C, 2 h + RT overnight)	SPE (mixed mode anion exchange)	LC–MS/MS (ESI−) Synergy polar RP column, 30 °C MP: 0.1% FA/ACN Run time: 10 min	2	8	23
Alcohol	EtG	30	Water Acetone	Pulverization	Water (ultrasound, 2 h/RT)	SPE (mixed mode anion exchange)	GC–MS (CI-) HP-5 MS capillary column Run time: 9 min Derivatization (HFBA)	Not evaluated	2	24
Alcohol	EtG	50	None	Pulverization	ACN/water (50 : 50, v/v) (ultrasound, 25 °C, 2 h)	Direct injection	LC–MS/MS (ESI−) Zorbax HILIC Pluc column, 25 °C MP: water/ACN Run time: 10 min	0.48 ppb	1.61 ppb	25
Alcohol	EtG	30	DCM MeOH	Cut	Water (ultrasound, RT, 2 h)	Direct injection	LC–MS/MS (ESI−) Hypersil Gold aQ C18 column, 42 °C MP: 0.1% FA/ACN Run time: 10 min	3	10	26
Tobacco	Nicotine	10–30	DCM	Cut	1 mM NaOH (50 °C, overnight)	LL (DEE)	LC-EC Prodigy ODS(3) column MP: 50 mM KH_2PO_4/MeOH/ACN (93.5 : 2.5 : 4, v/v) pH 4.8 Run time ≥20 min	10	Not evaluated	27
Tobacco	Nicotine, cotinine, NNAL	20–80	DCM (×2)	None	1 N NaOH (50 °C, overnight)	– LL (DCM, acidic pH) for nicotine/cotinine – SPE (1st diatomeas earth; 2nd mixed mode	– GC–MS (EI) for nicotine, cotinine HP-5 capillary column Run time ≥9 min	0.02–12	Not evaluated	28

						cation exchange) for NNAL	– LC–MS/MS (ESI+) for NNAL ZORBAX SB C18 column, 25 °C MP: MeOH/water (35 : 65, v/v) Run time: 15 min			
Tobacco	NNN	40–100	DCM (×2)	None	1 N NaOH (50 °C, overnight)	SPE (1st diatomeas earth; 2nd mixed mode cation exchange; 3rd normal-phase, basic pH)	LC–MS/MS (ESI+) ZORBAX SB C18, 25 °C MP: MeOH/water (35 : 65, v/v) Run time ≥20 min	0.02	Not evaluated	31
Tobacco	Nicotine, cotinine, myosmine	20–30	DCM	Cut	1 N NaOH (50 °C, overnight)	LL (DCM)	GC–MS (EI) DBS-MS capillary column Run time ≥30 min	10–35	30–105	29
Tobacco	Cotinine, 3-OH-cotinine	10	0.1% SDS	Cut	1 N NaOH (RT, overnight)	Acetonitrile + QuECHERS ($MgSO_4$ + NaCl + PSA)	LC–MS/MS (ESI+) Kinetex C18, 30 °C MP: water + MeOH/ACN (50 : 50, v/v) Run time: 15 min	Not evaluated	10	30
Cannabis	THC, THCCOOH	>10	0.1% SDS (ultrasound) Water (ultrasound, ×3) MeOH (ultrasound, ×3)	None	1 M NaOH (95 °C, 30 min)	LL (EtOAc, basic pH for THC; EtOAc, acidic pH for THCCOOH)	GC–MS (EI) HP-1 capillary column Run time: 25 min Derivatization: 1% TMCS in BSTFA (95 °C/20 min)	100	Not evaluated	32
Cannabis	THCCOOH	10–50	DCM	Cut	1 M NaOH (80 °C, 1 h)	SPE (mixed mode anion exchange)	2D-GC–MS/MS (CI-) DB-1MS column DB-17MS column Run time: 37.4 min Derivatization: HFIP + PFPA (70 °C/20 min)	0.01	0.02	34

Table 10.2 (*Continued*)

Drug group	Analytes	Sample weight (mg)	Washing procedure	Homogenization	Digestion	Clean-up	Analytical technique	LOD (pg mg^{-1})	LOQ (pg mg^{-1})	Ref.
Cannabis	THC, THCCOOH, OHTHC, diOHTHC, CBD, CBN	30	DCM (×5)	Pulverization	1 M NaOH (95 °C, 15 min)	SPE (mixed mode anion exchange)	LC–MS/MS (ESI+) Kinetex C18, 30 °C MP: 0.1% FA/ACN Run time: 10 min	10–100	20–100	35
Cannabis, amphetamine	THC, THCCOOH, AMP, MAMP, MDMA, MDA	30	Water MeOH (×3)	None	1 M NaOH (95 °C, 30 min)	LL (1st EtOAc; 2nd hexane/EtOAC, 9 : 1, v/v, acidic pH) both extracts mixed	GC–MS (EI) DB-5 MS capillary column Run time: 30 min Derivatization: MSTFA (70 °C/15 min)	16–5	100–200	33
Amphetamines	MAMP, AMP	Not indicated	MeOH/water (50 : 50, v/v, ultrasound)	None	2.5 N NaOH (80 °C, 30 min)	LL (*n*-pentane)	GC–MS (CI) 2% Thermon 3000 on Chromosorb W (80–100 mesh) glass column Derivatization: TFA (80 °C/30 min)	Not evaluated	Not evaluated	36
Amphetamines	MAMP, AMP, *p*-OH-MAMP	10–30	MeOH (×5) Water (×5)	Crushed (mortar)	0.6 N HCl	LL ($CHCl_3$/2-propanol, 3 : 1, v/v)	GC–MS (EI) 3% OV-17 on Chromosorb W AW DMCS glass column Derivatization: TFA (55 °C/20 min)	Not evaluated	Not evaluated	37
Amphetamines	AMP, MDA, MDMA	14	DCM	None	1 N NaOH	LL (1st EtOAc, acidic pH; 2nd EtOAc, basic pH)	GC–MS (EI) HP-5MS capillary column Derivatization: PFPA + PFP-OH Run time: 18.5 min	Not evaluated	Not evaluated	38

Amphetamines	AMP, MAMP	Not indicated	MeOH	None	2 N NaOH (80 °C, 1 h)	LL (1st EtOAc; 2nd EtOAc, acidic pH; 3rd aqueous layer back extracted, EtOAc, basic pH)	GC–MS (EI) HP-1MS capillary column Run time: 18.5 min Derivatization: HFBA (55 °C/20 min)	2000	2000	39
Amphetamines	*R*-MDMA, *S*-MDMA, *R*-MDA, *S*-MDA	5	Water	Pulverization	MeOH (ultrasound, 50–60 °C, 4 h)	Direct injection after filtration	LC–MS/MS (ESI+) Kinetex C18 column, 35 °C MP: 0.1% FA in 25 mM ammonium formate/ 0.1% FA in ACN Run time: 12 min Chiral derivatization: 0.3% DNPV (50 °C/30 min)	0.5	2	42
Amphetamines, ketamine	AMP, MAMP, MDA, MDMA, ketamine, norketamine	20	Water MeOH (×2)	None	1 M NaOH (95 °C, 30 min)	LL (EtOAc)	GC–MS (EI) DB-5MS capillary column Run time: 17 min Derivatization: HFBA (60 °C/30 min)	15–94	50–314	40
Amphetamines, ketamine	AMP, MAMP, MDA, MDMA, norketamine	20	Water MeOH (×3)	Cut + pulverization	MeOH (ultrasound, 50 °C, 1 h)	Direct injection after high-speed centrifugation	GC–MS (EI) DB-5MS capillary column Run time: 17 min Derivatization: HFBA (60 °C/30 min)	12–24	50–80	41
Cocaine	COC, BE, CE, EEE, EME, norBE, NorCOC, AEME, *m*-OHBE	25	MeOH	Cut	MeOH (40 °C, 16 h)	SPE (mixed mode cation exchange)	GC–MS (EI) HP-5MS capillary column Derivatization: 1% TMCS in BSTFA (65 °C/30 min)	100–250	250–500	43

Table 10.2 (*Continued*)

Drug group	Analytes	Sample weight (mg)	Washing procedure	Homogenization	Digestion	Clean-up	Analytical technique	LOD ($pg\,mg^{-1}$)	LOQ ($pg\,mg^{-1}$)	Ref.
Cocaine	COC, BE, CE, NorCOC	5	MeOH (×3)	None	MeOH (40 °C, 16 h)	SPE (mixed mode cation exchange)	GC–MS (EI) HP-5MS capillary column Run time: 17.8 min Derivatization: PFPA/PFP (70 °C/30 min)	3000–3500	Not evaluated	44
Cocaine	COC, BE, EME, AEME	50	DCM (×5)	Pulverization	0.1 M HCl (56 °C, overnight)	LL (1st $CHCl_3$/2-propanol/*n*-hexane, 50 : 17 : 33, v/v, basic pH; 2nd $CHCl_3$, basic pH)	GC–MS (EI) HP-5MS capillary column Run time: 14 min Derivatization: 1% TMCS in BSTFA (70 °C/20 min)	50–500	Not evaluated	46
Cocaine	COC, BE, norCOC, CE	5	Water MeOH	Cut + pulverization	MeOH (ultrasound, 7 h)	Direct injection after centrifugation, evaporation and reconstitution	LC–MS/MS (ESI+) Kinetex PFP column, 35 °C MP: 25 mM ammonium formate/0.1% FA in ACN Run time: 12 min	2–5	10	45
Opiates	6-AM, MORPH, COD, acetylCOD, heroin	Not indicated	Water ACN (×2)	Frozen pulverizarion	Borate buffer, pH 9 (ultrasound, RT, 30 min)	LL ($CHCl_3$/2-propanol, 9 : 1, v/v)	LC–MS/MS (ESI+) Allure PFP propyl column, 45 °C MP: ammonium acetate/ACN Run time ≥10 min	10–30	50	47
Opiates	6-AM, MORPH, COD, MTD, EDDP	60	Water	Cut	0.1% FA in water/0.1% FA in MeOH (9 : 1, v/v) (55 °C, 72 h)	Direct injection after high-speed centrifugation	LC–MS/MS (ESI+) Kinetex biphenyl column, 25 °C MP: 0.1% FA in water/0.1% FA in MeOH Run time ≥15 min	20	50	48

Opiates	MTD, EDDP, EMDP	30	0.1% SDS (×3) Water (×3) MeOH (×3)	Cut	1 M NaOH (90 °C, 40–60 min)	LL (MTBE) SPE (mixed mode cation exchange)	GC–MS (EI) VF-5MS capillary column (30 m × 0.25 mm, 0.25 μm) Run time: 28 min Derivatization: MSTFA (80 °C/30 min)	3300–6000	10 400–20 800	49
Opiates	BUP, norBUP, naloxone	20	Water (×2) CH_3Cl (×2) MeOH (×2)	None	1 M NaOH (55 °C, 3 h)	LL ($CHCl_3$/2-propanol, 9 : 1, v/v)	LC–MS (ESI+) Discovery C18 column, 30 °C MP: 0.1% FA in water/ACN Run time ≥16 min	2.3–8.1	8.9–27	50
Cocaine, opiates	COC, BE, CE, EEE, EME, norBE, NorCOC, AEME 6-AM, COD, MORPH, norCOD, norMORPH	15	2-Propanol Phosphate buffer pH 6 (×3)	None	Protease XI + DDT in Tris buffer pH 7.4 (40 °C, overnight)	SPE (mixed mode cation exchange)	GC–MS (EI) HP-1 capillary Runt time: 16 min Derivatization: 1% TMCS in BSTFA (65 °C/30 min)	Not evaluated	130–1000	51
Cocaine, opiates	COC, BE, CE, NorCOC, EME, EG 6-AM, MORPH, COD, hydromorphone, hydrocodone, nalorphine	100	MeOH (×3)	Cut	0.1 M phosphate buffer pH 5 (ultrasound 1 h, soaked 72 h)	SPE (mixed mode cation exchange)	GC–MS (EI) DB-5 capillary column Run time: 14 min Derivatization: MSTFA (75 °C/45 min)	Not evaluated	Not evaluated	52
Cocaine, opiates	COC, MORPH, 6-AM	50	DCM (×2)	Cut	HCl 37% (100 °C/30 min + 12 min RT)	LL ($CHCl_3$/2-propanol, 3 : 1, v/v), basic pH SPE (mixed mode cation exchange)	GC–MS (EI) DB-5 Run time ≥11.6 min Derivatization: propionic anhydride (90 °C/1 h)	Not evaluated	Not evaluated	54
Cocaine, opiates	COC, BE, MORPH, MTD, caffeine, nicotine, cotinine	20	MeOH	Cut	0.1 N HCl (overnight)	SPE (mixed mode cation exchange)	GC–MS (EI) MS-5 capillary column Run time ≥14.75 min Derivatization: 1% TMCS in BSTFA	25	25–50	55

Table 10.2 (*Continued*)

Drug group	Analytes	Sample weight (mg)	Washing procedure	Homogenization	Digestion	Clean-up	Analytical technique	LOD ($pg\,mg^{-1}$)	LOQ ($pg\,mg^{-1}$)	Ref.
GHB	GHB, GHB glucuronide	25	DCM	None	Comedical VMA-TM3® (acidic aqueous buffer) (Comedical s.r.l., Trento, Italy)	SPE (reversed-phase)	LC–MS/MS (ESI−) HSS T3 C18 column MP: 0.1% FA/ACN (95 : 5, v/v) Run time: 10 min	20–100	70–300	68
Antidepressants	Fluoxetine, paroxetine, duloxetine	50	Water Acetone Water	None	MeOH (ultrasound, 50 °C, 2 h)	Direct injection after evaporation and reconstitution	HPLC-UV-Vis Hichrom Kromasil 100-5C18 column, RT MP: ammonium acetate/ACN (59 : 41) Run time: 18 min	100–600	300–1800	59
Antidepressants, benzodiazepines	6 antidepressants 6 benzodiazepines	30	DCM (×3)	Pulverization	Water/ACN (50 : 50) (horizontal agitation, 90 min)	SPE (mixed mode cation exchange)	LC–MS/MS (ESI+) XBridge column, 30 °C MP: 0.1% FA in ammonium formate/ACN Run time: 14 min	5–10	10	56
Benzodiazepines	14 benzodiazepines	50	Water EtOH (×2)	Cut	TFA/MeOH (1 : 50) (RT, 16–18 h)	LL (DCM)	LC–MS/MS (ESI+) Luna phenyl hexyl column MP: 2 mM ammonium formate/ACN Run time: 8 min	Not evaluated	Not evaluated	57
Benzodiazepines, zolpidem	31 benzodiazepines zolpidem	10	DCM MeOH	Pulverization	MeOH (ultrasound, 2 h)	Direct injection	LC–MS/MS (ESI+) Kinetex C18 column, 25 °C MP: 0.1% FA/ACN Run time: 20 min	0.03–13.1	0.1–43.7	58

Hypnotic	Zolpidem	5–10	Water (×2)	Pulverization	MeOH (ultrasound, 6 h)	Direct injection after evaporation and reconstitution	LC–MS/MS (ESI+) Kinetex C18 column, 40 °C MP: 5 mM ammonium formate/1 mM ammonium formate in ACN Run time: 8.3 min	Not evaluated	0.1	22
Antipsychotic	Clozapine	5	Water (×2) Acetone	Pulverization	1% FA in 20 mM ammonium acetate/ACN (30 : 70, v/v) (Ultrasound RT, 1 h)	Direct injection after dilution	LC–MS/MS (ESI+) Ultra IBD column, 25 °C MP: 0.1% FA in 20 mM ammonium acetate/ ACN Run time: 4.5 min	50	500	60
Antipsychotic	Asenapine	50	Water Acetone Water	Cut	MeOH (overnight) MeOH + MP (ultrasound, 10 min)	Direct injection after evaporation and reconstitution	HPLC-UV-Vis Inertil C18 column MP: (0.2 M AA + 0.2 M sodium acetate) + ACN + MeOH Run time ≥6 min	2700	8000	61
Antipsychotics	Clozapine, haloperidol, levomepromazine, olanzapine, quetiapine	30	DCM (×3)	Pulverization	Water/ACN (50 : 50) (horizontal agitation, 90 min)	SPE (mixed mode cation exchange)	LC–MS/MS (ESI+) XBridge column, 30 °C MP: 0.1% FA in ammonium formate/ ACN Run time: 8 min	2.5	10	62
GUS (∼3000 substances)	—	20	Water Acetone (×2)	Cut + pulverized	MeOH/ACN/ ammonium formate (25 : 25 : 50, v/v) (40 °C/ 800 rpm/ 18 h) (×2)	Direct injection after centrifugation and partial evaporation	LC-QTOF-MS (ESI+) Zorba Eclipse plus C18 column, 30 °C MP: 10 mM ammonium acetate/MeOH Run time: 23 min	Not evaluated	Not evaluated	65

Table 10.2 (*Continued*)

Drug group	Analytes	Sample weight (mg)	Washing procedure	Homogenization	Digestion	Clean-up	Analytical technique	LOD ($pg\,mg^{-1}$)	LOQ ($pg\,mg^{-1}$)	Ref.
Cocaine, opiates, amphetamines, ketamine, zolpidem, benzodiazepines, antipsychotics, antidepressant, antihistamins, antihypertensive, analgesic	76 drugs and metabolites	20	Water Acetone (×2)	Pulverized	MeOH/ACN/ammonium formate (25 : 25 : 50, v/v) (40 °C/800 rpm/18 h) (×2)	Direct injection after filtration	LC–MS/MS (ESI+) Kinetex C18 column MP: 0.1% FA in water/ACN Run time: 11 min (antipsychotics) Run time: 16 min (other drugs)	0.37–29.7	0.95–46.7	64
Cocaine, opiates, amphetamines	COC, BE, EME, 6-AM, MORPH, COD, MTD, EDDP, AMP, MAMP, MDMA, MDEA	20	Water Acetone	Pulverization	MeOH/ACN/ammonium formate (25 : 25 : 50, v/v) (500 rpm/18 h)	Direct injection after ultracentrifugation and filtration	LC–MS/MS (ESI+) EZ: C18 column MP: 0.01% FA in 2 mM ammonium formate/0.01% FA in ACN Run time: 21 min	Not evaluated	50	63
Cocaine, cannabis, tobacco, amphetamines, ketamine, GHB, synthetic cathinones, synthetic cannabinoids	COC, BE, CE, THC, cotinine, AMP, MAMP, MDMA, ketamine, norketamine, GHB, GBL, cathinone, methcathinone, 4-MMC, methylone, butylone, 4-MEC, JWH-398, JWH-122, JWH-081	25	DCM	None	VMA-T M3 reagent (Comedical s.r.l.)	SPE (reversed-phase)	LC–MS/MS – Cocaine, amphetamines, ketamine, THC: ESI+ BEH C18 column MP: 0.1% FA in 5 mM ammonium formate/ACN Run time >5 min	0.3–100	1–300	66

							– Synthetic cathinones and cannabinoids: ESI+ BEH C18 column MP: 0.1% FA/MeOH Run time >10 min – GHB/GBL: ESI– HSS T3 C18 column Run time >7 min			
Cocaine, cannabis, opioids, synthetic cathinones, synthetic cannabinoids, fentanyls, phenethylamines, phencyclidines, tryptamines, piperazines	COC, BE, THC, MORP, 6-AM, COD, MTD, EDDP, tramadol and 2 other opioids, 34 synthetic cathinones, 18 synthetic cannabinoids, 26 fentanyls, 10 phenethylamines, 3 phecyclidines, 2 trytamines, 2 piperazines	20	Water/ACN (×2)	Cryogenic ground	ACN/0.075% FA in 5 mM ammonium acetate (50 : 50, v/v) (ground at 3000/25 s × 4, RT)	Direct injection after filtration	LC–MS/MS (ESI+) HSS C18 column MP: 0.075% FA in 5 mM ammonium acetate/ 0.1% FA in ACN Run time: 15 min	2.5–25	5–50	67

[a] 3-OH-cotinine: 3-hydroxy-cotinine; 4-MEC: 4-methylethcathinone; 4-MMC: mephedrone; 6-AM: 6-acetylmorphine; ACN: acetonitrile; AEME: anhydro ecgonine methyl ester; AMP: amphetamine; BE: benzoylecgonine; BSTFA: *N*,*O*-bis(trimethylsilyl)trifluoroacetamide; BUP: buprenorphine; CBD: cannabidiol; CBN: cannabinol; CE: cocaethylene; CI: chemical ionization; COC: cocaine; COD: codeine; DCM: dichloromethane; DEE: diethylether; diOHTHC: 8-β-11-dihydroxy-THC; DNPV: *N*-(2,4-dinitro-5-fluorophenyl)-L-valinamide; EC: electron capture; EDDP: 2-ethylidene-1,5-dimethyl-3,3-diphenylpyrrolidine; EEE: ecgonine ethyl ester; EG: ecgonine; EI: electron impact; EMDP: 2-ethyl-5-methyl-3,3-diphenylpyrrolidine; EME: ecgonine methyl ester, ESI: electrospray; ETG: ethyl glucuronide; EtOAc: ethyl acetate; EtOH: ethanol; FA: formic acid; GHB: γ-hydroxybutyric acid; GBL: γ-butyrolactone; GUS: general unknown screening; HFBA: heptafluorobutyric anhydride; HFIP: hexafluoroisopropanol; LL: liquid–liquid extraction; LOD: limit of detection; LOQ: limit of quantification; *m*-OH-BE: *m*-bydroxybenzoylecgonine; MAMP: methamphetamine; MeOH: methanol; MP: mobil phase; MORPH: morphine; MSTFA: *N*-methyl-*N*-trimethylsilyltrifluoroacetamide; MTD: methadone; NNAL: 4-(methylnitrosamino)-1-(3-pyridyl)-1-butanol); NNN: *N*′-nitrosonornicotine; OHTHC: 11-hydroxy-THC; PFP-OH: 2,2,3,3,3-pentafluoro-L-propanol; PFPA: pentafluoropropionic anhydride; *p*-OH-MAMP: *p*-hydroxymethamphetamine; PSA: primary secondary amine; RT: room temperature; SDS: sodium dodecyl sulfate; SPE: solid phase extraction; TFA: trifluoroacetic anhydride; THC: Δ9-tetrahydrocannabinol; THCCOOH: 11-nor-9-carboxy-Δ9-tetrahydrocannabinol: TMCS: trimethylchlorosilane.

Post-column infusion of 2-propanol has been used to increase sensitivity.[26] The use of GC–MS/MS with negative chemical ionization has also been described, although it required the derivatization of EtG with HFBA. A non-polar analytical column was employed, achieving a LOQ of 2 pg mg^{-1}.[25]

10.2.2 Tobacco

Most methods for the identification of tobacco exposure included the analysis of nicotine[27] and/or the metabolites cotinine[28,29] and 3-hydroxy-cotinine (OH-cotinine).[30] Some authors analyzed other minor tobacco biomarkers, including the specific tobacco nitrosamines 4-(methylnitrosamino)-1-(3-pyridyl)-1-butanol) (NNAL)[28] or *N′*-nitrosonornicotine (NNN),[31] or myosmine,[29] to help discriminate between active and passive smokers. In all the methods, toenail instead of fingernail was analyzed as it is less exposed to external contamination.

10.2.2.1 Pre-analytical Steps

The amount of nail employed ranged between 10 and 30 mg of nail clippings for the analysis of nicotine and its metabolites, and myosmine. However, determination of NNAL or NNN required a higher amount of nail (40–100 mg) due to the lower concentrations of these tobacco biomarkers in the real samples from smokers. Decontamination was performed with DCM in most cases.[27–29,31] Kim *et al.*[30] found similar efficiency in the external contamination elimination with 0.1% sodium dodecyl sulfate (SDS) and methanol, although extraction of cotinine and OH-cotinine from inside the sample was much higher with methanol (up to 30%). Incubation was performed in all cases with 1 N NaOH, usually at 50 °C overnight. For nicotine and metabolites, clean-up was accomplished by liquid–liquid extraction with DCM[27–29] or QuECHERS extraction.[30] Determination of NNAL[28] and NNN[31] required a more intensive sample clean-up, including extraction with two and three cartridges, respectively, with different sorbents (diatomaceous earth, mixed mode cation exchange and/or normal-phase cartridges).

10.2.2.2 Analytical Technique

GC–MS with electron impact (EI),[28,29] LC with electron capture detection[27] or LC–MS/MS with ESI+[29] were employed for myosmine, nicotine, and metabolites, although for the more polar NNAL and NNN, LC–MS/MS with ESI+ was required.[28,31] LOD were 10 to 35 pg mg^{-1} for myosmine, nicotine and metabolites, and 0.02 pg mg^{-1} for NNAL and NNN.

10.2.3 Cannabis

Four methods were developed for the detection of cannabis in nails, including the identification of tetrahydrocannabinol (THC) and 11-nor-9-carboxy-THC

(THCCOOH),[32,33] only THCCOOH,[34] or THC, cannabidiol (CBD) and cannabinol (CBN), and the metabolites THCCOOH, 11-hydroxy-THC (OH-THC) and 8-β-11-diOH-THC.[35]

10.2.3.1 Pre-analytical Steps

The amount of nail employed ranged between 10 and 50 mg of nail clippings. An extensive decontamination procedure with four to seven solvent steps was usually applied due to the high possibility of external contamination of fingernails from cannabis manipulation. The solvents employed were SDS, water and/or methanol,[32,33] or DCM.[34,35] After drying the samples, some authors cut[34] or pulverized the nail clippings[35] before subjecting them to incubation with NaOH with heating for 15 to 60 min. Clean-up of the digested extracts was performed by SPE with mixed-mode anion exchange cartridges, or by liquid–liquid extraction with ethylacetate (EtOAc). In order to extract THCCOOH, the liquid–liquid extraction must be performed in acidic conditions, although at this pH THC was not extracted.[32] To extract both analytes, Kim *et al.*[33] re-extracted the sample with hexane/EtOAc (9 : 1, v/v) at acidic pH after EtOAc extraction in basic conditions, and mixed both extracts.

10.2.3.2 Analytical Technique

GC–MS with EI has been used for the determination of THC and THCCOOH, achieving a LOD from 35 to 100 pg mg^{-1} after analyte derivatization with 1% TMCS in BSTFA or with MSTFA.[32,33] Shorter run times (from 25–30 min to 10 min) were possible using LC–MS/MS with ESI+, using acetonitrile and 0.1% formic acid in water as the mobile phase, and a reversed-phase C18 column.[35] This method also included CBD, CBN and additional metabolites, achieving similar LOD.[35] Considering the low sensitivity required for THCCOOH, Jones *et al.*[34] developed a method with 2D-GC–MS/MS with chemical impact, reaching a LOD of 0.01 pg mg^{-1} after derivatization with HFIP and PFPA.

10.2.4 Amphetamines

Several methods were developed for the determination of amphetamine, methamphetamine, and/or MDMA,[36–42] in one case in conjunction with cannabis.[33] Some methods also included the metabolites *p*-hydroxy-methamphetamine[37] or MDA.[33,38,40–42] Kim *et al.* additionally included ketamine[40] and its metabolite norketamine,[40,41] as this drug is usually concomitantly consumed with MDMA.

10.2.4.1 Pre-analytical Steps

The amount of nail employed ranged between 5 and 30 mg, usually from nail clippings. Madry *et al.*[42] compared concentrations in nail clippings and nail scrapings for *R,S*-MDMA and *R,S*-MDA in a controlled MDMA administration study. Although no significant differences were observed in the sum of enantiomers, lower concentrations were detected in scrapings, probably due to the extraction of the analytes from the nail surface during daily hygiene.

Decontamination was performed with DCM,[38] or methanol,[39] water[42] or both solvents.[36,37,40,41] Some authors analyzed the entire washed nail sample, although the lowest LOQ was achieved when the nails were pulverized.[42] Samples were then subjected to incubation usually with NaOH (at 80 to 95 °C for 30 to 60 min)[36,38–40] or with MeOH (at 50 °C for 1–4 h).[41,42] For clean-up, liquid–liquid extraction was mainly performed with EtOAc, although a simple high-speed centrifugation[41] or filtration[42] was possible when the sample was pulverized and digested with methanol.

10.2.4.2 Analytical Technique

Most of the published methods were developed using GC–MS with EI, after derivatization of the analytes with different acetylating agents.[37–41] Kim *et al.*[40] compared results after derivatization with TFAA, PFPA, HFBA, observing the best results with HFBA. The lowest reported LOD range was 12–24 pg mg^{-1}.[41] Finally, Madry *et al.*[42] performed a chiral derivatization that allowed the determination of MDMA and MDA enantiomers by LC–MS/MS with ESI+ analysis, achieving a LOD of 0.5 pg mg^{-1}.

10.2.5 Cocaine and Opioids

Although some of the published methods allow the identification of cocaine[43–46] or opioids,[47–50] several methodologies were developed for the simultaneous detection of both types of drug.[51–55] For cocaine analysis, its main metabolite benzoylecgonine (BE) was included in all the methods. As BE could be present in the sample due to spontaneous hydrolysis of cocaine, identification of other metabolites could avoid a false positive result due to external contamination. Madry *et al.*[45] suggested the identification of norcocaine, less prone to external contamination, as an effective tool to discriminate between cocaine consumption and external contamination. Other cocaine metabolites present in some of the published methods were ecgonine methyl ester (EME), ecgonine ethyl ester, cocaethylene (for concomitant use of cocaine and ethanol), norbenzoylecgonine, *m*-hydroxybenzoylecgonine or anhydro ecgonine methyl ester (for smoked cocaine identification). For opiates, the main analytes included in the methods were morphine, 6-acetylmorphine (6-AM) and codeine. Other metabolites were acetylcodeine, norcodeine or normorphine. Analytical methods for pharmaceutical opiates such as methadone (and its metabolites EDDP[48] and EMDP[49]), buprenorphine (and its metabolite norbuprenorphine),[50] hydrocodone or hydromorphone[52] were also developed in nail samples.

10.2.5.1 Pre-analytical Steps

Varying amounts of nail (mostly clippings), ranging from 5 to 100 mg, were used in the different published methods. Madry *et al.*[42] found 6-fold higher concentrations in nail scrapings compared to nail clippings. As scrapings represent only the surface of the nail, these higher concentrations were mainly attributed to drug deposition from external contamination.

Typical decontamination solvents such as methanol[43,44,52] or DCM,[54,55] or combinations with water and methanol or acetonitrile[45,47–49,55] were usually employed. For these drugs, a critical step in sample pretreatment is incubation. The use of acid or alkaline solvents has been described for this purpose; however, these conditions facilitate cocaine or 6-AM hydrolysis to BE or morphine, respectively. For cocaine, soaking nails in phosphate buffer pH 5 for 72 h led to higher cocaine recovery than methanol,[52] but subsequent publications demonstrated BE formation with this solvent, while it was residual with methanol.[43,44] Finally, Shen *et al.*[47] combined pulverization of the nails at low temperatures (frozen pulverization) with sonication with phosphate buffer pH 9.4, which shortens digestion time while avoiding 6-AM hydrolysis. For sample clean-up, SPE with mixed mode cation exchange cartridges[43,44,51,52,55] or liquid–liquid extraction with methyl tert-butyl ether or chloroform with 2-propanol[46,47,50,54] were the most common procedures, although direct injection of the digested sample has also been described.[45,48]

10.2.5.2 Analytical Technique

GC–MS with EI was the most common analytical technique for the analysis of cocaine and opiates, requiring derivatization of the analytes, for which 1% TMCS in BSTFA, MSTFA, PFPA/PFP or propionic anhydride have been employed.[43,44,46,49,52,54,55] LOD ranged between 25 and 3500 pg mg^{-1}. The use of LC–MS/MS with ESI+ has also been used,[45,47,48] allowing lower LOD than those reported by GC–MS (2–20 pg mg^{-1}).

10.2.6 Benzodiazepines, Antidepressants, Antipsychotics

Three methods were published for the determination of one[22] or several benzodiazepines and/or hypnotics (zolpidem)[56–58] in nails; two methods for antidepressants;[56,59] and three methods for one[60,61] or several antipsychotics.[62]

10.2.6.1 Pre-analytical Steps

The amount of nail employed ranged between 5 and 50 mg of nail clippings. Solvents for decontamination varied from water,[61] to water and acetone,[59–61] water and ethanol,[57] DCM,[56,62] or DCM and methanol,[58] applying several washes (2 to 5 steps). In most cases, samples were pulverized before incubation.[22,56,58,60,62] The samples were incubated with methanol or water/

acetonitrile, generally ultrasonicating or shaking for 1.5–6 h. Irving *et al.*[57] added 1% TFA to the methanol to avoid zopiclone degradation. For antipsychotics, Cobo-Golpe *et al.*[62] evaluated the use of methanol, 2 M NaOH, acetonitrile, and water/acetonitrile in different proportions, submitting the samples to heating, agitation, and sonication at different timepoints (15, 30, 90, or 180 min), with the best results achieved with water/acetonitrile (50 : 50, v/v) with horizontal agitation for 90 min.

For clean-up, in some methods SPE with cation exchange columns[56,62] or liquid–liquid extraction with DCM was performed,[57] although in most cases samples were directly injected without previous sample extraction,[22,58–61] even for multianalyte methods.[58,59]

10.2.6.2 Analytical Technique

Although the use of HPLC-UV-Vis was reported,[59,61] in most cases LC–MS/MS with ESI+ was the technique of choice, using reversed-phase columns and formic acid or ammonium formate or acetate and acetonitrile as the mobile phase, allowing greater sensitivity. LOD ranged from 0.03–5 pg mg^{-1} for benzodiazepines, from 5–100 pg mg^{-1} for antidepressants, and from 2.5–2700 pg mg^{-1} for antipsychotics.

10.2.7 Multianalyte Methods

Most methods for the determination of drugs of abuse and psychoactive medicines in nail have been developed for the identification of a few analytes belonging to one or two different drug classes. However, considering that polydrug use and polymedication is a common practice, and that the amount of sample available could limit the number of analyses to be done, the development of multianalyte methods allowing the identification of substances from different drug groups represents an important advantage.

Currently, five methods have been published for the identification of analytes from three or more drug classes. Cappelle *et al.*[63] developed a method for identification of cocaine, amphetamines, and heroin and morphine consumption. The list of drugs was extended in the method published by Krumbiegel *et al.*,[64] which includes 76 drugs and metabolites (cocaine, opiates, amphetamines, ketamine, antipsychotics, antidepressant, BZD, zolpidem, antihistamines, metoprolol, and gabapentin). The same authors also described the first method to perform a general unknown screening in nails.[65] Finally, two authors developed methods which allowed the detection of new psychoactive substances (NPS).[66,67] The method developed by Busardò *et al.*,[66] apart from the classical drugs of abuse cocaine, cannabis and amphetamines, allowed the identification of six synthetic cathinones, three synthetic cannabinoids, GHB and GBL, and ketamine; and Liu *et al.*'s[67] method included cocaine, cannabis, classical opiates, and almost 100 analytes belonging to different NPS classes (synthetic cathinones and cannabinoids, fentanyl derivatives, phenethylamines, phencyclidines, tryptamines or piperazines).

10.2.7.1 Pre-analytical Steps

The amount of sample employed in all these multianalyte methods was 20–25 mg of nail clippings.

Decontamination was performed with DCM,[66] water, and acetone[63–65] or water/acetonitrile.[67] For incubation, Busardò *et al.*[66] used the commercial acidic aqueous buffer VMA-T M3® (Comedical s.r.l.). Although the process was fast (1 h), it required incubation at 100 °C, and could affect the stability of thermolabile compounds. For the other methods, a mixture of organic and aqueous solvents such as methanol/acetonitrile/ammonium formate[63–65] or acetonitrile/0.075% formic acid in 5 mM ammonium acetate[67] were employed. Incubation for 18 h at 500–800 rpm was further applied, except for Liu *et al.*[67] These authors significantly reduce the incubation time by applying an initial pulverization of the nail at low temperature using a liquid nitrogen freezer mill, obtaining small nail particles that increase the surface area. This procedure allowed a faster penetration of the solvent and, therefore, digestion with the organic/aqueous solvent could be further performed by grinding the nails with zirconia balls for just 1 min at room temperature. Except for Busardò *et al.*,[66] who performed SPE with reversed-phase columns for sample clean-up, in the remaining methods the incubation solvent was ultracentrifuged and/or filtered before direct injection into the analytical instrument.

10.2.7.2 Analytical Technique

All the published multi-analyte methods employed LC–MS/MS with ESI+ for the analysis. The mass spectrometer analyzer was a triple quadrupole, except for the general unknown screening, for which a QTOF was used.[65] Total run time for the analysis of all the compounds ranged from 15 to 27 min; however, Krumbiegel *et al.*[64] applied two different LC gradients, one for the analysis of antipsychotics and another for the remaining analytes; and Busardò *et al.*[66] applied three different LC gradients; one for GHB and GBL, other for synthetic cathinones and cannabinoids, and other for cocaine, amphetamines, ketamine and THC. Considering the amount and variety of compounds included in these methods, and the different sensitivity required in each case, the methods' LOD ranged from 0.3 to 100 $pg\,mg^{-1}$.

10.3 Considerations Regarding Nail Analytical Methods

Although the analytical steps applied for hair and nails are the same, some aspects should be considered when developing a method for the analysis of nail samples:

10.3.1 Use of Nail Clippings or Nail Scrapings

Although nail scraping has been used in some of the methods developed for nail analysis, it just represents the external layer of this matrix. For this reason, nail scrapings are more susceptible to environmental contamination than nail clippings. In addition, drugs present in the nail scrapings are more easily removed by daily hand washing, and also by the applied decontamination procedure, as will be further detailed in Section 10.4.2. For these reasons, it is preferable to use nail clippings for method development and for real sample analysis.

10.3.2 Nail Decontamination Procedure

As mentioned before, the incorporation of drugs into hair and nails from environmental exposure is well known and, therefore, the analytical methodologies must include a decontamination procedure. Similar procedures are usually carried out for hair and nail decontamination; however, unlike what occurs in the nails, the inner layer of the hair is externally protected by the cuticle, which acts as a barrier, hindering the entrance of drugs inside the hair due to environmental exposure. Therefore, nails are more susceptible to being affected by external contamination, as would happen with damaged hair. On the other hand, the presence of the hair cuticle also limits the loss of drug that has been incorporated into the hair through the blood supply (which represents active exposure to the drug) through the decontamination procedure. Therefore, in the absence of this protective barrier, the application of extended washing procedures for nail samples decontamination must be avoided to limit the washout of the drug incorporated inside the matrix.

10.3.3 Nail is a Harder Matrix Than Hair

As for hair, an important step for nail analysis is the extraction of the drugs from the inner part of the matrix. Penetration of the incubation solvents into the matrix required the use of strong alkali solutions, which could compromise the stability of some compounds (*e.g.*, cocaine or 6-AM), or soaking the sample in organic and/or aqueous buffer solutions, usually heating and/or sonicating the mixture. Extraction of the analytes could be accelerated with a previous pulverization of the matrix, which is usually performed with a ball mill. However, it should be noted that, due to the hardness of the nail, to obtain the same matrix particle size and, therefore, similar extraction efficiency, nails should be pulverized longer and/or at a higher speed than hair, which increases the temperature achieved.[56,62] The use of liquid nitrogen cryogenic grinding has also been described,[67] which allows pulverization before digestion at low temperature, ensuring that the analytes will not degrade, followed by a short incubation (1 min) in buffer/organic solvent and simultaneous pulverization with zirconia beds.

10.4 Usefulness of Nail Samples

10.4.1 Fingernails or Toenails?

Choosing fingernails or toenails for the analysis can lead to different results. One difference between the two is the growth rate, with toenails growing at half the speed of fingernails, which will lead to different detection windows between these two nail samples and when compared to hair. Furthermore, the accumulation of substances for a longer time can lead to the amplification of the concentrations in toenails. In addition, fingernails are more susceptible to contamination *via* manipulation of the drug, and toenails are more affected by sweat contamination. Moreover, there is a possibility of drug loss in fingernails due to frequent handwashing.

The comparison of finger and toenail results should be performed intra-individually because of all the variables that can affect drug concentrations. A general comparison of concentrations using the means or medians of a certain studied cohort does not clarify the relationship between the two matrices.

There are few studies comparing concentrations in fingernail and toenail samples from the same individuals. Some drugs appear to be always in higher concentration in fingernails or in toenails. This is the case of cocaine and its metabolites,[43,53,69] morphine and 6-AM,[53,69] cannabinoids (THC, CBN, CBD),[35] EDDP[69] and PCP.[70] On the other hand, some studies have found higher concentrations in toenails for amphetamines,[36,69] methadone,[69] caffeine and amlodipine,[71] fluoxetine and sertraline,[56] and clozapine, levomepromazine and haloperidol.[62] For other drugs including venlafaxine, zolpidem, trazodone, olanzapine, quetiapine, or codeine,[53,56,62,69] the distribution between matrices is not so clear, with some differences between individuals that make it difficult to reach a conclusion.

A larger study was published by Shu *et al.*,[72] which involved the analysis of around 10 000 nail samples from a high-risk population. Although the authors did not compare fingernails and toenails results on a case-by-case basis, looking at the mean and median concentrations, fingernail concentrations were higher for cocaine, opioids and EtG in fingernails, and lower for amphetamine and methamphetamine. Concentrations of THCCOOH were similar in fingernails and toenails, and for other drugs no comparison was available.

In the end, it is difficult to reach a conclusion about the distribution of drugs in fingernails and toenails since most of the available studies analyzed a too limited number of samples to be representative and in cases where an appropriate number of samples was analyzed, the samples were not compared in a pairwise manner. Furthermore, the interpretation is complicated by the different and wide windows of detection of these two nail samples, and the difficulty of controlling the variability in drug consumption before sample collection.

10.4.2 Nail Clippings or Nail Scrapings?

Nail scrapings are collected by scraping the surface of the nail with a scalpel. They have been rarely described as another method of collecting nail samples, as opposed to[38] or to complement nail clipping collection.[42,45]

However, nail scrapings are considered a less suitable sample for two main reasons: environmental contamination is higher, as scrapings are usually collected from the external layers of the nail, and the investigated drugs could be more easily washed out by daily hygiene and by the analytical decontamination procedure, which could even lead to the drugs being undetectable.[42,51]

Supporting these statements, Madry *et al.* published two studies were drug concentrations in nail scrapings and clippings were compared.[42,45] In the first work, the authors analyzed toenail scrapings and clippings from cocaine users.[45] Nail clippings were washed twice with water/acetone; however, for scraping samples, the surface of the toenails was just wiped with acetone before sample collection to avoid cocaine extraction from the pulverized nail matrix during the wash procedure. Higher cocaine concentrations were found in nail scrapings, which, together with a much lower BE/cocaine ratio, was attributed to the greater environmental contamination of the scrapings. Furthermore, a postmortem fingernail from a cocaine user was scraped in nine different layers, with the highest cocaine concentration found in the dorsal layer, attributed to external contamination, with decreasing concentrations until the last two layers at the ventral side, where concentrations increased again, coming from the nail blood supply. In a second study, MDMA and MDA were determined in toenails clippings and scrapings from subjects participating in a controlled MDMA administration study.[42] Higher MDA and MDMA concentrations were observed in clippings than scrapings in most cases, which was explained by daily hygiene extracting more drug from the nail surface represented by nail scrapings.

10.4.3 How Do Nails Compare to Hair?

To consider the usefulness of nails as an alternative matrix to hair it is necessary to collect empirical information about the concentrations found in both matrices to determine if the results are comparable or correlated, and thus, if the nails could be used interchangeably or as a complement to hair analysis.

To date there are some studies comparing concentrations in hair and nails. However, conclusions from these data required to consider some questions:

- The study design:
  - What is the chosen cohort: is it a retrospective study of a big population, a smaller study in a selected cohort of drug users/treated patients, or a case report of one or a few cases, in living or autopsy cases?
  - What is the selected nail sample being compared to hair? Fingernails, toenails, both; or is it not specified? What is the length of the hair and nail samples analyzed? Are the samples segmented?

    ○ Are the drugs administered in a controlled study, is it an uncontrolled study of drug users with self-reported dosage, or no data about consumption habits are reported?
- The description of the results:
    ○ Is the comparison between drug concentrations in the two matrices performed on a case-by-case basis?
    ○ Is the concentration data detailed for each subject or are results expressed as mean/median and concentration ranges for the whole population of the study?

For a general comparison of hair and nail concentrations, the ideal is to study a cohort big enough to be statistically significant while allowing for the description of the results for each individual. A controlled administration with a long study period or having good retrospective data about drug consumption is important due to the long detection windows of these matrices, and choosing fingernails or toenails to compare with hair can affect the comparison. Case reports, while not so useful to draw general conclusions, are interesting to describe specific circumstances that could alter the results.

10.4.3.1 Single Dose Administration

Studies designed to determine the concentration of a certain drug in nails after a single administration are a good way of avoiding variability due to changes in the uptake across the months previous to sample collection. Four studies have been published comparing nail and hair concentrations after single dose drug administration. Madry *et al.*[22] and Hang *et al.*[21] evaluated the distribution of zolpidem, providing information about the three routes of drug incorporation into the nail for the first time. In both studies, when comparing hair and fingernail results in the segments corresponding to the time of consumption, higher concentrations were observed in hair, which was attributed to the high affinity of zolpidem for binding to melanin.

Hang *et al.*[21] also obtained paired toenails and found average concentrations higher in toenails than in fingernails, although still much lower than in hair.

More recently, Kuwayama *et al.*[20,73,74] published three papers measuring concentrations in nails after a single dose administration of different drugs, and analyzed a paired hair sample in two of them. The first study[74] included 12 different drugs and metabolites: ibuprofen, hydroxyibuprofen, 2-carboxy-ibuprofen, fexofenadine, chlorpheniramine, desmethylchlorpheniramine, dihydrocodeine, ephedrine, methylephedrine, acetaminophen, allyl isopropyl acetylurea, and epinastine. Acidic compounds were detected in nails (except ibuprofen and its metabolites) but not in hair, or at very low concentrations (weakly acidic compounds). Basic compounds, except methylephedrine and the zwitterionic compound fexofenadine, were quantified at higher concentrations in hair, attributed to melanin binding.

In another study, Kuwayama *et al.*[73] studied chlorpheniramine concentrations in hair and nail clippings from one volunteer collected 35 days after drug administration. The hair sample was cut lengthwise in 0.4 mm segments, and the central 2 mm portion of the nail was cut lengthwise in 0.2 mm segments. Concentrations in the hair were higher than those observed in the nail. Furthermore, while two peaks could be observed in the concentration-segment curve of the hair sample, attributed to separate incorporation from blood to the hair matrix and from sweat and sebum to the formed hair, a homogeneous distribution of the drug was observed along the nail sample, attributed to incorporation from both the nail matrix and the nail bed.

Finally, Madry *et al.*[42] published a controlled study of MDA and MDMA after two doses of MDMA in 15 subjects. Hair samples of 1–2 cm in length contained concentrations of MDA and MDMA significantly higher than the corresponding nail clippings and scrapings. Enantiomeric ratios (*R*/*S*) and metabolite ratios (MDA/MDMA) were in the same range in hair and nail samples, indicating that incorporation for these drugs in both matrices may be comparable, while the difference in concentrations would be explained by their shifted detection windows.

10.4.3.2 Chronic Drug Use

Tables 10.3 and 10.4 show a summary of the comparison between hair and nail concentrations, and between finger and toenail concentrations, respectively, for different drugs and medicines consumed chronically. Some drugs of abuse or pharmaceuticals showed the same behavior in all subjects studied with respect to hair and nail concentrations. For drugs such as the antipsychotics haloperidol, levomepromazine, clozapine, and quetiapine,[60,62,75–77] the antifungal terbinafine,[78] and cotinine,[30] concentrations were always higher in hair. A probable explanation is the higher affinity of these drugs for melanin, which implies a higher binding to hair. Conversely, other analytes like EtG[23,79–81] and morphine[47,54,69] have been found to be more incorporated to nails. These substances do not show a preference for melanin, and since nails grow slower, the drugs are accumulated for a longer time, amplifying the detected concentrations. For cannabinoids, THCCOOH, THC, CBN and CBD were found at higher concentrations in fingernails than in hair,[34,35] but lower concentrations for THC, CBN and CBD were found in toenails.[35]

Unfortunately, results for all the other studied drugs are equivocal due to a high interindividual variability and differences in study designs. The highest variability was found for amphetamine-type stimulants, with some studies showing different patterns in different subjects for amphetamine and methamphetamine.[36,37] However, comparing the sum of the enantiomers in fingernails or toenails to 5 cm of hair, the concentrations were higher in the hair samples.[36,37] Lin *et al.*[39] also found higher concentrations in 1.5 cm hair samples than in fingernails. Conversely, Cappelle *et al.*[69] found consistently higher concentrations of amphetamine, MDMA and MDEA in fingernails than in hair, but lower concentrations in toenails.

Table 10.3 Summary of published methods comparing hair and nail concentrations for different drugs.[a]

Class	Analyte	Comparison	Samples	Ref.
Hair > nails				
Antipsychotics	Haloperidol		20	76
			20	77
			13	62
	Clozapine		16	60
			1	75
			1	62
	Haloperidol, quetiapine, levomepromazine		13	62
Antidepressants and benzodiazepines	Zolpidem	Hair > nails	7	21
			9	22
			2	56
	Trazodone, sertraline, alprazolam, venlafaxine, paroxetine, fluoxetine		16	56
Other	Terbinafine		12	78
	Cotinine, 3-hydroxycotinine		26	30
Nails > hair				
Alcohol	EtG		529	23
			45	79
			40	80
			21	81
Cannabinoids	THCCOOH	Nails > hair	60	34
Opioids	Morphine		18	54
			18	47
			26	69
No clear pattern				
Cannabinoids	THC, CBD, CBN	Fingernails > hair > toenails	17	35
Amphetamines	Amphetamine, methamphetamine	Similar	7	36
		High variability	15	37
		Hair > fingernails	6	39
		Fingernails, toenails > hair	26	69
	Amphetamine	Similar	1	38
	MDA, MDMA	Hair > nail	13	42
		Similar	1	38
	MDMA, MDEA	Fingernails, toenails > hair	26	69
Cocaine and opioids	Cocaine	Toenails > hair	18	54
	6-AM	Similar	18	54
	6-AM, codein, acetylcodeine	Hair > nails	18	47

Table 10.3 (*Continued*)

Class	Analyte	Comparison	Samples	Ref.
	Cocaine, BE, norcocaine, cocaethylene	Hair > toenails	20	45
	Cocaine	Hair > fingernails, toenails	26	69
	6-AM, methadone, EDDP, BE, EME	Fingernails, toenails > hair	26	69
	Buprenorphine	Nails > hair	46	50
Antipsychotics	Olanzapine	High variability	13	62

[a] BE: benzoylecgonine; CBD: cannabidiol; CBN: cannabinol; EDDP: 2-ethylidene-1,5-dimethyl-3,3-diphenylpyrrolidine; EME: ecgonine methyl ester; EtG: ethylglucuronide; MDA: 3,4-methylendioxiamphetamine; MDEA: 3,4-methylenedioxyethylamphetamine; MDMA: 3,4-methylendioximethamphetamine; THCCOOH: 11-nor-9-carboxy-Δ9-tetrahydrocannabinol; THC: Δ9-tetrahydrocannabinol; 6-AM: 6-acetylmorphine.

Cocaine has been detected in higher concentrations in hair than in fingernails or toenails in three studies,[45,54,69] although a high variability was observed in postmortem samples.[54] Regarding the metabolites BE, norcocaine, EME and cocaethylene, higher concentrations were detected in hair compared to toenails,[45,69] but lower concentrations when compared to fingernails.[69]

While morphine concentrations were consistently higher in nails, the same cannot be said for 6-AM and codeine, which were found in higher concentrations in nails in one study[69] and in hair in another.[47] Hair and nail concentrations of buprenorphine and its metabolite norbuprenorphine were evaluated in one study, in which median concentrations were compared. The median concentration of buprenorphine was higher in nails, while for norbuprenorphine the medians were similar in both matrices.[50]

Finally, for some antidepressants and benzodiazepines, three publications reported the analysis of hair and nail samples. Cobo-Golpe *et al.*[56] found higher concentrations for trazodone, sertraline, alprazolam, zolpidem, venlafaxine, paroxetine, and fluoxetine in the hair from patients under treatment with these pharmaceuticals. The other two studies[57,65] are reports of specific clinical and postmortem cases, from which definitive conclusions could not be drawn.

10.4.4 Cut-offs in Nails

Due to the differences between hair, fingernail and toenail concentrations, the cut-offs established for hair are not valid to interpret the concentrations found in nails. Few studies proposed cut-offs in nail samples for some analytes, including EtG, THC, amphetamine, cocaine, BE and EME.

All published studies comparing EtG in hair and fingernails found higher concentrations in fingernail samples, and thus proposed that the cut-off for alcohol abstinence or excessive consumption should be higher than the cut-off in hair.[24,80,82] Berger *et al.*[82] suggested a cut-off of 56 pg mg^{-1} for excessive alcohol consumption, while Cappelle *et al.*[24] recommended a value of

Table 10.4 Summary of published methods comparing fingernail and toenail concentrations for different drugs.[a]

Class	Analyte	Comparison	Samples	Ref.
Fingernails > toenails				
Cocaine and opioids	Cocaine, BE,		14	43
	EME, norcocaine, CE		17	53
			18	69
	Morphine, 6-AM,		17	53
	EDDP	Fingernails > toenails	18	69
Cannabinoids	THC, CBD, CBN		19	35
Other	PCP		3	70
Toenails > fingernails				
Amphetamines	Amphetamine, methamphetamine,		4	36
	MDA, MDEA		18	69
Antidepressants	Fluoxetine, sertraline		3	56
Antipsychotics	Clozapine, levomepromazine, haloperidol	Toenails > fingernails	6	62
Other	Caffeine, amlodipine		1	71
	Methadone		18	69
No clear pattern				
Antidepressants and benzodiazepines	Zolpidem	Toenails > fingernails	4	21
		Similar	2	56
	Venlafaxine	Similar	3	56
	Alprazolam, trazodone	Variable	9	56
Codeine	Codeine	Similar	17	53
		Toenails > fingernails	18	69

[a] BE: benzoylecgonine; CBD: cannabidiol; CBN: cannabinol; CE: cocaethylene; EDDP: 2-ethylidene-1,5-dimethyl-3,3-diphenylpyrrolidine; EME: ecgonine methyl ester; KET: ketamine; MDA: 3,4-methylendioxiamphetamine; MDEA: 3,4-methylenedioxyethylamphetamine; NCOC: norcocaine; NKET: norketamine; PCP: phencyclidine; THC: Δ9-tetrahydrocannabinol; 6-AM: 6-acetylmorphine.

123 pg mg^{-1} as a cut-off for chronic excessive consumption and of 59 pg mg^{-1} for alcohol abstinence. Most recently, a study by Vermeulen *et al.*[83] proposed 7.6 pg mg^{-1} as the cut-off for alcohol abstinence, as this value was the 97.5% percentile of concentrations in fingernails of teetotalers, with a maximum detected concentration of 23 pg mg^{-1}.

For THC, Cobo-Golpe *et al.*[35] found that the concentrations in toenails correlated better with hair concentrations than fingernail concentrations, and suggested a cut-off of 16.5 pg mg^{-1} for chronic cannabis use.

Finally, Cappelle *et al.*[69] found higher concentrations in hair for cocaine, and lower for BE, EME and amphetamine when compared to finger and toenail, and proposed the following cut-offs in fingernail and toenail respectively: 440 and 150 pg mg^{-1} for cocaine, 175 and 105 pg mg^{-1} for BE, 80 and 40 pg mg^{-1} for EME, and 485 and 505 pg mg^{-1} for amphetamine.

10.4.5 Considerations for the Interpretation of Nail Results

Hair and nails are matrices known to accumulate drugs over long periods of time and from which retrospective information about drug consumption can be obtained. Since both are keratinized matrices, they share common characteristics but differ in some aspects, which will affect drug results interpretation.

One of them is the mechanism of drug incorporation, which in nails is double through the nail bed and nail matrix, while in hair is only performed through the blood capillaries in the follicle. Another difference is the growth rate: while in head hair a rate of 1 cm per month is accepted, nails grow more slowly (about 3 mm per month for fingernails and 1.1 mm per month for toenails). Furthermore, nail growth is constant, unlike hair's cyclical growth with different stages of active growth and rest. All this means that substances are incorporated constantly all along the nail and during a longer period of time, which can be translated into an amplification of the concentrations and a lengthening of the window of detection.

Differences in the structure include the presence of melanin in hair, known to influence the incorporation of certain substances depending on their physicochemical properties. Nails of Caucasians only contain melanin in pathological conditions (melanonychia), although in Black, Asian, Hispanic, and Middle Eastern individuals, the presence of melanin is not pathological and is quite common, increasing with age.[10] In general, the absence of melanin is regarded as an advantage because it mitigates inter-individual variability in substance incorporation, although this can also lead to cases where the concentrations incorporated to nails are too low to be detected.

Furthermore, nails do not possess the cuticle layer that protects hair strands from contamination and extraction, increasing the risk of incorporation into nails due to external or sweat contamination, and also increasing the wash out effect of the analytes during decontamination procedures[51] and even normal hygiene.

The effect of cosmetic treatment in hair drug concentrations is well documented,[84] but the effect of treatments such as nail polish, nail polish remover, or treatment for onychomycosis has barely been considered. Most studies opt for excluding participants with nail treatment. Only Cobo-Golpe *et al.*[35] documented a few cases, observing no discernable difference in concentrations after cosmetic treatment, but a marked decrease in concentrations compared to what was expected in a case of onychomycosis treatment.

Acknowledgements

The authors would like to acknowledge the Xunta de Galicia (Galicia, Spain) for the Consolidation and Structuring of Competitive Research Units Help (2021-PG054). María Cobo-Golpe would like to thank her "Margarita Salas" postdoctoral contract (Ministerio de Universidades, Spanish Government).

References

1. R. Baran, R. P. R. Dawber and E. Haneke, *Skinmed*, 2005, **4**, 18.
2. S. Yaemsiri, N. Hou, M. M. Slining and K. He, *J. Eur. Acad. Dermatol. Venereol.*, 2010, **24**, 420.
3. A. Palmeri, S. Pichini, R. Pacifici, P. Zuccaro and A. Lopez, *Clin. Pharmacokinet.*, 2000, **38**, 95.
4. D. Cappelle, M. Yegles, H. Neels, A. L. N. van Nuijs, M. De Doncker, K. Maudens, A. Covaci and C. L. Crunelle, *Forensic Toxicol.*, 2015, **33**, 12.
5. A. Rossi, L. Barbieri, G. Pistola, P. Bonaccorsi and S. Calvieri, *J. Appl. Cosmetol.*, 2002, **21**, 1.
6. D. A. R. de Berker, J. André and R. Baran, *Int. J. Cosmet. Sci.*, 2007, **29**, 241.
7. N. Zaias, *Arch. Dermatol.*, 1963, **87**, 37.
8. M. Johnson and S. Shuster, *Br. J. Dermatol.*, 1993, **128**, 277.
9. M. R. Baumgartner, *Bioanalysis*, 2014, **6**, 2189.
10. V. A. Hill, G. N. Stowe, R. B. Paulsen and M. Schaffer, *J. Anal. Toxicol.*, 2018, **42**, 425.
11. J. Jefferson and P. Rich, *Dermatol. Res. Pract.*, 2012, **2012**, 952186.
12. A. W. Kopf and E. Waldo, *Australas. J. Dermatol.*, 1980, **21**, 59.
13. A. S. Geyer, N. Onumah, H. Uyttendaele and R. K. Scher, *J. Am. Acad. Dermatol.*, 2004, **50**, 229.
14. L. F. Edwards and R. G. Schott, *Ohio J. Sci.*, 1937, **37**, 91.
15. W. Clark and L. Buxton, *Br. J. Dermatol. Syph.*, 1938, **50**, 221.
16. J. B. Hamilton, H. Terada and G. E. Mestler, *J. Gerontol.*, 1955, **10**, 401.
17. B. Geoghegan, D. F. Roberts and M. R. Sampford, *J. Appl. Physiol.*, 1958, **13**, 135.
18. K. M. Donovan, *Br. J. Dermatol.*, 1977, **96**, 507.
19. A. Abraham, G. Roga and N. Thomas, *Indian J. Dermatol.*, 2015, **60**, 562.
20. K. Kuwayama, H. Miyaguchi, T. Kanamori, K. Tsujikawa, T. Yamamuro, H. Segawa, Y. Okada and Y. T. Iwata, *Anal. Chim. Acta*, 2020, **1108**, 89.
21. C. Hang, X. Ping and S. Min, *Anal. Bioanal. Chem.*, 2013, **405**, 7281.
22. M. M. Madry, A. E. Steuer, T. M. Binz, M. R. Baumgartner and T. Kraemer, *Drug Test. Anal.*, 2014, **6**, 533.
23. J. Jones, M. Jones, C. Plate, D. Lewis, M. Fendrich, L. Berger and D. Fuhrmann, *Am. J. Anal. Chem.*, 2012, **3**, 83.
24. D. Cappelle, H. Neels, M. Yegles, E. Fransen, K. Dueffels, S. Bremenfeld, K. E. Maudens, A. L. N. van Nuijs, A. Covaci and C. L. Crunelle, *Forensic Toxicol.*, 2016, **34**, 158.

25. A. Keten, C. Zeren, M. M. Arslan, N. Daglioglu, R. Karanfil and B. B. Sen, *Rom. J. Leg. Med.*, 2013, **21**, 67.
26. L. Morini, M. Colucci, M. G. Ruberto and A. Groppi, *Anal. Bioanal. Chem.*, 2012, **402**, 1865.
27. W. K. Al-Delaimy, G. N. Mahoney, F. E. Speizer and W. C. Willett, *Cancer Epidemiol., Biomarkers Prev.*, 2002, **11**, 1400.
28. I. Stepanov, R. Feuer, J. Jensen, D. Hatsukami and S. S. Hecht, *Cancer Epidemiol., Biomarkers Prev.*, 2006, **15**, 2378.
29. K. Schütte-Borkovec, C. W. Heppel, A. Heling and E. Richter, *Biomarkers*, 2009, **14**, 278.
30. J. Kim, H. Cho, J. H. Suh, J. Lee, E. Lee, C. H. Jin, Y. Wang, S. Cha, H. Im and S. B. Han, *Molecules*, 2020, **25**, 1763.
31. I. Stepanov and S. S. Hecht, *Cancer Epidemiol., Biomarkers Prev.*, 2008, **17**, 945–948.
32. N. P. Lemos, R. A. Anderson and J. R. Robertson, *J. Anal. Toxicol.*, 1999, **23**, 147.
33. J. Y. Kim, J. C. Cheong, M. K. Kim, J. I. Lee and M. K. In, *Arch. Pharm. Res.*, 2008, **31**, 805.
34. J. Jones, M. Jones, C. Plate and D. Lewis, *Am. J. Anal. Chem.*, 2013, **4**, 1.
35. M. Cobo-Golpe, A. de-Castro-Ríos, A. Cruz, M. López-Rivadulla and E. Lendoiro, *J. Anal. Toxicol.*, 2021, **45**, 969.
36. O. Suzuki, H. Hattori and M. Asano, *Forensic Sci. Int.*, 1984, **24**, 9.
37. S. Suzuki, T. Inoue, H. Hori and S. Inayama, *J. Anal. Toxicol.*, 1989, **13**, 176.
38. V. Cirimele, P. Kintz and P. Mangin, *Arch. Toxicol.*, 1995, **70**, 68.
39. D. Lin, R. Yin, H. Liu, C. Wang and R. H. Liu, *J. Anal. Toxicol.*, 2004, **28**, 411.
40. J. Y. Kim, S. H. Shin and M. K. In, *Forensic Sci. Int.*, 2010, **194**, 108.
41. J. Y. Kim, J. C. Cheong, J. I. Lee, J. H. Son and M. K. In, *J. Forensic. Sci.*, 2012, **57**, 228.
42. M. M. Madry, A. E. Steuer, C. M. Hysek, M. E. Liechti, M. R. Baumgartner and T. Kraemer, *Anal. Bioanal. Chem.*, 2016, **408**, 545.
43. D. Garside, J. D. Ropero-Miller, B. A. Goldberger, W. F. Hamilton and W. R. Maples, *J. Forensic Sci.*, 1998, **43**, 974.
44. S. Valente-Campos, M. Yonamine, R. L. de Moraes Moreau and O. A. Silva, *Forensic Sci. Int.*, 2006, **159**, 218.
45. M. M. Madry, A. E. Steuer, M. Vonmoos, B. B. Quednow, M. R. Baumgartner and T. Kraemer, *Bioanalysis*, 2014, **6**, 3183.
46. C. Ragoucy-Sengler and P. Kintz, *J. Anal. Toxicol.*, 2005, **29**, 765.
47. M. Shen, H. Chen and P. Xiang, *J. Chromatogr. B*, 2014, **967**, 84.
48. R. Scendoni, E. Bury, E. Buratti, R. Froldi, M. Cippitelli, G. Mietti and M. Cingolani, *Toxics*, 2022, **10**, 420.
49. T. P. Magalhães, S. Cravo, D. D. da Silva, R. J. Dinis-Oliveira, C. Afonso, M. L. Bastos and H. Carmo, *J. Anal. Toxicol.*, 2018, **42**, 192.
50. M. N. Tzatzarakis, E. Vakonaki, L. Kovatsi, S. Belivanis, M. Mantsi, A. Alegakis, J. Liesivuori and A. M. Tsatsakis, *J. Anal. Toxicol.*, 2015, **39**, 313.

51. J. D. Ropero-Miller, B. A. Goldberger, E. J. Cone and R. E. Joseph, *J. Anal. Toxicol.*, 2000, **24**, 496.
52. D. A. Engelhart, E. S. Lavins and C. A. Sutheimer, *J. Anal. Toxicol.*, 1998, **22**, 314.
53. D. A. Engelhart and A. J. Jenkins, *J. Anal. Toxicol.*, 2002, **26**, 489.
54. M. Cingolani, S. Scavella, R. Mencarelli, D. Mirtella, R. Froldi and D. Rodriguez, *J. Anal. Toxicol.*, 2004, **28**, 128.
55. F. Mari, L. Politi and E. Bertol, *Forensic Sci. Int.*, 2008, **179**, 176.
56. M. Cobo-Golpe, A. de-Castro-Ríos, A. Cruz, M. Páramo, M. López-Rivadulla and E. Lendoiro, *Forensic Sci. Int.*, 2021, **326**, 110935.
57. R. C. Irving and S. J. Dickson, *Forensic Sci. Int.*, 2007, **166**, 58.
58. M. Moretti, L. Andrello, S. Visonà, C. Vignali, A. Groppi, F. Freni, A. Osculati, L. Tajana and L. Morini, *J. Pharm. Biomed. Anal.*, 2018, **152**, 137.
59. V. Samanidou, K. Pantazidou, L. Kovatsi, S. Njau and A. Livanos, *J. Sep. Sci.*, 2012, **35**, 839.
60. H. Chen, P. Xiang, Q. Sun and M. Shen, *Yao Xue Xue Bao*, 2012, **47**, 1193.
61. L. Kovatsi, A. Titopoulou, A. Tsakalof and V. Samanidou, *J. Liq. Chromatogr. Relat. Technol.*, 2015, **38**, 1666.
62. M. Cobo-Golpe, A. de-Castro-Ríos, A. Cruz, M. Páramo, M. López-Rivadulla and E. Lendoiro, *J. Pharm. Biomed. Anal.*, 2020, **189**, 113443.
63. D. Cappelle, M. De Doncker, C. Gys, K. Krysiak, S. De Keukeleire, W. Maho, C. L. Crunelle, G. Dom, A. Covaci, A. L. N. van Nuijs and H. Neels, *Anal. Chim. Acta*, 2017, **960**, 101.
64. F. Krumbiegel, M. Hastedt, L. Westendorf, A. Niebel, M. Methling, M. K. Parr and M. Tsokos, *Forensic Sci. Med. Pathol.*, 2016, **12**, 416.
65. F. Krumbiegel, M. Hastedt and M. Tsokos, *Forensic Sci. Med. Pathol.*, 2014, **10**, 496.
66. F. P. Busardò, M. Gottardi, R. Pacifici, M. R. Varì, A. Tini, A. R. Volpe, R. Giorgetti and S. Pichini, *J. Anal. Toxicol.*, 2020, **44**, 69.
67. P. Liu, W. Liu, H. Qiao, S. Jiang, Y. Wang, J. Chen, M. Su and B. Di, *Anal. Chim. Acta*, 2022, **1226**, 340170.
68. F. P. Busardò, M. Gottardi, A. Tini, C. Mortali, R. Giorgetti and S. Pichini, *Molecules*, 2018, **23**, 2686.
69. D. Cappelle, S. De Keukeleire, H. Neels, F. Been, M. De Doncker, G. Dom, C. L. Crunelle, A. Covaci and A. L. N. van Nuijs, *Drug Test. Anal.*, 2018, **10**, 1110.
70. A. J. Jenkins and D. A. Engelhart, *J. Anal. Toxicol.*, 2006, **30**, 643.
71. F. Takahashi, M. Kobayashi, A. Kobayashi, K. Kobayashi and H. Asamura, *Molecules*, 2018, **23**, 3231.
72. I. Shu, J. Jones, M. Jones, D. Lewis and A. Negrusz, *J. Anal. Toxicol.*, 2015, **39**, 624.
73. K. Kuwayama, H. Miyaguchi, Y. T. Iwata, T. Kanamori, K. Tsujikawa, T. Yamamuro, H. Segawa and H. Inoue, *Anal. Chim. Acta*, 2016, **948**, 40.
74. K. Kuwayama, H. Miyaguchi, Y. T. Iwata, T. Kanamori, K. Tsujikawa, T. Yamamuro, H. Segawa and H. Inoue, *Drug Test. Anal.*, 2017, **9**, 571.

75. H. Chen, P. Xiang and M. Shen, *J. Forensic Leg. Med.*, 2014, **22**, 62.
76. T. Uematsu, R. Sato, K. Suzuki, S. Yamaguchi and M. Nakashima, *Eur. J. Clin. Pharmacol.*, 1989, **37**, 239.
77. T. Uematsu, R. Sato, O. Fujimori and M. Nakashima, *Arch. Dermatol. Res.*, 1990, **282**, 120.
78. J. Faergemann, H. Zehender, J. Denouël and L. Millerioux, *Acta Derm. Venereol.*, 1993, **73**, 305.
79. D. Cappelle, H. Neels, S. De Keukeleire, E. Fransen, G. Dom, A. Vermassen, A. Covaci, C. L. Crunelle and A. L. N. van Nuijs, *Forensic Sci. Int.*, 2017, **279**, 187.
80. J. T. Fosen, L. Morini, C. Sempio, N. Giarratana, A. Enger, J. Mørland and G. Høiseth, *Alcohol Alcohol.*, 2017, **52**, 580.
81. R. Paul, L. Tsanaclis, C. Murray, R. Boroujerdi, L. Facer and A. Corbin, *Alcohol Alcohol.*, 2019, **54**, 402.
82. L. Berger, M. Fendrich, J. Jones, D. Fuhrmann, C. Plate and D. Lewis, *Addiction*, 2014, **109**, 425.
83. L. Vermeulen, A. L. N. van Nuijs, C. L. Crunelle, W. Jacobs and H. Neels, *Forensic Sci. Int.*, 2022, **335**, 111278.
84. G. Skopp and L. Pötsch, *Ther. Drug Monit.*, 1997, **19**, 386.

CHAPTER 11

Hair Analysis in Asia

EUNMI KIM

National Forensic Service, Wonju, Republic of Korea
Email: emkim4725@korea.kr

11.1 Introduction

Asia is the largest continent on earth, with an area of about 44 000 000 km^2, which accounts for 30% of the total area of the earth. Its population is about 4.2 billion, or about 60% of the world's population. Regionally, Asia is divided into East Asia, Southeast Asia, South Asia, Central Asia, and West Asia, and there are a total of 47 countries. Korea,[†] Japan, and China, which will be the main focus of this chapter, are all located in East Asia.

Although many countries in Asia each have their own long history, the forensic science network between Asian countries was organized only relatively recently. The Asian Forensic Science Network (AFSN)[1] was formed in 2008 with the purpose of having a platform for forensic scientists in Asia to come together to discuss, share, and advance forensic science. AFSN has grown from six member institutes in six countries in 2008, to 62 member institutes from 17 countries (as of September 2021). AFSN has nine technical workgroups (Crime Scene Investigation, Digital Forensics, DNA, Fingerprint, Forensic Medicine, Illicit Drugs, Toxicology, Trace Evidence, and Questioned Document) and a Quality Assurance and Standard Committee. In addition, the Toxicology Work Group (TXWG) deals with hair analysis related to drug abuse.

Due to the large population in Asia, it is difficult to outline the extent of hair analysis in all of these countries. Therefore, this chapter will focus on the current state of hair analysis in Korea. In addition, commonly abused

[†]Korea, South Korea, and Republic of Korea are all described as Korea in this chapter.

Issues in Toxicology No. 47
Perspectives and Challenges of Hair Analysis
Edited by Pascal Kintz, Alberto Salomone and Marco Vincenti

Published by the Royal Society of Chemistry, www.rsc.org

drugs in Asia will be selected and their hair analytical methods summarized. In other words, methamphetamine (MA), cannabis, and new psychoactive substances (ketamine and synthetic cannabinoids) are reviewed here, and finally, propofol, which was specifically designated as a controlled drug in Korea.

11.2 Methamphetamine Analysis

MA is a potent and highly addictive psychostimulant that was first synthesized in Japan in 1893. It is the most frequently abused drug and represents an increasing trend among youths in East Asian countries, such as Korea, Japan, and China.[2] Many East Asian and Pacific countries have ranked MA as their primary drug of concern. According to the drug report by UNODC,[3] MA continues to be seized mainly in North America and in East and Southeast Asia, which accounted for 49% and 43%, respectively, of the global quantity of MA seized in the period 2015–2019. Several countries including Brunei Darussalam, Cambodia, Malaysia, Singapore, and Korea have seen an rising trend in the use of crystalline MA.

Hair analysis of narcotics was first introduced in Korea in 1993, and before then, urine was the main specimen used in drug testing.[4] When the analytical method was established, the only drug that could be detected by analysis in hair was MA. The established method was to detect MA and its metabolite, amphetamine (AM) in hair, using gas chromatography/mass spectrometry (GC/MS). In 2003, an analytical method was established for methylenedioxymethamphetamine (MDMA), a new psychoactive substance (NPS) with a similar structure to MA, enabling simultaneous analysis of MA and MDMA in hair.[5] In 2006, a simultaneous analytical method for amphetamine-type stimulants (ATS) was established. These drugs were amfepramone, fenfluramine, phentermine (PTM), phendimetrazine, MA, MDMA, and their metabolites norephedrine, norfenfluramine, hydroxyphentermine, phenmetrazine, amphetamine (AM), and methylenedioxyamphetamine (MDA).[6] In 2005, an automated system for analyzing MA and MDMA in hair was developed.[7] By modifying the Chemstation® and applying dynamic data exchange (DDE) technology to the components, this system was designed to automatically collect and process the analytical data obtained from GC/MS. A series of processes such as identifying peaks, extracting ions, and quantifying target drugs were calculated simultaneously, and fully automated quantification reports were generated (Figure 11.1).

The analytical method currently implemented in Korea is GC/MS analysis based on agitation extraction. Briefly, 10–20 mg hair samples are cut into tiny pieces of less than 1 mm. Target drugs from hair are extracted with 1% HCl in methanol in the presence of MA-d_5 and AM-d_5 as internal standards. After derivatization with trifluoroacetic anhydride (TFAA), identification and quantification of MA and AM are performed by GC/MS–selected ion monitoring (SIM) mode. In this method, both the LODs and LOQs of MA and AM are 0.25 and 0.125 ng mg^{-1}, respectively.[8–10]

In a study on the MA concentrations and AM/MA ratios in hair, Han *et al.*[11] from Korea determined the AM/MA ratios in the hair of 2444 MA abusers, and reported the AM/MA ratios were in the range 0.004–1.16 (mean 0.09). When they were divided into six groups according to the concentration of MA, there was a statistically significant difference among the six range groups, namely, in groups of higher MA concentrations, lower ratios of

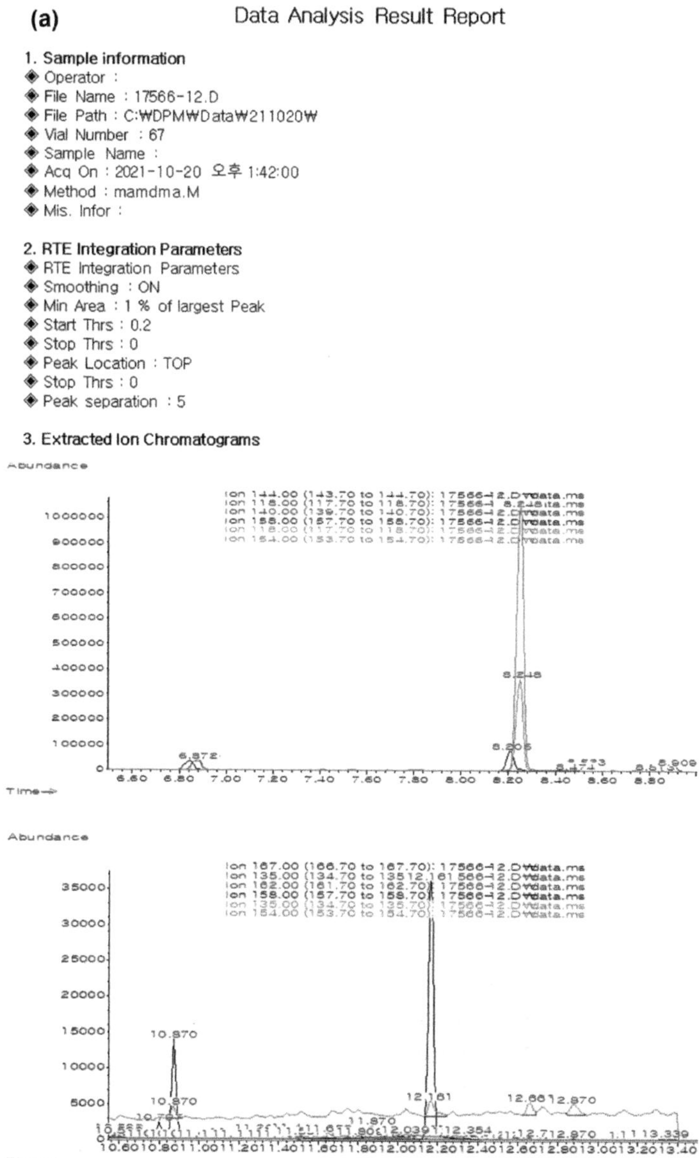

Figure 11.1 Automated quantification report of MA and MDMA.

(b)

4. Target Ion Results

Name	Ion	Ref. R.T.	Margin	R.T.	Corr. Area	Peak Height
AM-d5	144	6.85	0.10	6.84	97532	36462
AM	118	6.87	0.10	6.87	107134	36145
AM	140	6.87	0.10	6.89	108603	41312
MA-d5	158	8.16	0.10	8.21	173412	77534
MA	118	8.24	0.10	8.25	754090	363904
MA	154	8.24	0.10	8.25	2176900	1076160
MDA-d5	167	10.86	0.05	10.87	22281	13679
MDA	135	10.89	0.05	10.87	3054	1792
MDA	162	10.89	0.05	10.93	33	29
MDMA-d5	158	12.17	0.05	12.16	73470	37488
MDMA	162	12.21	0.05	12.23	317	170
MDMA	154	12.21	0.05			

5. Quantification & Qualification Results

Calculation Area	Quan Name	Ratio Ion	Ion Ratio	Con. Spot	211015(Tot	Amount
Quantification	AM/AM-d5	140/144	1.113525,50,100		55.5255	5.5525
Quantification	MA/MA-d5	154/158	12.553325,50,100		707.1915	70.7191
Quantification	MDA/MDA-d5	162/167	0.001525,50,100		0.0000	0.0000
Quantification	MDMA/MDMA-d5	154/158	0.000025,50,100		0.0000	0.0000

6. Data Validation

Calculation Area2	Qual Name	Ratio Ion	STDs ion Ratio	Allowable Range	Min Value	Max Value	Sample Ion Ratio
Qualification	AM/AM	118/140	0.7306±10%	(1.0)	0.6575	0.8036	0.9865
Qualification	MA/MA	118/154	0.4282±20%	(0.3)	0.3426	0.5138	0.3464
Qualification	MDA/MDA	162/135	0.3279±20%	(0.4)	0.2623	0.3934	0.0108
Qualification	MDMA/MDMA	162/154	0.7115±10%	(0.6)	0.6404	0.7827	0.0000

7. Reference

◆ Data Processing date : 2021-10-25 오후 5:03:08
◆ Data Processed by : 국립과학수사연구원 > 독성학과 > 과장 > 보건연구관

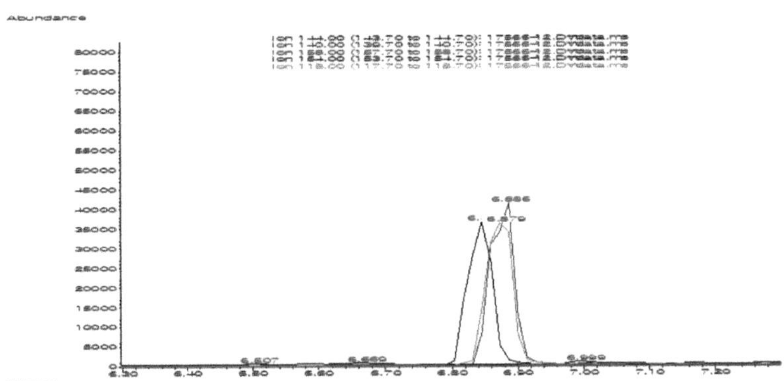

Figure 11.1 (Continued)

AM/MA were found. Meanwhile, the average ratio of MDA/MDMA in 53 hair samples was 0.12, which was significantly higher than that of AM/MA. In another study, MA concentrations in the oral fluid, urine, and hair of 12 drug abusers were compared.[10] It showed a similar pattern to the former study, MA concentrations were reciprocally proportional to the ratios of AM/MA. The average ratios of AM/MA in the urine, oral fluid, and hair samples for 11 drug abusers were 0.12, 0.11, and 0.06, respectively. By statistical analysis, the AM/MA ratio of hair was significantly lower than urine and oral fluid.

In 2011, the National Forensic Service (NFS) of Korea succeeded in developing certified reference material (CRM) for MA using MA abusers' hair,[8,12–14] based on the international standard (ISO 17034). A brief explanation of this CRM preparation process is as follows. (I) Create a hair pool: hair samples were collected from MA abusers' hair specimens, which were submitted to the NFS. All the subjects were Korean with naturally black hair. The roots of the hair were removed, and when it was long enough up to about 12 cm was taken. The specimens, where the MA concentrations ranged from 10 to 20 ng mg^{-1}, were mixed to create a pool. (II) Preparation of candidate hair CRM: the hair was washed in dichloromethane and homogenized in distilled water. After that, it was cut into about 1 mm pieces, sieved, blended, and finally bottled, *ca.* 100 mg per bottle. (III) Quantification of MA and AM in hair by instrumental analysis: the hair sample was accurately weighed (*ca.* 10 mg) and internal standards (MA-d_5 and AM-d_5) were added. Extraction, derivatization, and GC/MS analysis were the same as described above. (IV) Securing traceability using a CRM: Standard Reference Material® 2379 (the National Institute of Standards and Technology, USA) was used to compare the extraction method and to verify the analytical method. (V) Evaluation of the homogeneity and the stability: the homogeneity test was performed by two different methods, one was by GC/MS, which was mentioned above, and the other was by high performance liquid chromatography–tandem mass spectrometry (HPLC–MS/MS) after micropulverized extraction.[15] The stability of MA and AM in hair was evaluated for three months, under two different storage conditions (at room temperature and at 4 °C). (VI) Characterization: the uncertainties in the homogeneity test by two methods were calculated, and the expanded uncertainties of MA and AM were calculated. Finally, the certified values of MA and AM were obtained (Figure 11.2).

This achievement brought financial benefits. The prime example was saving the cost of purchasing expensive standard products, which was

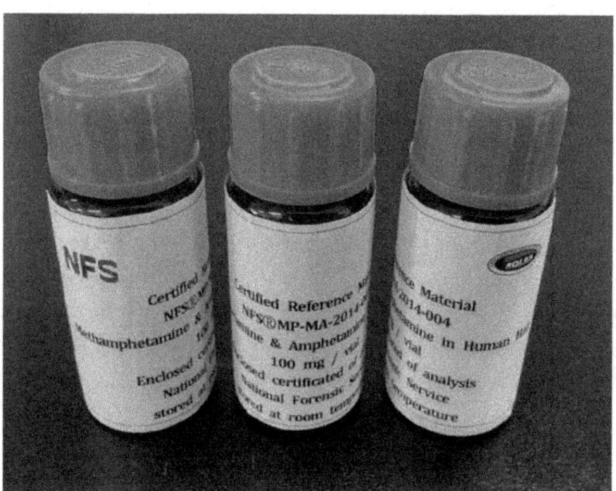

Figure 11.2 Certified reference material (CRM) for MA.

estimated to have saved $60 000 over the last 10 years. It also contributed to the development of forensic science by providing free distribution to developing Asian countries.

Meanwhile, in Japan, the first country with synthetic MA, hair analysis was studied relatively early. As a sensitive method for the analysis of MA and AM in human hair, the GC/chemical ionization (CI)/MS method was presented in 1984.[16] Hair samples were extracted with an organic solvent and derivatized with TFAA before the GC/MS analysis. The detection limit was about 10 pg in an injected volume. This method enabled the detection of MA in a single human hair with high sensitivity. Nakahara[17] discussed pretreatment, extraction, and analytical methods for amphetamines in hair using immunoassay, HPLC, and GC/MS. When compared with three extraction methods, *i.e.*, digestion (alkaline agent), acid extraction, and enzymatic treatment, the acidic method was the most efficient. In addition, Nakahara *et al.*[18] showed the usefulness of stable-isotope dilution GC/MS for precise determination of MA and AM in hair by using MA-d_4 and AM-d_4 in 1991. When a 5–10 mg hair sample was used, the detection limit was regarded as 0.5 ng mg^{-1} using GC/MS–SIM mode.

Even though the GC/MS method is the most popular method for analyzing a very small amount of drugs in hair, recent advances in analytical techniques such as high performance liquid chromatography–tandem mass spectrometry (HPLC–MS/MS) have made it possible to detect extremely small amounts of drugs in hair. A novel micropulverized extraction (MPE) method for HPLC–MS/MS determination of MA and AM in human hair was developed.[15] When MPE was compared with two conventional methods (*i.e.*, acidic methanol extraction and alkaline digestion), no significant differences were observed. They also reported that MPE was suitable for rapid drug screening because it took less than 30 minutes to obtain an HPLC–MS/MS result from hair samples. Moreover, it needed only a small amount of hair (2 mg). LLOQ was estimated at 0.040 ng mg^{-1} for both MA and AM in this method.

In China, hair has become a new biological matrix for the identification of drug addiction since 2017. The Ministry of Public Security of Drug Addiction clearly stipulated that hair samples should be used for testing, providing a scientific basis for law enforcement to identify drug abuse. In this regard, the combined method of high-speed grinding, solvent extraction, and solid-phase microextraction (SPME) was developed.[19] Using LC–MS, three drugs (MA, MDMA, and ketamine) and their metabolites were well determined with the LODs and the LLOQs as 0.067 and 0.2 ng mg^{-1}, respectively. In Korea, a method based on hybrid solid-phase extraction (hybridSPE) was recently developed.[20] This technique minimized the matrix effects, improving GC–MS analysis of hair, the LODs were 0.016 ng mg^{-1} for AM and 0.031 ng mg^{-1} for MA.

Recently, Chen *et al.*[21] from Taiwan developed a sensitive LC–MS/MS assay using microwave-assisted extraction and dansyl chloride derivatization in a single hair. A total of eight illicit drugs: MA, AM, MDMA, MDA, morphine,

6-acetylmorphine, ketamine (KT), and norketamine (NK) were simultaneously analyzed in a micro-segment of a single hair (0.4 mm). The LODs were in the range 0.015–0.050 ng mg^{-1} and LOQs were in the range 0.045–0.125 ng mg^{-1}.

11.3 Cannabinoid Analysis

Cannabis is the second most commonly abused illicit drug in Korea after MA, and its use has been controlled since 1976. 11-Nor-9-carboxy-tetrahydrocannabinol (THC-COOH) is a metabolite of tetrahydrocannabinol (THC); because of its acidic property, THC-COOH is rarely incorporated into hair.[22] Therefore, THC-COOH is found only in very low concentrations in hair, and highly sensitive analytical techniques are required to detect it.

In 2011, two different forensic laboratories in Korea simultaneously developed new methods for the determination of THC-COOH in hair using GC–MS/MS with negative ion chemical ionization (NCI) mode.[23,24] Since the limit of quantification (LOQ) of THC-COOH with these methods was 0.00005 ng mg^{-1} (0.05 pg mg^{-1}), Korean forensic labs have set this as the cut-off value for THC-COOH detection and have been performing hair analysis of cannabis abusers with this system. The cut-off value of 0.00005 ng mg^{-1} for THC-COOH in the hair is in agreement with that of the regulations outlined by the Substance Abuse and Mental Health Services Administration (SAMHSA) of the United States,[25] and with the recommended value from the Society of Toxicological and Forensic Chemistry (GTFCh).[26] However, a four times higher cut-off value of 0.0002 ng mg^{-1} (0.2 pg mg^{-1}) is recommended by the Society of Hair Testing (SoHT) for THC-COOH in the hair.[27]

Cho *et al.*[28] from Korea, developed a LC–MS/MS analysis method in 2019 with a multiple reaction monitoring (MRM) mode for detection of THC-COOH in the hair of drug abusers. In order to eliminate the high background level of the hair matrix lipids produced from alkaline hydrolysis of the hair, a column-switching valve system was used, which was previously introduced.[29] These samples were analyzed in two steps by the LC–MS system using two different acquisition modes, where MRM was first used for identification and quantitative analysis, and MS^3 was used for reconfirmation of THC-COOH detection.

Figure 11.3 shows representative chromatograms of THC-COOH and THC-COOH-d_3 with two acquisition modes of an authentic hair sample (Figure 11.3A, MRM mode; Figure 11.3B, MS^3 mode). The LOD and LOQ of the method was 0.0001 ng mg^{-1} (0.1 pg mg^{-1}). The concentrations of THC-COOH in the authentic Korean hair samples (766 hair segments of 586 hair samples) were in the range 0.10–27.30 pg mg^{-1}, and its concentrations were classified as low, medium, or high ranges, *i.e.*, 0.10–0.39 pg mg^{-1}, 0.39–1.99 pg mg^{-1}, or 1.99–27.30 pg mg^{-1}, respectively, according to statistical evaluation (Table 11.1).

In addition, Cho *et al.*[30] developed a simultaneous detection method for THC, cannabidiol (CBD), cannabinol (CBN), and THC-COOH in hair using LC–MS/MS. When this method was applied to authentic hair, THC, CBD,

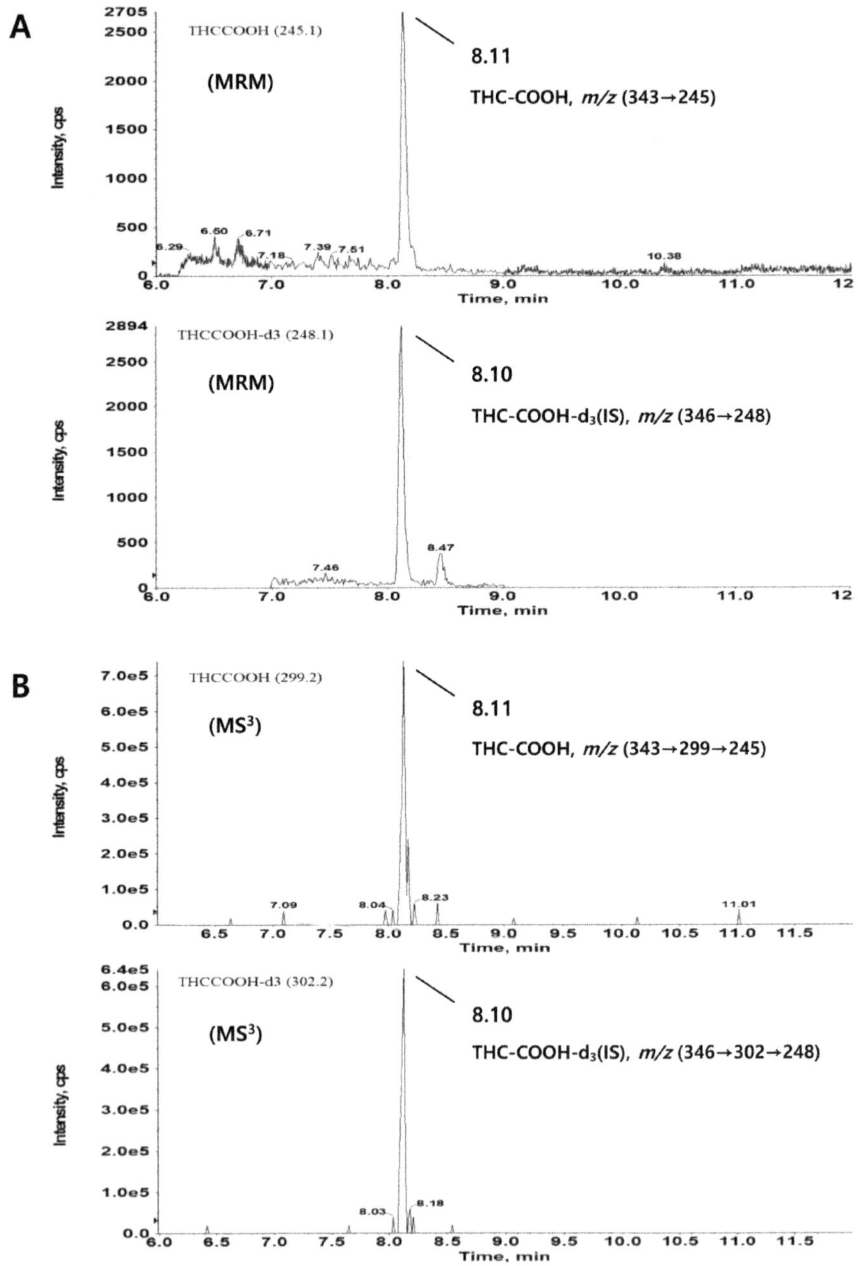

Figure 11.3 Representative chromatograms of THC-COOH and THC-COOH-d_3 in an authentic hair sample of MRM (A), and MS^3 (B) modes. Reproduced from ref. 28 with permission from Elsevier, Copyright 2019.

Table 11.1 THC-COOH concentration in authentic hair samples ($n = 766$ segments). Data from ref. 28.

	THC-COOH (pg mg^{-1})		Interpretation
Cut-off	0.10		
Mean	1.84		
Standard error	0.10		
Minimum	0.10	⇕	Low range
Percentile 25	0.39	↑	
Median	0.87	↓	Medium range
Percentile 75	1.99	⇕	High range
Maximum	27.30		

and CBN, as well as THC-COOH were well detected in cannabis abuse cases (Figure 11.4). The LOQ of THC-COOH was 0.1 pg mg^{-1}, while the LOQs of THC, CBD, and CBN were 5 pg mg^{-1}. Of the total 32 cannabis-positive samples, concentrations of THC-COOH, THC, CBD, and CBN in hair were in the ranges 0.10–5.80, 6.2–8600.0, 8.0–332.5, and 7.6–175.6 pg mg^{-1}, respectively.

Similarly, a simultaneous analytical method for three cannabinoids (THC, CBD, and CBN), two THC metabolites (THC-COOH and THC-COOH-glucuronide), and the six amphetamine-type stimulants (ATS) MA, AM, PTM, MDMA, MDA, and methylenedioxyethylamphetamine (MDEA) was developed using liquid chromatography–high resolution mass spectrometry (LC–HRMS).[31] The LOQs were 0.1 pg mg^{-1} for THC-COOH and THC-COOH-glucuronide, 4 pg mg^{-1} for THC, CBD, and CBN, and 10 pg mg^{-1} for the six ATS.

In Japan, Kuwayama *et al.*[32] developed a simple and highly sensitive analytical method for THC-COOH in spiked hair using micropulverized extraction (MPE) and LC–MS/MS in 2015. The cut-off level of THC-COOH was 0.2 pg mg^{-1}, which is recommended by the SoHT. Furthermore, they detected concentrations of unconjugated THC-COOH and conjugated THC-COOH (*i.e.*, THC-COOH-glucuronide) from authentic hair.[33] The percentage of THC-COOH-glucuronide in a total THC-COOH measured by LC–MS/MS following MPE with and without hydrolyzation was approximately 26%.

11.4 Ketamine Analysis

Ketamine (KT), one of the new psychoactive substances (NPS), is used for medical purposes as a sedative for animal anesthesia. However, structurally and pharmacologically it is related to phencyclidine and has been abused as a recreational drug for some time.[34] The Drug Enforcement Administration (DEA) of the USA registered KT as a drug of abuse in 1995 and classified it as

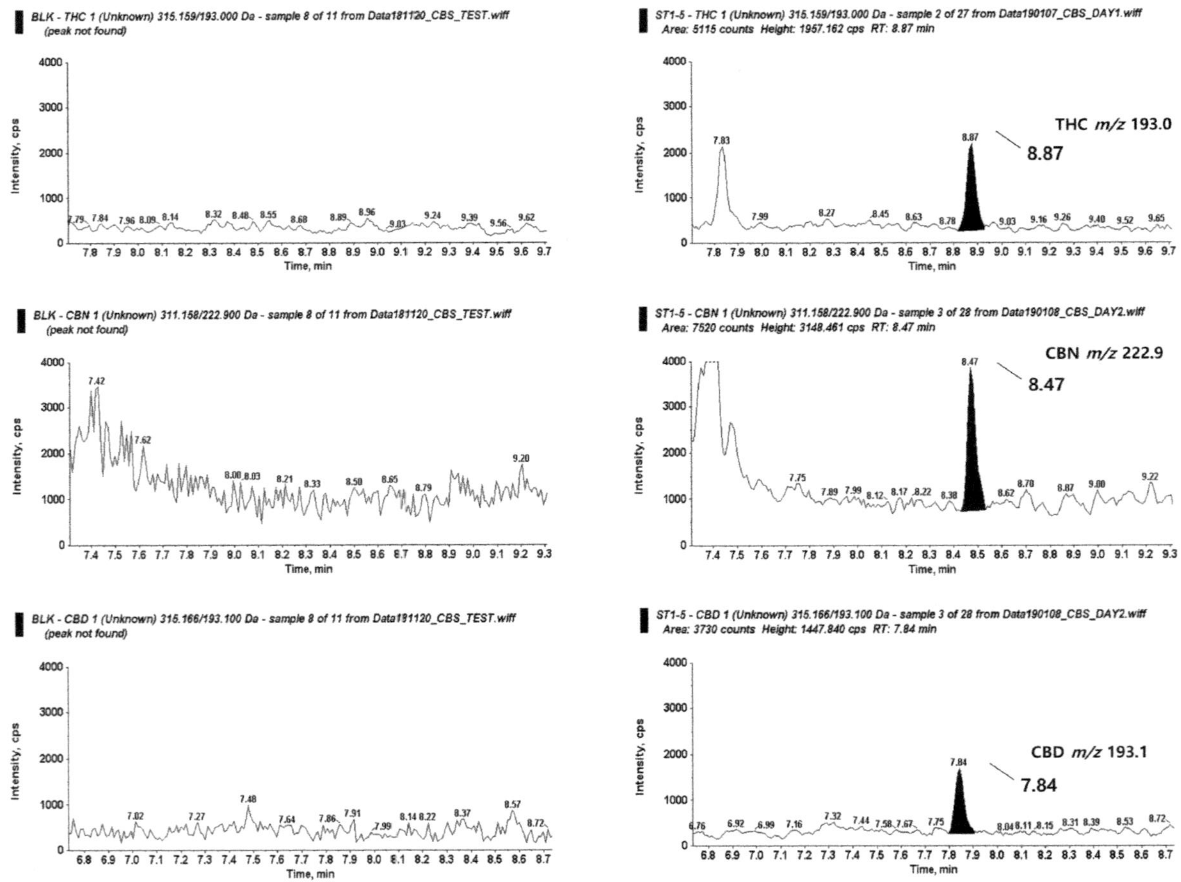

Figure 11.4 Representative chromatograms of THC, CBN and CBD in an authentic hair sample.

Schedule II in 1999. In the case of Korea, KT has been regulated as a controlled drug since November 2005.

Recently, KT has been abused a lot in Asian countries such as China, Taiwan, and Vietnam, and the reason seems to be related to punishment regulations. For example, in Taiwan, KT is classified as Class III and there is no punishment for abuse, and the Taiwanese government is also considering upgrading the regulatory rating of KT to Class II, which is punishable by three years in prison for violations.[35] On the other hand, in relation to punishment, Korea classifies KT as a controlled drug at the same level as MA in the Narcotics Control Act. Penalties are imprisonment of up to 10 years for illegal use, life imprisonment or 5–10 years in prison for possession, and life imprisonment or more than 5 years in prison for smuggling and manufacturing.[36]

Table 11.2 The number of detection of narcotics from seized materials and urine samples at NFS in the metropolitan area[a] of Korea for four years (2017–2020).[b]

Matrix	Subjects/year	2017	2018	2019	2020	Total	Percent (%)
Seized material	Methamphetamine	767	661	1062	1186	3676	50.3
	Cannabis	274	252	446	496	1468	20.1
	Opium (morphine)	180	220	181	370	951	13.0
	Synthetic opioids	8	9	9	10	36	0.5
	MDMA	18	15	52	80	165	2.3
	Cocaine	4	8	28	17	57	0.8
	Benzodiazepines	107	106	169	132	514	7.0
	Ketamine	9	21	48	88	166	2.3
	LSD	4	10	10	12	36	0.5
	Propofol	11	6	13	26	56	0.8
	GHB	4	5	13	2	24	0.3
	Synthetic cannabinoids	4	1	55	104	164	2.2
	Total	1390	1314	2086	2523	7313	100.0
Urine	Methamphetamine	1086	999	1162	1380	4627	54.3
	Cannabis	364	390	656	641	2051	24.1
	Morphine	3	4	9	2	18	0.2
	Synthetic opioids	10	7	15	15	47	0.6
	MDMA	40	36	66	68	210	2.5
	Cocaine	5	22	13	6	46	0.5
	Benzodiazepines	253	301	397	322	1273	15.0
	Ketamine	14	19	32	50	115	1.4
	LSD	11	14	13	7	45	0.5
	Propofol	12	16	17	7	52	0.6
	GHB	1	2	7	1	11	0.1
	Synthetic cannabinoids	0	0	9	11	20	0.2
	Total	1799	1810	2396	2510	8515	100.0

[a] The data is limited to the metropolitan area (Seoul, Incheon, and Gyeongi-do) of Korea.
[b] MDMA: methylenedioxymethamphetamine; LSD: lysergicaciddiethylamide; GHB: gamma-hydroxybutyricacid.

The number of detections of narcotics from seized materials and urine samples in the metropolitan area of Korea from 2017 to 2020 is shown in Table 11.2. MA was the most detected drug in both seized materials and urine samples, as it comprised 50% of the total. The second was cannabis, which accounted for 20 and 24% of seized materials and urine samples, respectively. The third was opiates in seized materials (13%) and benzodiazepines in urine samples (15%), showing different results. Meanwhile, the detection rates of KT from 2017 to 2020 were relatively low at 2.3 and 1.4% in seized material and urine samples, respectively, but the detection rate tended to increase year by year.

Park *et al.*[37] investigated concentrations of KT and NK in hair samples from 61 subjects, which detected KT at the National Forensic Service (NFS) in Korea from January to December 2020. The results showed KT was more abused by men than women, and the average age was 26.5 years, which was estimated to be preferred by young people in their 20s. In 104 KT-positive hair or hair segments obtained from 61 KT abusers, the KT concentration was found in the range 0.50–88.33 ng mg^{-1} (mean 9.07; median 4.21), and the NK concentration was found in the range 0.02–73.64 ng mg^{-1} (mean 2.52; median 0.88) (Figure 11.5). The average ratio of NK to KT concentrations in the hair was 0.47 (median 0.22).

Regarding the analysis of KT in biological samples, an analytical method by Positive Chemical Ionization–Gas Chromatography/Mass Spectrometry (PCI–GC/MS) after extraction with a solid phase extraction (SPE) from urine was developed in 2008.[38] In addition, an analytical method by GC/MS was established after SPE.[39] For KT analysis in hair, a simultaneous analysis for amphetamine (AM), methamphetamine (MA), methylenedioxyamphetamine (MDA), methylenedioxymethamphetamine (MDMA), and norketamine (NK) by GC/MS using micro-pulverized extraction was established.[40] In 2015, simultaneous analysis of paramethoxyamphetamine (PMA), KT, and NK in hair from dead bodies was developed.[41] Methods using liquid chromatography–tandem mass spectrometry (LC–MS/MS) were also reported.[42,43]

Of the 61 cases, only three cases detected KT solely, while the remaining 58 cases showed abuse with other drugs along with KT, accounting for 95% of the total. MDMA was detected in 52 cases of total, showing a high

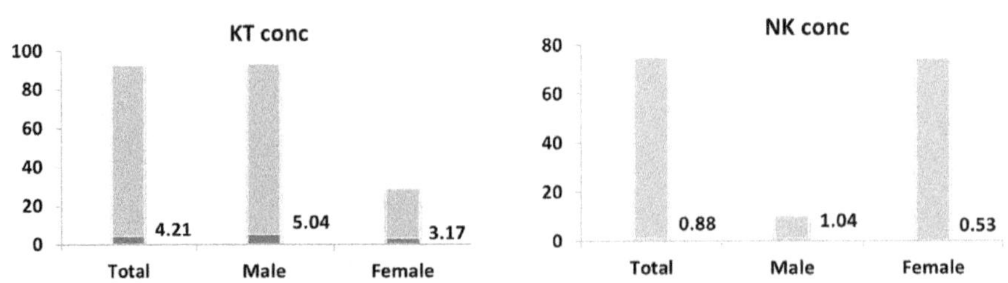

Figure 11.5 Hair concentration data for KT and NK showing concentration ranges and median values in ng mg^{-1}.

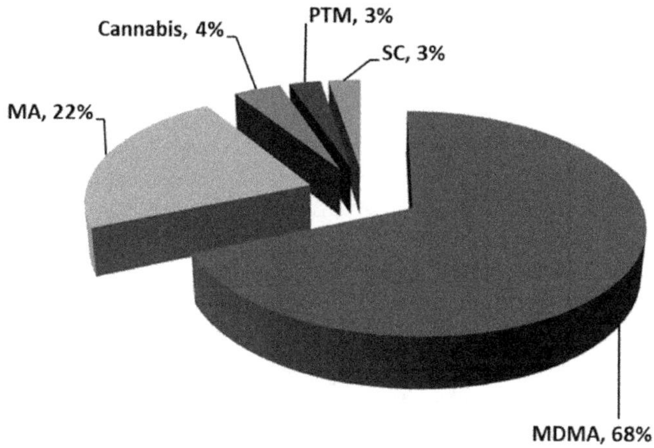

Figure 11.6 Percentage of each drug detected along with KT.

detection rate of 68%, followed by MA with 22% ($n = 17$), cannabis 4% ($n = 3$), phentermine (PTM) and synthetic cannabinoids (SC) with 3% ($n = 2$), respectively (Figure 11.6). In addition, amphetamine-type stimulants (ATS) such as MDMA, MA, and PTM accounted for 93% of the total, indicating that KT was abused with these ATS rather than alone. When comparing past research data by Rhee *et al.*,[43] from 2016 to recently, KT abuse in Korea showed a pattern of co-administration with other drugs rather than a single administration.

Meanwhile, the Academy of Forensic Science (AFS) of Shanghai in China reviewed the prevalence of drugs of abuse by hair analysis from August 2018 to July 2019.[44] Among the 5610 cases, 1713 (30.5%) were positive for drugs of abuse, ATS (57%) was the most frequently detected drug, and the detection rate of KT was 7%. There were 132 cases (7.7%) of the 1713 positive cases detected with one type of drug. The most common combination with KT was MDMA ($n = 51$), followed by amphetamines ($n = 19$), which showed a similar pattern to that of Korea.

11.5 New Psychoactive Substances: Synthetic Cannabinoid Analysis

Since the first detection of JWH-018 in Korea in 2009, NPS abuse has rapidly increased. According to data from the National Forensic Service (NFS), the number of detected cases gradually increased from 8 in 2009 to 22 in 2011. However, the number was 184 in 2019, a 24 times increase compared to a decade ago (Figure 11.7). The detection of NPS by their types between 2016 and 2019 was collated. From a total of 426 cases, 315 cases of synthetic cannabinoids (SC) were detected, accounting for 74% of the total cases (Figure 11.8).[45]

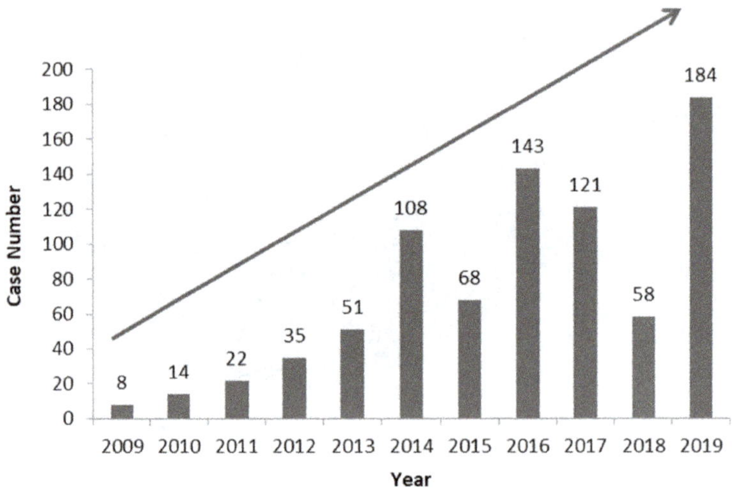

Figure 11.7 The detection of seized material in cases containing NPS at NFS in Korea (2009–2019). Reproduced from ref. 45 with permission from TIAFT, Copyright 2023.

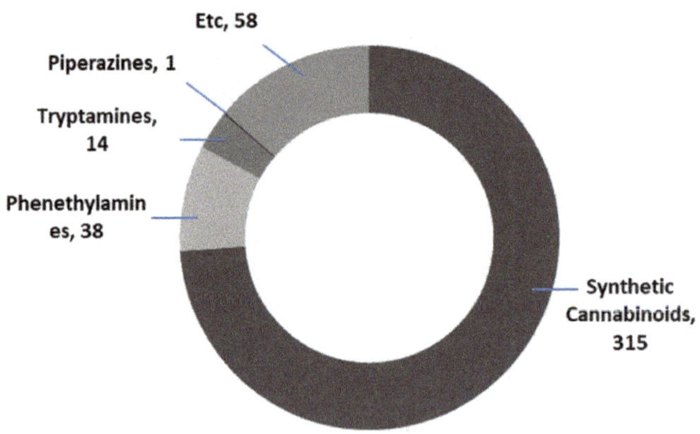

Figure 11.8 The detection of NPS by their types from seized material in cases at NFS in Korea (2016–2019). Reproduced from ref. 45 with permission from TIAFT, Copyright 2023.

The first study on SC analysis in Korea commenced in 2012. A simultaneous analytical method detecting JWH-018, JWH-073 and associated metabolites in hair samples was established in 2013.[46] When this method was applied to the hair samples of 18 abusers, only JWH-018 and its metabolites JWH-018 N-5-OH and JWH-073 were detected. The metabolite of JWH-073 was not detected, showing a different outcome from the analysis using urine samples.[47] Meanwhile, there were no significant differences in SC concentrations between black and white hair samples of experimental animals. In the other study,[48] AM-2201 did not show a significant distribution of concentrations in black and white hair samples of the experimental animals. Considering the relatively high binding power of hair pigments to basic drugs and no effect on acidic or neutral drugs, it can be predicted that hair pigmentation has no influence on the distribution of synthetic cannabinoids. In 2015, a case of XLR-11 and its metabolites in the hair samples of abusers using liquid chromatography–tandem mass spectrometry (LC–MS/MS) was detected.[49] It was applied to 14 hair samples from abusers suspected of XLR-11 use, XLR-11, and four out of the five metabolites tested were detected. In 2017, an analytical method to detect AB-CHMINACA and six of its metabolites (M2, M3A, M4, M5A, M6, and M7) in hair samples of abusers was established.[50] Hair samples from a total of 37 abusers were found to contain AB-CHMINACA and its metabolites M2 and M4, and the results of the quantitative analysis showed AB-CHMINACA as the most commonly detected substance, followed by the metabolite M2. As of June 2021, Cho *et al.*[51] established a simultaneous analytical method using LC–MS/MS for 25 SC and 27 of their metabolites in hair samples, which was based on their previous research.[52]

In 2021, a total of 1127 NPS cases were detected at the NFS in Korea, and among them, 465 cases were detected in seized materials, accounting for 41% of the total. Next, 417 cases of hair samples, and 245 cases of urine were detected, accounting for 37 and 22%, respectively (Figure 11.9).

From 2014 to 2021, a total of 21 SC were detected in the hair of drug abusers in Korea and the highest in 2020 with 64 cases (Table 11.3). Among the detected NPS, AB-CHMINACA was the most detected with 43 cases, followed by, 4F-MDMB-BUTINACA ($n = 39$), MDMB-4en-PINACA ($n = 30$), 5F-ADF ($n = 27$), JWH-210 ($n = 20$), 5F-MDMB-PICA ($n = 20$), and AMB-FUBINACA ($n = 19$) (Figure 11.10).[53] In particular, MDMB-4en-PINACA first appeared in Korea in 2021 and is estimated to be a major abused NPS in the future.

It was found that one type of NPS used to be (solely, mainly) abused. However, unlike in the past it is usual that two or more NPS are being abused

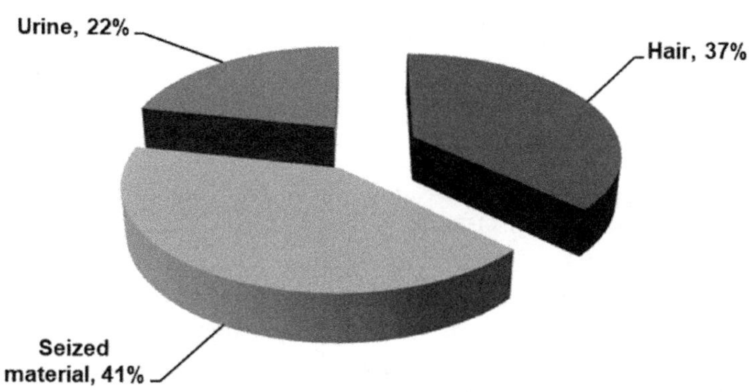

Figure 11.9 The detection number of NPS at NFS in Korea (2021).

Table 11.3 Summary of detected synthetic cannabinoids in authentic hair samples at NFS in Korea (2014–2021).

	Subjects/year	2014	2015	2016	2017	2018	2019	2020	2021	Total
1	AB-CHMINACA			17	12	3	11			**43**
2	JWH-210	5	3	3	8	1				**20**
3	AKB-48			2						**2**
4	XLR-11	2	1	1	1					**5**
5	5F-AKB-48			1	1					**2**
6	PB-22		1	1						**2**
7	AB-PINACA				1					**1**
8	AB-FUBINACA				1					**1**
9	UR-144	11								**11**
10	MAM-2201	4								**4**
11	AM-2201	2								**2**
12	JWH-122	5								**5**
13	JWH-018	5								**5**
14	5F-AB-PINACA		1					1		**2**
15	5F-ADB						18	8	1	**27**
16	5F-MDMB-PICA							19	1	**20**
17	AMB-FUBINACA							13	6	**19**
18	4F-MDMB-BUTINACA							23	16	**39**
19	MDMB-4en-PINACA								30	**30**
20	ADB-BUTINACA								1	**1**
21	4F-ABUTINACA								1	**1**
	Total	**34**	**6**	**25**	**24**	**4**	**29**	**64**	**56**	**242**

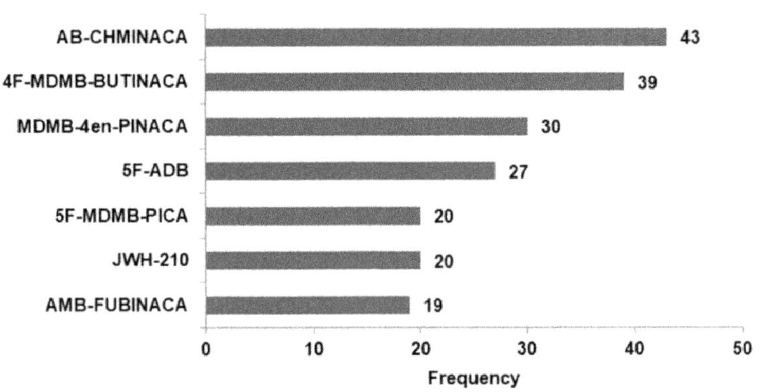

Figure 11.10 Top seven most detected synthetic cannabinoids in hair samples in Korea from 2014–2021.

in a mixed form in recent years. In addition, it used to take at least a year to bring another new drug to the market but new drugs have gained popularity more quickly and changed more frequently in recent years.

Meanwhile, in May 2021, China announced new legislation to control SC as a class all using commonly encountered structural backbones. This has resulted in the emergence of new generations of SC with core components that were previously not encountered and/or not well characterized.

Zhou *et al.*[54] from China developed a UPLC–MS/MS method for quantitative analysis of 29 SC and their metabolites in human hair. A total of 59 actual positive hair samples were collected from 2020 to 2021. Among them, the most frequently detected SC was MDMB-4en-PINACA ($n = 45$), followed by ADB-BUTINACA ($n = 35$). These results show that MDMB-4en-PINACA and ADB-BUTINACA might show a trend of abuse in China. As shown in Table 11.3, the two most detected SC in Korea between 2020 to 2021 were 4F-MDMB-BUTINACA ($n = 39$) and MDMB-4en-PINACA ($n = 30$). This is a prime example of the recent NPS trend similarities between Korea and China.

In order to prove the abuse of NPS, it is important to develop an analytical method from biological samples, which is usually carried out in four steps: information collection, reference standard material (RSM) and metabolites obtainment, method development and validation, and application to authentic samples. There is a long lag time to obtain RSM, making it difficult to develop analytical methods. Considering the characteristics of NPS being abused at an exceptionally fast rate, the government needs to develop a system to shorten the time required to secure RSM and provide budgets to secure research personnel and high-resolution equipment.

11.6 Propofol Analysis

Propofol is a sedative anesthetic used for the induction, maintenance, and sedation of critically ill patients undergoing artificial respiration. However, propofol has become a social problem as it has been abused by Korean celebrities for the purpose of inducing euphoria and hallucinations rather than for medical purposes. In February 2011, Korea was the only country in the world to designate propofol as a controlled drug and began to regulate its use. Therefore, it became urgent to establish an analytical method that can prove propofol administration from abusers. As a result of intensive research conducted in the National Forensic Service (NFS), in January 2013, the detection method of propofol in hair was established for the first time.[55]

This was the first study where propofol-glucuronide, a metabolite of propofol, in hair was proposed as a marker to prove chronic propofol use. An analytical method for the determination of propofol-glucuronide in hair was developed by methanol extraction of hair followed by LC–MS/MS analysis with ESI in negative mode. propofol-glucuronide in hair is confirmed by multiple reaction monitoring (MRM) transitions (*m/z* 353.2/177.2-quantifier), qualifier ion (*m/z* 353.2/175.0-qualifier) of the target drug, and the retention time (Table 11.4). At this time, the detection and quantification limits of propofol were 2 pg mg^{-1} and 5 pg mg^{-1}, respectively, and this method was applied to the hair of 23 propofol abusers in Korea. Among 89 hair segments of 23 hair samples, the concentration of propofol-glucuronide was in the range 5–1410 pg mg^{-1} (mean 321; median 187).

Table 11.4 MRM transitions, retention times of propofol-glucuronide and testosterone-glucuronide-d_3. Reproduced from ref. 57 with permission from the author, Copyright 2014.[a]

Compound name	Retention time (min)	Precursor ion (*m/z*)	Product ion (*m/z*)	
propofol-glucuronide	4.0	353.2	**177.2**	175.0
Testosterone-glucuronide-d_3	3.7	466.0	113.0	85.0

[a] Transitions presented in bold were used for quantification.

Table 11.5 Various cut-off values in specifications for the detection of drugs in three standards from SoHT guidelines, China, and Korea in human hair samples.[a]

		Cut-off (ng mg^{-1})		
Group	Target analyte	SoHT	Shanghai Key Lab (China)	National Forensic Service (Korea)
Amphetamines	Amphetamine	0.2	0.2	0.125
	Methamphetamine	0.2	0.2	0.5
	MDA	0.2	0.2	0.125
	MDMA	0.2	0.2	0.5
Cannabinoids	THC	0.05	0.05	NA
	THC-COOH	0.0002	NA	0.00005 (GC–MS/MS) 0.0001 (LC–MS/MS)
Cocaine	Cocaine	0.5	0.5	0.5
	BZE, EME, CE, NC	0.05	0.05 (BZE)	0.05
Opiates	Morphine	0.2	0.2	0.05
	Codeine	0.2	NA	0.05
	6-monoacetylmorphine	0.2	0.2	0.1

[a] MDA: methylenedioxyamphetamine; MDMA: methylenedioxymethamphetamine; THC: tetrahydrocannabinol; THC-COOH: carboxytetrahydrocannabinol; BZE: benzoylecgonine; EME: ecgoninemethylesther; CE: cocaethylene; NC: norcocaine; NA: not available; GC–MS/MS: gas-chromatography–tandem mass spectromery; LC–MS/MS: liquid chromatography–tandem mass spectrometry.

11.7 Cut-off Values for Drugs in Hair

Although there is a cut-off value for each drug in hair recommended by the SoHT, each country has varied values due to the different analytical conditions such as method, equipment, and even cultural characteristics.

Table 11.5 shows the cut-off values for each drug in hair recommended by the SoHT,[27] Shanghai Key Lab (China),[56] and the National Forensic Service (Korea).[57]

In the case of China, the cut-off value is consistent with the recommendations of SoHT, for ensuring comparability between different labs and consistency with international rules. In Korea, the cut-off value of THC-COOH is set at 0.00005 ng mg^{-1}, when analyzed by GC–MS/MS, and when analyzed by LC–MS/MS, set to 0.0001 ng mg^{-1}. These values are lower than those of SoHT (0.0002 ng mg^{-1}). Meanwhile, both MA and MDMA cut-off values in Korea are 0.5 ng mg^{-1}, which is higher than those of SoHT (Table 11.5).

11.8 Future Technology

Research on hair metabolomics using animals is being actively conducted.[58] One of the findings was that metabolic alterations in hair could provide insight into metabolic perturbation over a longer period than other conventional biological samples (*i.e.*, plasma, urine). Kim *et al.*[59] investigated human hair following heavy MA abuse using mass spectrometry and integrated network analysis. Considering the data from the targeted and untargeted metabolomics, the biosynthesis or metabolism of lipids as well as the metabolism of amino acids was significantly affected by heavy MA abuse. These new analytical techniques could be a groundbreaking approach with significant potential for the evaluation of human pathological conditions and the determination of drug addiction.

Meanwhile, Ueki *et al.*[60] reviewed the development of the mass spectrometry imaging (MSI) technique. MSI enables the visualization of the history of drug use in hair samples, thus the imaging analytical technology is expected to be used in a wider range of fields, including the monitoring of the history of health conditions.

Recently, Kuwayama *et al.*[61] investigated drug distribution in scalp hair using diphenhydramine (DP) and lidocaine (LD) by micro-segmental hair analysis combined with LC–MS/MS, which visualizes drug distribution at 0.4 mm intervals in individual hairs. As the results showed, DP and LD were predominantly incorporated into scalp hair *via* the bloodstream and *via* sweat/sebum, respectively, showing double-peak profiles. Through the micro-segmental analysis, the distribution profile clearly distinguished drug administration from external drug adsorption (contamination). This technology would be useful to elucidate how drugs are incorporated into hair and to estimate the circumstances of drug use in drug-related crimes.

11.9 Conclusions

Asian countries have been using their own analytical methods by referring to international guidelines. Korea is also continuously developing new analytical methods, such as simultaneous analysis by increasing the number of abusers and drugs.

Since the outlined hair analysis of substances is limited to the countries described here, it only partially underlines the current situation of hair analysis in Asia. Other drugs not covered in this chapter are expected to be discussed in the future.

Recently, new analytical technologies in hair such as metabolomics, imaging analysis, and single hair analysis are being actively developed in Asia. This research is expected to be extremely useful, not only in forensic toxicology, but also in clinical medicine.

As He and Li[62] pointed out in their review, to minimize the risk of error, forensic laboratories must continuously develop standard methods and establish processes to ensure comparability, consistency, and traceability of forensic outcomes. Although it will be time consuming, the forensic laboratories of the world will need a process of standardization of analytical methods and to coordinate with each other with a common rule. International meetings provide a place for countries to share new information and cooperate, and Asian countries have to actively participate in expanding and strengthening their knowledge of forensic science.

References

1. https://www.asianforensic.net
2. N. Kwon and E. Han, A commentary on the effects of methamphetamine and the status of methamphetamine abuse among youths in South Korea, Japan, and China, *Forensic Sci. Int.*, 2018, **286**, 81.
3. World Drug Report 2021: Booklet 4 – Drug market trends: Cocaine, Amphetamine-type stimulants. By United Nations Office on Drugs and Crime (UNODC), https://www.unodc.org/unodc/en/data-and-analysis/wdr-2021_booklet-4.html
4. Y. Yoo, H. Chung, H. Choi, H. Lee and C. Kim, Determination of methamphetamine in the hair of Korean abuser by GC/MS, *Annu. Rep. N.I.S.I.*, 1994, **26**, 155, (Korean, an in-house publication).
5. M. Park, E. Kim, M. Lim and H. Chung, Analysis of 3,4-Methylenedioxymethamphetamine (MDMA) in Human Hair by Gas Chromatography/Mass Spectrometry, *Yakhak Hoeji*, 2003, **47**(4), 195, (Korean).
6. E. Han, S. Lee, J. Lee, S. Back, D. Kim, S. Woo, C. Kim, Y. Park, M. Lim and H. Chung, Simultaneous Analysis of Amphetamine Type Stimulants in Hair by GC/MS, *Annu. Rep. N.I.S.I.*, 2007, **39**, 240, (Korean, an in-house publication).
7. S. Choi, S. Kim, J. Lee, E. Han, E. Kim, Y. Park, S. Baeck, H. Lee and M. Lim, Development of Raw data Management system for hair Analysis, *Annu. Rep. N.I.S.I.*, 2006, **38**, 282, (Korean, an in-house publication).
8. S. Lee, Y. Park, W. Yang, E. Han, S. Choe, S. In, M. Lim and H. Chung, Development of a reference material using methamphetamine abusers' hair samples for the determination of methamphetamine and amphetamine in hair, *J. Chromatogr. B*, 2008, **865**, 33.

9. E. Han, W. Yang, J. Lee, Y. Park, E. Kim, M. Lim and H. Chung, Correlation of methamphetamine results and concentrations between head, axillary, and pubic hair, *Forensic Sci. Int.*, 2005, **147**, 21.
10. E. Kim, J. Lee, H. Choi, E. Han, Y. Park, H. Choi and H. Chung, Comparison of methampheteamine concentrations in oral fluid, urine and hair of twelve drug abusers using solid-phase extraction and GC–MS, *Ann. Toxicol. Anal.*, 2008, **20**(3), 145.
11. E. Han, Y. Park, W. Yang, J. Lee, S. Lee, E. Kim, M. Lim and H. Chung, The study of metabolite-to-parent drug ratios of methamphetamine and methylenedioxymethampheteamine in hair, *Forensic Sci. Int.*, 2006, **161**, 124.
12. S. Lee, Y. Park, E. Han, S. Choe, M. Lim and H. Chung, Preparation and application of a fortified hair reference material for the determination of methamphetamine and amphetamine, *Forensic Sci. Int.*, 2008, **178**, 207.
13. S. Lee, Y. Park, W. Yang, E. Han, S. Choe, M. Lim and H. Chung, Estimation of the measurement uncertainty of methamphetamine and amphetamine in hair analysis, *Forensic Sci. Int.*, 2009, **185**, 59.
14. S. Lee, H. Miyaguchi, E. Han, E. Kim, Y. Park, H. Choi, H. Chung, S. Oh and K. Chung, Homogeneity and stability of a candidate certified reference material for the determination of methamphetamine and amphetamine in hair, *J. Pharm. Biomed. Anal.*, 2010, **53**, 1037.
15. H. Miyaguchi, M. Kakuta, Y. T. Iwata, H. Matsuda, H. Tazawa, H. Kimura and H. Inoue, Development of a micropulverized extraction method for rapid toxicological analysis of methamphetamine in hair, *J. Chromatogr. A*, 2007, **1163**, 43.
16. O. Suzuki, H. Hattori and M. Asano, Detection of methamphetamine and amphetamine in a single human hair by gas chromatography/chemical ionization mass spectrometry, *J. Forensic Sci.*, 1984, **29**(2), 611.
17. Y. Nakahara, Detection and diagnostic interpretation of amphetamines in hair, *Forensic Sci. Int.*, 1995, **70**, 135.
18. Y. Nakahara, K. Takahashi, M. Shimamine and Y. Takeda, Hair analysis for drug abuse, I. Determination of methamphetamine and amphetamine in hair by stable isotope dilution gas chromatography/mass spectrometry method, *J. Forensic Sci.*, 1991, **36**, 70.
19. L. Meng, Y. Dai, C. Chen and J. Zhang, Determination of amphetamines, ketamine and their metabolites in hair with high-speed grinding and solid-phase microextraction followed by LC–MS, *Forensic Sci. Res.*, 2021, **6**(3), 273.
20. N. Kwon, Y. Lee, H. Kim, J. Cheong and J. Kim, Hybrid Solid-Phase Extraction for Selective Determination of Methamphetamine and Amphetamine in dyed Hair by Using Gas Chromatography–Mass Spectrometry, *Molecules*, 2019, **24**, 2501.
21. H. C. Chen, P. D. Lee and Y. Z. Chang, Development of a rapid and sensitive LC–MS/MS assay for the quantification of commonly abused drugs in Asia in a micro-segment of a single hair using microwave-

assisted extraction and dansyl chloride derivatization, *J. Pharm. Biomed. Anal.*, 2022, **213**, 114678.

22. F. Pragst and M. A. Balikova, State of the art in hair analysis for detection of drug and alcohol abuse, *Clin. Chim. Acta*, 2006, **370**, 17.
23. J. Y. Kim, J. C. Cheong, J. I. Lee and M. K. In, Improved gas chromatography–negative ion chemical ionization tandem mass spectrometric method for determination of 11–nor–Δ^9–tetrahydrocannabinol–9–carboxylic acid in hair using mechanical pulverization and bead–assisted liquid–liquid extraction, *Forensic Sci. Int.*, 2011, **206**, e99.
24. E. Han, Y. Park, E. Kim, S. In, W. Yang, S. Lee, H. Choi, S. Lee, H. Chung and J. Song, Simultaneous analysis of Δ^9–tetrahydrocannabinol and 11–nor–9–carboxy–tetrahydrocannabinol in hair without different sample preparation and derivatization by gas chromatography-tandem mass spectrometry, *J. Pharm. Biomed. Anal.*, 2011, **55**, 1096.
25. D. Bush, The U.S. Mandatory Guidelines for Federal Workplace Drug Testing Programs: Current status and future considerations, *Forensic Sci. Int.*, 2008, **174**, 111.
26. F. Mußhoff, H. Sachs and D. Thieme, Anlage zu den Richtlinien der GTFCh zur Qualitätssicherung bei forensisch–toxikologischen Untersuchungen. Anhang B: Qualitätsstandards für spezielle Analyte: 2. Untersuchung von Haarproben, *Toxichem + Krimtech*, 2004, **71**, 140.
27. G. A. A. Cooper, R. Kronstrand and P. Kintz, Society of Hair Testing guidelines for drug testing in hair, *Forensic Sci. Int.*, 2012, **218**, 20.
28. H. S. Cho, B. Cho, J. Sim, S. K. Baeck, S. In and E. Kim, Detection of 11–nor–9–carboxy–terahydrocannabinol in the hair of drug abusers by LC–MS/MS analysis, *Forensic Sci. Int.*, 2019, **295**, 219.
29. M. Park, J. Kim, Y. Park, S. In, E. Kim and Y. Park, Quantitative determination of 11–nor–9–carboxy–tetrahydrocannabinol in hair by column switching LC–ESI–MS^3, *J. Chromatogr. B*, 2014, **947**, 179.
30. B. Cho, H. S. Cho, S. In and E. Kim, Poster presentation at the AAFS, Baltimore, USA, Feb, 2019.
31. Y. Shin, J. Kim, J. Cheong, J. Kim, J. Kim and H. Kim, Liquid chromatography-high resolution mass spectrometry for the determination of three cannabinoids, two (−)-*trans*–Δ^9–tetrahydrocannabinol metabolites, and six amphetamine-type stimulants in human hair, *J. Chromatogr. B*, 2020, **1149**, 122157.
32. K. Kuwayama, H. Miyaguchi, T. Yamamuro, K. Tsujikawa, T. Kanamori, Y. Iwata and H. Inoue, Micro-pulverized extraction pretreatment for highly sensitive analysis of 11–nor–9–carboxy–Δ^9–tetrahydrocannabinol in hair by liquid chromatography/tandem mass spectrometry, *Rapid Commun. Mass Spectrom.*, 2015, **29**, 2158.
33. K. Kuwayama, H. Miyaguchi, T. Yamamuro, K. Tsujikawa, T. Kanamori, Y. Iwata and H. Inoue, Highly sensitive quantification of unconjugated 11–nor–9–carboxy–Δ^9–tetrahydrocannabinol in a cannabis user's hair using micropulverized extraction, *Forensic Sci. Int.*, 2016, **262**, e34.

34. A. Mozayani, Ketamine–Effects on Human Performance and Behavior, *Forensic Sci. Rev.*, 2002, **14**(1–2), 124.
35. L. Y. Hsu, Ketamine use in Taiwan: moral panic, civilizing process, and democratization, *Int. J. Drug Policy*, 2014, **25**, 819.
36. E. Han, N. J. Kwon, L. Y. Feng, J. H. Li and H. Chung, Illegal use patterns, side effects, and analytical methods for ketamine, *Forensic Sci. Int.*, 2016, **268**, 25.
37. Y. Park, H. Baik, J. Rhee and E. Kim, Poster presentation at TIAFT, Virtual meeting, Feb, 2022.
38. E. Kim, J. Lee, S. Choi, M. Lim and H. Chung, Analysis of ketamine and norketamine in urine by automatic solid-phase extraction (SPE) and positive ion chemical ionization–gas chromatography-mass spectrometry (PCI–GC–MS), *Forensic Sci. Int.*, 2008, **174**, 197.
39. SOP of National Forensic Service, QI-NM-11: Detection of ketamine and norketamine in biological samples, 2016 In house documentation, (Korean).
40. J. Y. Kim, S. H. Shin, J. I. Lee and M. K. In, Rapid and simple determination of psychotropic phenylalkylamine derivatives in human hair by gas chromatography–mass spectrometry using micro-pulverized extraction, *Forensic Sci. Int.*, 2010, **196**, 43.
41. M. Jang, W. Yang, S. Jeong, S. Park and J. Kim, A fatal case of para-methoxyamphetamine poisoning and its detection in hair, *Forensic Sci. Int.*, 2016, **266**, e27.
42. H. Choi, S. Kim, S. Ahn, H. Chang, S. Lee and Y. Lee, Development of an analytical method for simultaneous detection of psychotropic phenylalkylamines in hair by LC–MS/MS with a multi-mode reversed-phase column using pH gradient elution, *Forensic Sci. Int.*, 2016, **259**, 69.
43. J. Rhee, J. Lee, J. Kim, M. Jang, I. Shin and S. Lee, Determination of Ketamine and Norketamine in Hair and Evaluation of Polydrug Use in Ketamine Abusers Using Hair Analysis in Korea, *J. Anal. Toxicol.*, 2021, **45**, 993.
44. X. Wang, J. Cui, Y. Zhuo, B. Shen, S. Zhang, W. Liu, M. Shen and P. Xiang, A Retrospective of Prevalence of Drug of Abuse by Hair Analysis in Shanghai using LC–MS–MS, *J. Anal. Toxicol.*, 2020, **44**, 482.
45. B. Cho, J. Kim, H. Baik, J. Rhee and E. Kim, Detection of synthetic cannabinoids in drug abusers' hair, *TIAFT Bull.*, 2021, **51**(3), 11.
46. J. Kim, Y. Park, M. Park, E. Kim and S. Lee, Deposition of JWH-018, JWH-073 and their metabolites in hair and effect of hair pigmentation, *Anal. Bioanal. Chem.*, 2013, **405**, 9769.
47. M. Jang, W. Yang, H. Choi, H. Chang, S. Lee, E. Kim and H. Chung, Monitoring of urinary metabolites of JWH-018 and JWH-073 in legal cases, *Forensic Sci. Int.*, 2013, **231**, 13.
48. J. Kim, Y. Park, M. Park, E. Kim, W. Yang, S. Lee and S. Han, Simultaneous determination of five naphthoylindole-based synthetic cannabinoids and metabolites and their deposition in human and rat hair, *J. Pharm. Biomed. Anal.*, 2015, **102**, 162.

49. M. Park, S. Yeon, J. Lee and S. In, Determination of XRL–11 and its metabolites in hair by liquid chromatography–tandem mass spectrometry, *J. Pharm. Biomed. Anal.*, 2015, **114**, 184.
50. J. Sim, H. S. Cho, J. Lee, S. In and E. Kim, Determination of AB-CHMINACA and its metabolites in human hair and their deposition in hair of abusers, *J. Pharm. Biomed. Anal.*, 2017, **140**, 162.
51. B. Cho, J. Kim, H. Baik, Y. Park, J. Rhee and E. Kim, Poster presentation at the AAFS, Seattle, USA, Feb, 2022.
52. B. Cho, H. S. Cho, J. Kim, J. Sim, I. Seol, S. Baeck, S. In, D. H. Shin and E. Kim, Simultaneous determination of synthetic cannabinoids and their metabolites in human hair using LC–MS/MS and application to human hair, *Forensic Sci. Int.*, 2020, **306**, 110058.
53. J. Lee, W. Yang and E. Kim, Poster presentation at TIAFT, Versailles, France, Sep, 2022.
54. L. Zhou, M. Shen, B. Shen, H. Chen, X. Wang, H. Deng, P. Xiang and Y. Shi, Application of a UPLC–MS/MS method for quantitative analysis of 29 synthetic cannabinoids and their metabolites, such as ADB-BUTINACA and MDMB-4en-PINACA in human hair in real cases, *Forensic Sci. Int.*, 2022, **331**, 111139.
55. J. Kim, S. In, Y. Park, M. Park, E. Kim and S. Lee, Quantitative analysis of propofol-glucuronide in hair as a marker for propofol abuse, *Anal. Bioanal. Chem.*, 2013, **405**(21), 6807.
56. H. Yan, P. Xiang and M. Shen, Current status of hair analysis in forensic toxicology in China, *Forensic Sci. Res.*, 2021, **6**(3), 240.
57. E. Kim, *Understanding of Drug Analysis in Hair*, Design MIP, Seoul, 2014, (Korean).
58. W. Jang, J. Choi, B. Park, J. Seo, Y. Seo, S. Lee, C. Jeong and S. Lee, Hair Metabolomics in Animal Studies and Clinical Settings, *Molecules*, 2019, **24**, 2195.
59. S. Kim, W. Jang, H. Yu, J. Kim, S. Lee, C. Jeong and S. Lee, Revealing Metabolic Perturbation Following Heavy Methamphetamine Abuse by Human Hair Metabolomics and Network Analysis, *Int. J. Mol. Sci.*, 2020, **21**, 6041.
60. R. Ueki, E. Fukusaki and S. Shimma, History of hair analysis by mass spectrometry imaging, *J. Biosci. Bioeng.*, 2022, **133**, 89.
61. K. Kuwayama, H. Miyaguchi, T. Kanamori, K. Tsujikawa, T. Yamamuro, H. Segawa, Y. Okada and Y. Iwata, Distribution profiles of diphenhydramine and lidocaine in scalp, axillary, and pubic hairs measured by micro-segmental hair analysis: good indicator for discrimination between administration and external contamination of the drugs, *Forensic Toxicol.*, 2022, **40**, 64.
62. X. He and C. Li, Development of forensic standards in China: a review, *Forensic Sci. Res.*, 2022, **7**, 1.

CHAPTER 12

Metals in Hair

J. P. GOULLÉ

Laboratoire de Toxicologie, Faculté de Médecine et de Pharmacie, Université de Rouen, 22 Boulevard Gambetta, 76183 Rouen CEDEX 1, France
Email: jean-pierre.goulle@univ-rouen.fr

12.1 Introduction

In 1836, the British chemist James Marsh presented a method of separation which made it possible to measure arsenic (As), even in trace amounts, in organic matter.[1] It was applied to the hair in many famous court cases throughout the nineteenth and twentieth centuries in many countries. This method has also been used to measure As in the hair of exhumed corpses. The interest of analysing mineral substances in the hair has been somewhat forgotten with the development of assay techniques such as atomic absorption spectrometry (AAS) in traditional biological media (blood and urine) during the last century.

Surprisingly, at the end of the century the interest in hair analysis emerged first in forensic toxicology and secondly in clinical toxicology because it is usually the other way around. The first applications did not concern mineral analysis, but drugs. The very active Society of Hair Testing (SoHT) has published guidelines for hair drug testing that provide laboratories with recommended best practice guidelines whether they are currently offering drug testing in hair, or plan to offer a hair testing service in the future.[2] The guidelines include reference to recommended sample collection and storage procedures, through sample preparation, pre-treatment and analysis and the use of cut-offs. Unfortunately, there is no equivalent published for metals in hair at the SoHT. Metals in hair can be categorized into three main groups,

Issues in Toxicology No. 47
Perspectives and Challenges of Hair Analysis
Edited by Pascal Kintz, Alberto Salomone and Marco Vincenti

Published by the Royal Society of Chemistry, www.rsc.org

either they are essential, possibly essential, or toxic. Since most metals have a strong affinity for keratin and melanin, compared to blood, hair has the unique characteristic of concentrating most metals. A comparison of mean hair concentrations with mean whole blood concentrations for subjects living in Sweden shows that the ratio hair concentration ($\mu g\,g^{-1}$ or $ng\,mg^{-1}$) divided by blood concentration ($mg\,L^{-1}$ or $\mu g\,mL^{-1}$) is in the range: 1–10 for Mg, Ba, Br, Se, Li, Ca, I; 10–100 for Si, Be, Th, Tl, Re, Zn, Cu, Pt, Ir, Mo, Zr, As, Mn, Sc, Nb, Pb, B, Sb, Sr, Ga; 100–1000 for Ta, Hg, Co, W, Ni, Y, Al, Cr, Cd, Sn, V, REE, Ti; and over 1000 for Au, Bi, Ag, U,[3] which is in agreement with our findings.[4] Since hair growth is close to 1 cm per month, each cm represents both the internal exposure and the external exposure for a month. So, hair is a complementary medium with a larger window of detection that gives information on long-term exposure, which blood or urine cannot. Xenobiotics decrease rapidly in the blood, within some hours or a few days, and less rapidly in urine but only days, and hair is a good marker for much older exposures, depending on the length of the hair.

Furthermore, hair is a very attractive biological matrix as sampling is easy and non-invasive and it is very easy to store at room temperature. The major problem in hair analysis is the possible contamination with exogenous material, particularly minerals, that is not the case for molecules for which the exposure is only internal. This touches on one of the main difficulties in interpreting the data which is that decontamination, or the washing step is a major hurdle. However, considerable analytical progress has been made over the past two decades and the scope of hair analysis is now broad, covering many areas including archaeology, biology, environment, epidemiology, nutrition, clinical toxicology, and forensic toxicology. Over the past years there have been detailed reviews devoted to elemental hair analysis.[5–7] These papers highlight the superiority of the ICP-MS technique that is the leading quantitative ultra-trace elemental analysis for hair analysis over blood, urine, and tissue.[8]

In addition, this technique has undergone significant developments in recent years, for example the coupling of liquid chromatography to ICP-MS which allows speciation analyses in order to specify the nature of the metallic species, toxic or non-toxic. The interest of isotopic ratios measurements in ICP-MS to specify the origin of a metal will be evoked. Major progress has been made in the field of metallomic bioimaging, in particular with laser ablation coupled with ICP-MS which allows elementary longitudinal analysis of a single hair.[9] The ability to perform longitudinal or temporal analysis of individual hair strands shows a significant advantage of hair compared to other biological media such as blood or urine. With regard to hair metal concentrations, in all scientific data published it is not possible to fix "normal ranges" due to a large inter-individual variability of the results as shown by the reviews of the scientific literature.[5–7] In these reviews, large differences between various publications are observed for reference values in hair. This diversity reflects variations in numerous factors affecting metal content in hair such as dietary habits, lifestyle, smoking habits, and the geochemical environment. Moreover, the concentrations of some elements in hair may depend on ethnicity, gender, age, type of hair, hair colour, and melanin content. Finally, the results obtained are also closely linked to the decontamination/washing procedure and the analytical method.[5–7] Considering these elements, all of which hamper the possibility of establishing reference values, a transfer of normal values obtained in a given population cannot be used for another. As early as 2001, Siegel[10] and then several other authors drew attention to the fact that commercial laboratories have each established their own reference ranges which are not comparable.[11] Some articles even doubt that the results of hair mineral analyses determined by these laboratories are reliable and significant.[11]

12.2 Metal Incorporation in Hair

Incorporation into hair due to external metal pollution is a major problem. Under most conditions, external air metal pollution is so embedded that it cannot be adequately removed with washing procedures. Furthermore, elemental concentration in hair is the sum of biological incorporation through digestive, pulmonary, cutaneous exposure, and of external pollution (see Figure 12.1). Hair provides an effective medium for binding material such as dust that may contain a huge number of metals. Incorporated inorganics are located inside hair shafts, *i.e.*, in the cortex and in the medulla, while the external contamination likely remains mainly on the surface, *i.e.*, on the cuticle of hair scales. Thus, it is necessary to eliminate exogenous contamination resulting from the external deposition of air, water, and dust in order to specifically measure endogenous exposure. For metals, there are more sources of external contamination than for drugs, these include dust, sweat, sebaceous secretions, but also soaps, shampoos, conditioners, permanent wave products, bleaches, and hair spray. To study hair metal incorporation, Israleson *et al.* have examined the distribution of uranium (U) in single human hair shafts using two synchrotron radiation based micro X-ray fluorescence techniques (SR μ-XRF).[12] Two different groups have been studied, workers at a nuclear fuel fabrication factory, mainly exposed by inhalation, and owners of drilled bedrock wells exposed by ingestion of water. U at the fg level was detectable and the U distribution in single hair shafts was derived. The metal was located peripherally on the shafts in what seems to be a layer of approximately 10–15 μm thickness. Unwashed hair samples from the occupationally-exposed group also showed sites of very high U signal intensity which were identified as U bearing particles by complementary electron microscopy analysis. These particles were not present on washed hair, reinforcing the point that hair analysis requires very rigorous washing procedures to measure only hair metal incorporation by endogenous exposure.[12] The adsorption of many metals on hair is dependent of their acidity and/or the medium in which the hair sample is immersed, suggesting that hair is an ion exchanger.[13] The pK_a is estimated to be between 4.5 and 5.0. Most metals have a strong affinity for keratin due to sulfhydryl groups and for melanin that has negatively charged groups,

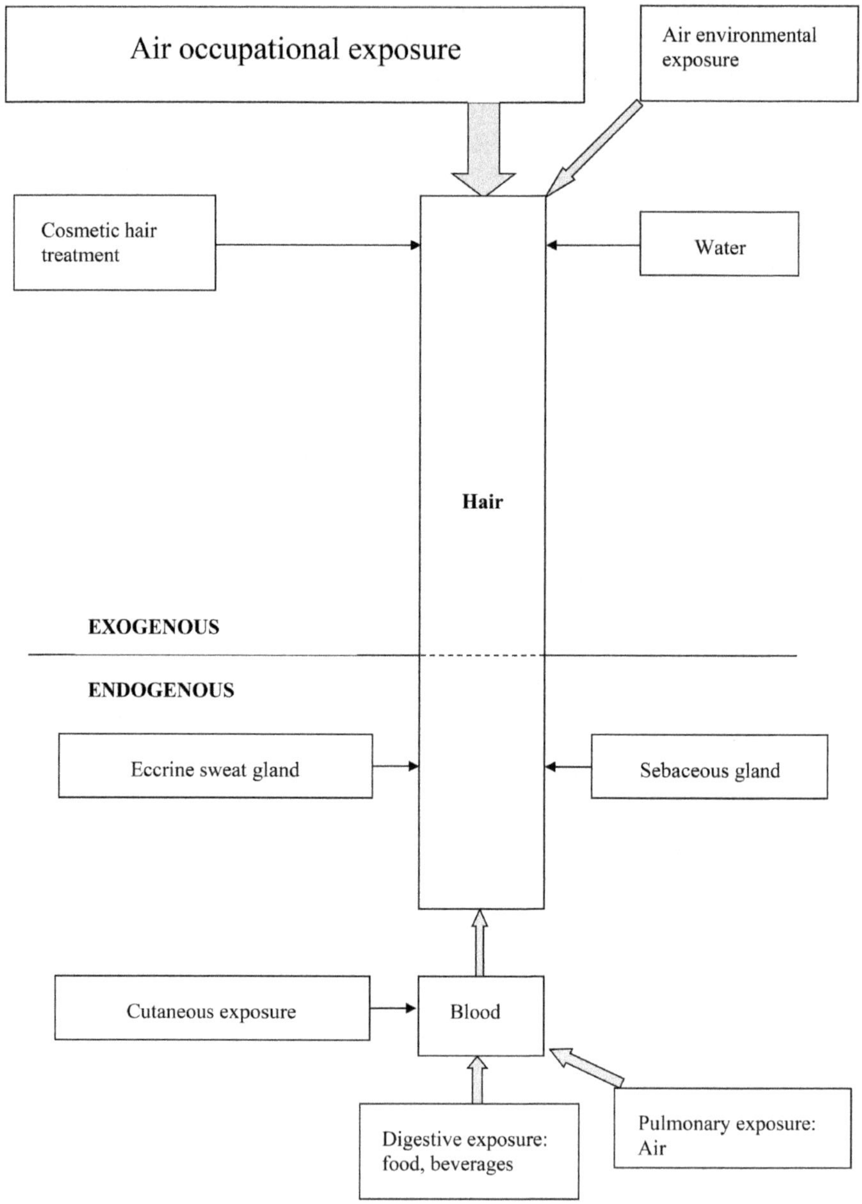

Figure 12.1 Flow-chart of metals in scalp hair sources.[80]

so they are easily incorporated into hair due to the high content of sulfhydryl groups in this matrix.[13,14] To measure exogenous metal incorporation, Abad *et al.* have performed both longitudinal analysis of human hair by laser ablation coupled to ICP-MS (LA-ICP-MS) and speciation analysis by gas chromatography coupled to ICP-MS (GC-ICP-MS) in single hair strands of five individuals which were occupationally exposed to high levels of mercury vapour three months previously and showed acute mercury poisoning symptoms.[15] On non-washed hair, LA-ICP-MS showed levels of mercury near the root of the hair *ca.* 1000 times lower compared to the part of the hair corresponding to the period of exposure. Additionally, speciation by GC-ICP-MS confirmed that Hg in all hair samples was present in inorganic form. These results point the direct adsorption of Hg vapour in hair at the time of exposure and the fact that Hg vapour can be irreversibly adsorbed on human hair.[15]

12.3 Hair Washing Procedure and Sample Preparation

12.3.1 Hair Washing Procedure

Washing procedure remain is one of the major problems for hair analysis as no consensus or recommended method exists. The difficulties with this procedure constitute a limitation in the use of hair analysis for human biomonitoring. But washing procedures are essential to remove exogenous contamination resulting from the external deposition by air, water, and dust in order to specifically measure only endogenous exposition. Despite a standard washing method being proposed by the International Atomic Energy Agency (IAEA), no common practice exists.[16] Kempson and Skinner tested the two most common decontamination procedures on hairs from the same individual: the procedure of the IAEA recommendation using polar and non-polar solvents and one based on the use of non-ionic detergent[17] (note: the IAEA recommended a procedure comprising a 10 min wash with acetone, three 10 min washes in distilled water followed by a further 10 min wash with acetone and a detergent-based procedure using a 2% Triton® X100 solution, washing for 30 min, followed by three 10 min rinses in distilled water). The analytical comparison used time of flight secondary ion mass spectrometry (ToF-SIMS). Sectioning hair longitudinally, qualitative information on elemental content was obtained from the outer surface and inner region of the hair from the identical longitudinal location. This was used to compare unwashed and washed strands comparing the internal and exterior surface elements. Both washing methods had similar effects. All elements appear to be removed from the surface of the hair as well as from inside the hair (except for Fe). This clearly shows that washing procedures can also have a significant effect on reducing the internal elemental signal levels in hair,[17] which was reported in a review as early as 1986 by Taylor.[18] In another paper, Tipple *et al.* investigated five washing protocols of varying aggressiveness to determine internal and external Sr concentrations in hair.[19] They measured $^{87}Sr/^{86}Sr$ isotope ratios in hair using multi-collector ICP-MS (MC-ICP-MS). The percent change was calculated using the difference between the average Sr content in unwashed hair, *i.e.*, control hair and hair treated with the five washing methods. The two most aggressive treatments removed 57% and 93% of the Sr, respectively. Accordingly, the $^{87}Sr/^{86}Sr$ isotope ratio of hair and hair leachate treated with the most aggressive washing methods reflect the isotope $^{87}Sr/^{86}Sr$ ratios of the interior and total exterior Sr signatures, respectively. From this experiment, it was concluded that the unwashed and the most aggressive washing procedure best allowed determination of external and internal Sr concentrations in hair. Pozebon *et al.* in their review reported about 50 different washing procedures for inorganic analysis of hair.[7] It is difficult to plan a method that would clear only external contamination whereas the binding of minerals in hair is rather complex and very far from being understood. Moreover, elements exhibit different physico-chemical properties. However, it should be noted that over-aggressive washing may not only remove surface contamination, but also extract elements from within the hair. In their contamination experiment, Muikku and Heikkinen tested hair samples from six adult volunteers which were cut and then exposed externally using washing water with highly elevated levels of U.[20] The samples were then washed using two commonly used washing procedures in order to remove the external contamination. U concentrations in hair measured by ICP-MS were finally about three orders of magnitude higher (10^3) and cannot be eliminated during the washing procedures. So, the possibility of external contamination should be kept in mind, especially if hair is considered to estimate the internal dose. So, it is necessary to know that in cases of environment or workplace exposure, metals can contribute to major contamination of the hair, but there is no known cleaning procedure that has been proven to be adequately effective.[21]

12.3.2 Hair Sample Preparation

There are many sample preparation processes. After the washing step, hair is usually mineralized as the decomposition of organic matter is an important part of the determination of elements for bulk hair analysis. To decompose the matrix, a mixture of nitric acid with hydrogen peroxide or nitric acid alone at 70 °C in a closed vessel is sufficient followed by a dilution with ultrapure water.[4,8,22–24] Open-beaker acid digestion is not recommended for ICP-MS due to contamination by airborne particles, loss of volatile elements and hazardousness.[25] In some cases, treatment of hair samples includes a closed-system microwave-assisted digestion with nitric acid as the matrix solubilization medium followed with appropriate dilution with ultrapure water.[26] On-line pre-concentration is often used to improve sensitivity to allow other elements to be determined by ICP-OES. For example, Anthemidis *et al.* have developed a time-based sequential dispensing on-line column preconcentration procedure for Hg determination at trace levels by cold vapour generation inductively coupled plasma atomic emission spectrometry (CV-ICP-AES), by means of a unified module of a pre-concentration column and a gas–liquid separator.[27] Montuori *et al.* determined methylmercury (MeHg) in hair by gas-chromatography cold-vapour atomic fluorescence spectrometry (GC-CV-AFS), after headspace solid phase microextraction.[28]

Moreda-Piñeiro *et al.* proposed a short-automated pretreatment procedure with liquid pressure extraction (LPE): hot water, after grinding, and mixing the hair with a dispersing agent combined with rapid simultaneous detection by ICP-AES.[29] They applied this LPE to As species determination in human hair.[30] Zheng and Hu introduced a fused silica capillary for capillary microextraction (CME) of traces of Cu, Zn, Ni, Hg, and Cd followed by on-line ICP-MS detection.[31] In a second paper, CME was performed with a new polymer coating for microextraction of traces of Co, Ni and Cd followed by on-line ICP-MS detection.[32] A simple, selective, and sensitive method using magnetic nanoparticles (MSPME) for SPE-ICP-AES was developed for heavy metal trace analysis by Cheng *et al.*[33] The same authors applied the MSPME to the speciation of Hg (MeHg and Hg^{2+}) by SPE-ICP-MS.[34] For miniaturization of extraction, microfluidic chips offer an interesting platform. Wang *et al.* fabricated an on-chip liquid-phase microextraction (on-chip LPME) device and combined it with electrothermal vaporization inductively coupled plasma mass spectrometry (ETV-ICP-MS) for trace determination of Cu, Zn, Cd, Hg, Pb, and Bi in the hair.[35] Recently, an interesting dispersive solid phase microextraction of Co and Ni in hair using a new nanomaterial, graphitic carbon nitride fixed on fibrous TiO_2, prior to their determination by ICP-MS was used by Chen *et al.*[36] The surface of this new adsorbent has many functional groups such as $-NH_2$, –NH– and =N, which can selectively bind target ions such as Co and Ni. The enrichment factor was 100. An original and very sensitive method for determination of total Hg in biological tissue and hair by isotope dilution ICP-MS after UV photochemical vapour generation was published by Liu *et al.*[37] A limit of detection (LOD) of 0.5 $pg\,g^{-1}$, based on an external calibration, provided 350-fold improvement over that obtained by utilizing conventional pneumatic nebulization sample introduction. As ICP-MS digestion step has many disadvantages Lam and Salin have proposed an alternative with an induction heating-electrothermal vaporizer with ICP-MS detection for direct multi-element determination of sub-milligram amounts of hair material.[38] As, Cd, Cu, Hg, Pb, and Zn absolute LODs were in the range 0.01–8 ng, values sufficiently low that they should allow the determination of metals at normal levels in a single human hair strand.

12.4 Analytical Methods for Elements in Hair

Many analytical methods have been described to analyse minerals in hair. For elemental measurements, significant progress has occurred, particularly over recent decades, that has progressively moved from the classic techniques of flame atomic absorption spectrometry (FAAS) and electrothermal atomic absorption spectrometry (EAAS) to the ICP techniques, ICP atomic emission spectrometry (ICP-AES) and ICP-MS.[8] There are two main groups of analytical techniques. Those devoted to metal bulk content, typically inductively coupled plasma (ICP) and those that explore longitudinal content of elements in hair, characteristically laser ablation (LA).[9]

12.4.1 ICP-AES and ICP-MS

ICP-AES and ICP-MS are the most popular methods. ICP-MS in particular is a fast and reliable metal and metalloid analytical method. For bulk analysis it is the preferred technique due to its multi-element analysis capability in a single run with flexibility, good reproducibility, and ability to measure a large linear range of concentrations. For hair it has high sensitivity, with a limit of detection (LOD) in the pg mg^{-1} level or lower.[8] Moreover, speciation analysis and isotopic analysis are possible. It has also been shown to be more sensitive at detecting the lower limits of trace elements and less likely to provide discrepant reference levels than ICP-AES.[36] Our research has developed and validated a multi-elementary procedure to simultaneous quantify 32 elements in 20 mg hair sample: Li, Be, B, Al, V, Cr, Mn, Co, Ni, Cu, Zn, Ga, Ge, As, Se, Rb, Sr, Mo, Pd, Ag, Cd, Sn, Sb, Te, Ba, W, Pt, Hg, Tl, Pb, Bi, U.[4,8] We use the following analytical procedure: a Thermo Elemental X7CCT bench top series with PlasmaLab® software and without a dynamic reaction cell (Thermo Optek, Courtaboeuf, France) was used for multi-elementary determinations. Plasma torch argon purity was higher than 99.999% (Linde Gas, Gargenville, France). Water was purified with a Milli-Q_{PLUS} 185 system (Millipore, St. Quentin-en-Yvelines, France). Suprapur® nitric acid 65%, triton X100 and multi-element standard solutions (30 elements) were obtained from Merck (Darmstadt, Germany). Seven other elements (W, Pd, Pt, Sn, Ge, Hg, Sb) and the rhodium internal standard solution were obtained from CPI international (Amsterdam, Holland). Global performance was assessed using a quality control program. Our laboratory is a registered participant of the Institut National de Santé Publique du Québec (Sainte-Foy, Canada) inter-laboratory comparison program for whole blood, urine, and beard hair, of non-occupationally exposed individuals, spiked with selected elements. After warm water and acetone decontamination, 20 mg of hair was mineralized with 0.20 mL nitric acid at 70 °C for one hour. A quantity of 0.1 mL of this solution was diluted into 3.9 mL (0.5% (v/v) butanol, 0.65% (w/v) nitric acid, 0.01% (v/v) triton with Rh (1 ppb) as internal standard). A calibration curve from limit of detection to 25 ng mL^{-1} or 250 ng mL^{-1} depending on the element was prepared. Intra-essay inaccuracy was in the range 0.4–6.7%. Inter-essay inaccuracy was also considered satisfactory. Quantification limits ranged from 0.2 pg mg^{-1} (Tl) to 0.5 ng mg^{-1} (B) for hair (see Table 12.1). High resolution with multi-collector ICP-MS (MC-ICP-MS) and sector field (ICP-SFMS) are also applied to determine the mass of an element with a higher precision (0.01 mu or less instead of 0.1 amu) in hair, as ICP-MS can be affected by polyatomic mass interferences.[8]

12.4.1.1 *Multi-elemental Determination by ICP-AES and ICP-MS*

As, Cd, Cu, Fe, Hg, Mn, Ni, Pb, Se, and Zn were determined by ICP-MS in microwave digested samples of hair from subjects with As intoxication in

Table 12.1 Hair multi-elementary ICP-MS analytical validation.[4] [a]

Compound	*r*	LOD	LOQ	CV% (1)	CV% (2)
Lithium	0.9999	0.002	0.007	6.5	6.1
Beryllium	0.9998	0.002	0.007	3.9	8.8
Boron	0.9991	0.14	0.46	3.6	8.9
Aluminium	0.9993	0.02	0.08	2.3	7.7
Vanadium	0.9998	0.001	0.003	1.7	9.0
Chromium	0.9999	0.06	0.20	3.5	9.3
Manganese	0.9996	0.001	0.004	1.7	6.6
Cobalt	0.9998	0.0003	0.001	2.3	7.9
Nickel	0.9998	0.01	0.05	1.8	6.4
Copper	0.9999	0.01	0.03	1.3	10.4
Zinc	0.9996	0.01	0.04	1.1	8.1
Gallium	0.9998	0.0003	0.0009	2.2	8.9
Germanium	0.9999	0.001	0.002	1.8	7.6
Arsenic	0.9997	0.01	0.02	3.5	6.4
Selenium	0.9997	0.02	0.06	2.6	7.8
Rubidium	0.9995	0.0003	0.001	2.0	5.8
Strontium	0.9995	0.0002	0.0007	1.0	7.0
Molybdenum	0.9998	0.0004	0.001	3.9	8.2
Palladium	0.9995	0.001	0.003	2.9	22.3
Silver	0.9998	0.0005	0.002	0.7	9.9
Cadmium	0.9998	0.0003	0.0009	0.7	5.9
Tin	0.9998	0.001	0.002	1.0	5.9
Antimony	0.9998	0.0003	0.001	1.0	5.2
Tellurium	0.9997	0.0006	0.002	6.7	6.1
Barium	0.9998	0.001	0.003	0.8	5.5
Tungsten	0.9998	0.0002	0.001	2.1	7.2
Platinum	0.9999	0.0001	0.0002	1.5	6.2
Mercury	0.9986	0.004	0.013	0.4	9.5
Thallium	0.9995	0.00005	0.0002	3.7	4.7
Lead	0.9997	0.0003	0.001	0.7	4.4
Bismuth	0.9997	0.0008	0.003	1.4	5.3
Uranium	0.9998	0.00004	0.0002	2.0	7.2

[a] *r*: correlation coefficient, LOD: limit of detection ($\mu g\,g^{-1}$), LOQ: limit of quantification ($\mu g\,g^{-1}$), CV% (1) = intra-assay imprecision, CV% (2) = inter-assay imprecision.

West Bengal, India. As well as As, Mn, Ni, and Pb levels were high compared with values from the literature.[22] Biomonitoring of Hg, Cd, and Pb exposure in Japanese children was performed by ICP-MS for Cd and Pb and by cold vapour atomic absorption spectrometry (CV-AAS) for Hg by Ilmiawati *et al.*[39] Thirteen elements were determined by ICP-MS in the hair of 196 children (aged 10–13 years), living near a hazardous waste incinerator in Catalonia, Spain.[40] Between 1998 (baseline survey) and 2012, significant decreases in the levels of Cd, Hg, Mn, Ni, Pb, and Sn were observed. Capillary microextraction combined with fluorinating assisted electrothermal vaporization was successfully applied to the ICP-AES determination of trace lanthanum, europium, dysprosium, and yttrium in human hair by Wu *et al.*[41]

12.4.2 Bioimaging Techniques

This section starts with laser ablation (LA) coupled to ICP-MS (LA-ICP-MS) followed by secondary ion mass spectrometry techniques (SIMS) that include time-of-flight (ToF-SIMS) and nano (Nano-SIMS), and finally synchrotron-based X-ray fluorescence (SXRF) with X-ray microanalysis (μ-XRF) (see Figure 12.2). The basic principles of these techniques and their applications for qualitative and quantitative imaging of elements are presented. These techniques have undergone significant development over the past few years and their great usefulness is demonstrated for hair.

12.4.2.1 LA-ICP-MS

The ICP-ion source is versatile to associate with different sample introduction systems, including LA. The LA-ICP-MS provides direct analysis of

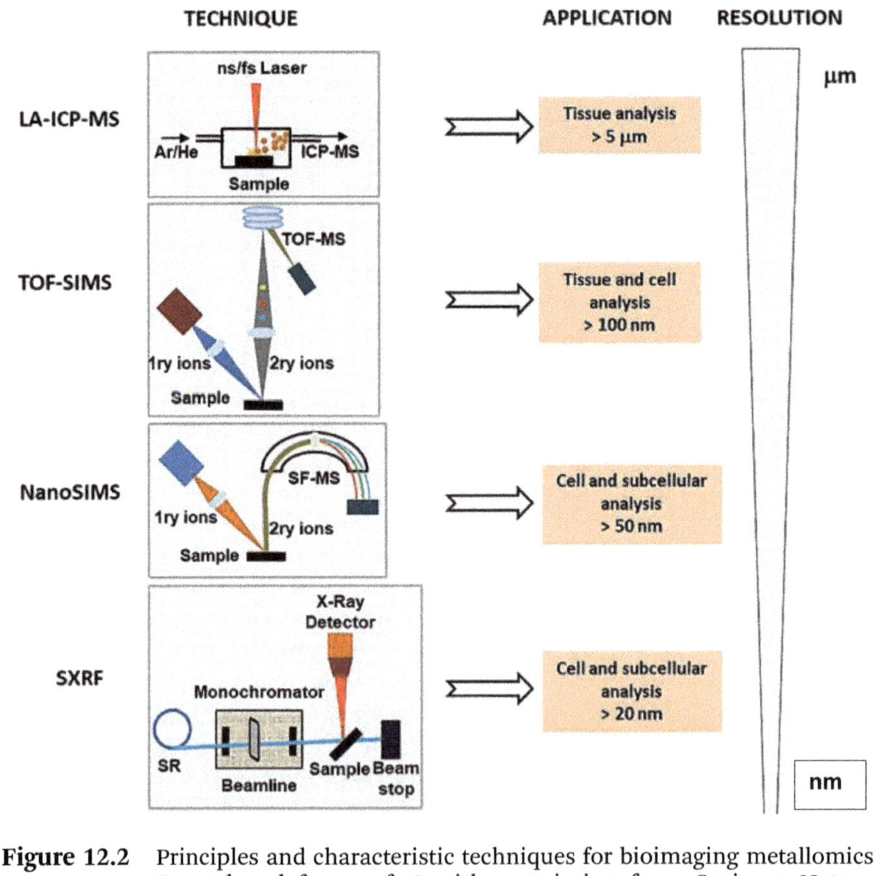

Figure 12.2 Principles and characteristic techniques for bioimaging metallomics. Reproduced from ref. 9 with permission from Springer Nature, Copyright 2018.

solids where the ablated material is carried to the ICP by a gas stream, usually Ar or Ar mixed with He or N_2.[9] The carrier gas passes through the LA cell which is connected to the injector tube in the plasma torch by a transfer line, typically a Tygon® or polytetrafluoroethylene (100 cm length × 0.6 cm internal diameter) tube.[9] Once the sample vapour is into the plasma, the analyte is ionized and then detected by ICP-MS (see Figure 12.2). LA-ICP-MS has been widely applied for both qualitative and quantitative analysis. This technique presents the huge potential to provide spatial resolution and time-track the monitoring of elements in hair strands, instead of conventional bulk analysis, which has many applications. Although images obtained by LA-ICP-MS do not have the fine spatial resolution of SIMS, or μ-XRF, it is a highly sensitive technique. Moreover, it can be used for isotopic analysis. Biomonitoring of essential and toxic metals in a single hair by LA-ICP-MS is possible with greater chronological resolution than other techniques as the sampling spot size is usually in the range 40–120 μm. So, LA-ICP-MS has many advantages:

- sampling hair strands is not invasive,
- analysis can be carried using only a single hair strand, which is a great advantage in forensic and archaeological investigations.

Even though the lack of sample representativeness might limit accuracy and precision, significative progress in the equipment overcomes the low sample amount ablated. Additional advantages of LA-ICP-MS are that digestion and grinding are not necessary since single hair strands can be analysed directly.[7] Single shot laser ablation ICP-MS has demonstrated the potential to detect Hg in single human hairs, with a sampling spot size of 50 μm, resolution corresponding to about 4 hours of growth, with a detection limit of approximately 0.2 $\mu g\,g^{-1}$.[42]

Variations in ablation efficiency, which is a problem with LA-ICP-MS, has been overcome by Cheajesadagul *et al.*[43] They have developed an innovative calibration strategy for this hair analysis: a lead-doped keratin film was used as a calibration standard for the determination of Pb in hair strands.

12.4.2.1.1 Multi-elemental Determination by LA-ICP-MS. Dressler *et al.* established that the LOD for Li, V, Mn, Ni, Co, Cu, Sr, Mo, Ag, Ba, Cd, I, Hg, Pb, Bi, and U in a single hair strand were in the range 0.001–0.90 $\mu g\,g^{-1}$, whereas those of Cr and Zn were 3.4 and 5.1 $\mu g\,g^{-1}$, respectively.[44] Calibration strategies for analysis of single hair strands by LA-ICP-MS were investigated by Kumtabtim *et al.* using either powdered hair certified reference material or human hair doped with known amounts of analytes, mounted on sticky tape.[45] The $^{34}S^{+}$ ion was used as the internal standard, thus improving the quantification of As and 16 other elements and the detection of exposure time profiles. The LOD was 0.6 $\mu g\,g^{-1}$ for As and was in the range 0.3–7.8 $\mu g\,g^{-1}$ for the other elements.

12.4.2.2 Secondary Ion Mass Spectrometry (SIMS)

The principle of the SIMS is the use of a primary ion source to sputter neutral particles, elements, and molecular ions from a sample surface. The negative or positive ions formed are then extracted and analysed by a mass spectrometer, whereas removal of the surface atoms or molecules through the energy loss of the primary ions occurs in the form of a collision cascade (see Figure 12.2).[7]

12.4.2.2.1 Time-of-flight Secondary Ion Mass Spectrometry (ToF-SIMS). This technique was used by Kempson and Skinner for the study of longitudinal, transverse, and localized distributions of elemental species in hair.[46] The authors studied elemental signals and spatial arrangements on the outer surface of hair as well as internal surfaces produced by the longitudinal sectioning of samples. The cuticle scales appear to play a major role in the external accumulation of exogenous elements. As the cuticle is continually exposed to environmental contamination the metal's greatest signal intensities were observed there. The majority of elements considered accumulated longitudinally on the outer surface of the hair above the scalp level. Internally, most elemental signals, and especially Al, decreased longitudinally once exposed to the environment with the exception of Si, which showed an increase. Moreover, they claim that bleaching hair can enhance the accumulation of contaminants. The complex patterns of concentration in hair make the interpretation of results complicated.

12.4.2.3 Nano Secondary Ion Mass Spectrometry (Nano-SIMS)

Nano-SIMS uses an ion source to produce a primary beam of ions. These primary ions erode the sample surface and produce atomic collisions, some of these collisions result in the release of secondary ion particles. These ions are transmitted through a mass spectrometer, where the masses are measured and identified. The primary ion beam is focused across the sample surface and a 'map' of the element and isotope distribution is created by counting the number of ions that originate from each pixel with at best a 50 nm resolution, that is 10–50 times greater than conventional SIMS.[47] Audinot and colleagues described a nano-SIMS 50 method for the determination of As in microtomed cross-sections of human hair.[48] The authors suggested that the method may be sufficiently sensitive to differentiate As in the medulla, cortex, and cuticle and so distinguish As intoxication by ingestion from surface contamination. The linearity of the relationship between the SIMS signal and As concentration was demonstrated. They applied this technique to the hair of Napoleon Bonaparte.[49]

12.4.2.4 *Synchrotron-based X-ray Fluorescence (SXRF and μ-XRF)*

SXRF spectrometry has a high sensitivity for most elements and does not destroy the sample. X-ray photons produced by X-ray tubes irradiate the sample atoms and eject a core electron with a binding energy lower than the incident photon energy. The resulting hole is then filled by an outer electron, thus producing an X-ray fluorescence emission. The sensitivity increases with atomic mass. Therefore, X-ray fluorescence is particularly appropriate for the study of heavy elements. The SXRF microprobes are composed of several key elements: the magnet source, the monochromator, the focusing optics, the X-ray detector (see Figure 12.2).[7] This technique was applied by Khudzari *et al.* to the determination of Mn, Fe, Ni, Cu, Zn, Se, and Sb in hair samples of sanitation workers and of a control group.[50] The mean concentration of sanitation workers showed elevation in comparison to the control group. Martin *et al.* analysed a single hair from a lead smelter worker using a synchrotron microprobe approach.[51] SXRF signals were detected from K, Ni, Zn, and Pb. The results suggest that the Pb originates both from ingestion and environmental exposure, however, direct deposition from the environment is the more important source of hair Pb. The persistence of this enrichment for a short distance up the hair shaft suggests that sweat and sebum may also be sources of these metals. Kondo *et al.* achieved a 3D imaging of trace elements in a single hair strand by a micro-X-ray fluorescence computed tomography (μXRF-CT) using synchrotron radiation.[52] A spatial resolution of around 2 μm determined by the beam size enabled detailed patterns in the cross-section and longitudinal direction of the hair strand to be visualized for Br, Cu, Fe, and Zn. The authors noted the potential interest of the application for dietary, archaeological, and forensic investigations.

12.4.2.5 *Proton Induced X-ray Emission Tomography (PIXE-T)*

Sela *et al.* have developed PIXE tomography that allowed the 3D quantitative mapping of the distribution of S, K, Cl, Ca, Fe, and Zn in a section of a strand of hair.[53]

12.4.3 Instrumental Neutron Activation Analysis (INAA)

NAA also was used for determining the concentrations of elements in hair. The technique allows discrete sampling of elements as it disregards the chemical form of a sample, and focuses solely on atomic nuclei.[54] The method is based on neutron activation and thus requires a source of neutrons. The sample is bombarded with neutrons, causing its constituent elements to form radioactive isotopes. The radioactive emissions and radioactive decay paths from each element are known. It is therefore possible to study the spectra of the emissions of the radioactive sample, and determine the concentrations of the various elements within it.[54] A particular advantage of this technique is that it does not destroy the sample, and thus has been used for the analysis of hair. INNA was applied to a suicide involving As_2O_3 to elucidate the discrepancy of ICP-MS As level of 26.4 $\mu g\,g^{-1}$ in hair collected at the autopsy and levels of As found by INAA in the range 0.16–0.26 $\mu g\,g^{-1}$ that excluded chronic poisoning, because the person died 14 h after the As_2O_3 ingestion.[55]

12.4.4 Atomic Fluorescence Spectrometry (AFS)

Gas chromatography coupled to cold-vapour AFS (GC-CV-AFS) was applied to the determination of MeHg in hair after SPE by Montori *et al.*[56] The digestion parameters were optimized to avoid MeHg degradation. In another AFS application Chen *et al.* used cellulose fibre as a sorption medium for Se preconcentration in hair followed by speciation with hydride generation before AFS.[57] Liu *et al.* have developed a slurry sample dilution followed by arsine generation which was then coupled with an *in situ* dielectric barrier discharge system to pre-concentrate As and reduce interferences prior to AFS.[58] The reported detection limit was in the ultra-trace range (14 pg). Recently, Deng *et al.* have proposed the simultaneous analysis of trace As, Sb, Se and Hg in hair certified refence material by chemical-vapour generation coupled non-dispersive atomic fluorescence spectrometry (CV-AFS).[59] The CV-AFS was compared by Astolfia *et al.*[60] to a frequently used advanced mercury analyser (AMA) to evaluate the total Hg level in human hair. The suitable condition for fast digestion of small size hair samples was obtained, 5–10 mg for AMA and 20 mg for CV-AFS. The validation parameters for both techniques show they are sensitive, precise, and reliable with a LOD of 0.4 $ng\,g^{-1}$ for AMA and 4 $ng\,g^{-1}$ for CV-AFS. The results show that the two methods are accurate and correlated.

12.4.5 Chemical Speciation Analysis

In their last review of advances in elemental speciation, Clough *et al.* observed that the most popular elements are still As, Hg, and Se,[61] but it is necessary to mention that sample preparation is critical in chemical speciation analysis as the interconversion of species generally occurs in this step.[7]

12.4.5.1 ICP and As Speciation Analysis

For As speciation analysis, ICP is generally coupled to HPLC. As there are many endemic areas around the world, the speciation analysis of the element proves to be a powerful tool for a complete analytical assessment in epidemiological studies covering the endemic areas. As species in water extracted from the hair of Bangladeshi people and controls were investigated as biomarkers of As exposure.[62] Water extraction from hair was carried out by high-performance liquid chromatography-inductively coupled to argon plasma mass spectrometry (HPLC-ICP-MS). Hair contained iAs(III) (58.9%), iAs(V) (34.8%), MMA(V) (2.9%), and DMA(V) (3.4%). As in hair was positively correlated with water As, suggesting that these measurements could be considered as a biomarker for As exposure. In a study of Chilean populations chronically exposed to iAs(V), total As and four As species, iAs(III), DMA(V), MMA(V) and iAs(V), were determined in water-extracts of hair samples by HPLC-HG (hybrid generation)-ICP-MS.[63] iAs(III) levels had the best correlation with chronic exposure to iA(V) in comparison with other species and total As. As(total), iAs(III) and iAs(V) in hair samples were correlated with the age of the exposed individuals ($R = 0.65$, 0.69, 0.57, respectively) and with the time of residence in the place ($R = 0.54$, 0.71, and 0.58, respectively). Sanz *et al.* described a new extraction procedure compared with established methods.[64] The extraction time was reduced from about 6 h to only a few minutes. They use an ultrasound probe sonication, an enzymatic treatment with lipase and protease in aqueous media and an HPLC column, placed into an oven. Total As was determined by ICP-MS and the species present in hair were quantified by HPLC-ICP-MS. The same authors determined after IAEA protocol washing, that concentrations found in hair samples from the arsenic-affected areas of the Middle and Lower Ganga plain by HPLC-ICP-MS were 39-fold higher than their background levels.[65] These samples contained mainly inorganic As in its tri- and pentavalent forms. Morado Piñeiro *et al.* have published a "green" automated speciation procedure by pressurized hot water extraction and HPLC-ICP-MS.[30] The extraction was carried out at 100 °C under high pressure (1500 psi) in a four-extraction step. Limits of quantification of 7.0, 6.3, and 50.3 ng g^{-1} for total As, iAs(III) and iAs(V), respectively were achieved.

12.4.5.2 ICP and Hg Speciation Analysis

The speciation of Hg in human hair was carried out by Gao *et al.* by combustion-atomic absorption spectrometry for total Hg with a LOD of 1.5 ng g^{-1} for a 20 mg sample, and by headspace-gas chromatography-atomic fluorescence spectrometry detection (HS-GC-AFS) for MeHg with a LOD of 0.04 ng g^{-1}.[66] Ma *et al.* developed a magnetic solid phase extraction (MSPE) with magnetic iron nanoparticles coupled with ICP-MS for the speciation of Hg in human hair samples.[67] As Hg^{2+} and MeHg exhibit different desorption behaviour on the magnetic particles, they can be separated. Queipo-Abad *et al.* validated GC-ICP-MS specific isotope dilution mass spectrometry for the simultaneous determination of MeHg, ethylmercury (EtHg), and iHg in hair using certified human hairs IAEA 085 and IAEA 086.[68] Analysis of Hg species in hair was also performed by a short two step procedure based on HPLC-ICP-MS with post-column volatile species generation (VSG).[69] The optimized post-column VSG step enhances sensitivity 30–40 times and improves the LOD for Hg species 4–5 times compared to HPLC-ICP-MS with eluent nebulization. Moreover, the post-column VSG step eliminates the effect of organic solvent gradient on individual species sensitivity. This HPLC-VSG-ICP-MS method was successfully applied to hair specimens and to human hair certified reference materials.

12.4.5.3 AFS and Se Speciation Analysis

Chen *et al.* have developed a sequential flow injection on-line Se separation and pre-concentration system with cellulose fibres coated as sorption medium in a micro-column, followed by hydride generation and measurement by AFS (FI-HG-AFS).[57] The procedure is validated by analysing Se in a reference material GBW 10 010 (rice) and human hair samples. When performing speciation, the Se(VI) was pre-reduced to Se(VI) under heating at 120 °C with HCl. Total inorganic Se is derived by pre-reduction of selenate and speciation is estimated by difference.

12.4.6 Isotopic Analysis

Isotopic ratios are increasingly investigated in many biomedical applications, together with elemental concentrations, to highlight small variations in the metabolic patterns of essential and toxic elements, but also to specify their origin; although the cost, time, skill, and equipment required for such measurements may limit their wider application. Isotopic ratios have been successfully applied to many elements and some examples are given in this section. The majority of the Hg in hair is MeHg; therefore, total Hg concentrations in hair are used for determining MeHg exposure levels from fish consumption. However, when people are exposed to significant levels of gaseous elemental mercury (GEM), total Hg concentrations in human hair may reflect the exposure routes of MeHg and GEM. Therefore, it is important to determine isotopic ratios, as they are different among the hair of gold miners and the hair of subjects exposed to MeHg oceanic fish. Yamakawa *et al.* determined the Hg isotopic ratios of National Institute for Environmental Studies (NIES) CRM no. 13.[70] Samples of human hair were analysed using cold vapour generation coupled to a multi-collector inductively coupled plasma mass spectrometer (CV-MC-ICP-MS) to meet the growing demand for better understanding of Hg exposure routes by using Hg isotopic compositions in human hair samples. In order to develop this isotope analysis, they measured the Hg isotopic compositions of four secondary reference materials—IAEA-085, IAEA-086, and CRPG-RL24H. As mentioned, Hg isotopic analysis in hair may be useful. Human exposure MeHg and elemental vapour Hg^0 are often estimated using total Hg concentrations in hair and urine, respectively. Sherman *et al.* investigated whether Hg stable isotopes could be used to better distinguish between these exposures.[71] They found that hair from North American dental professionals was characterized by high positive Δ^{199}Hg values (mean = 1.86‰). This confirms that among people who regularly consume fish, the total Hg concentrations in hair reflect exposure to MeHg. In contrast, they found that urine from the same individuals was characterized by a range of lower Δ^{199}Hg values (0.29 to 1.77‰) that were significantly correlated to the number of dental amalgams in each individual's mouth. So, they hypothesize that fish-derived MeHg is demethylated within the body, causing mass-dependent fractionation and the excretion of inorganic Hg in urine. Therefore, Hg in urine represents a mixture of demethylated fish-derived MeHg and amalgam-derived inorganic Hg. These data suggest that within populations that consume fish, urine total Hg concentrations may overestimate Hg exposure from personal dental amalgams.[71] For Zn, in order to provide information about transportation of metal elements *in vivo*, Ohno *et al.* have determined the precise Zn isotopic ratios of ^{66}Zn/^{64}Zn and ^{68}Zn/^{64}Zn of human hair samples by CV-MC-ICP-MS.[72] For U, the isotope study is of great interest. The process of enrichment of natural uranium (NU) produces depleted uranium (DU) as a by-product. NU contains 0.72% of ^{235}U, 99.27% of ^{238}U and DU 0.2% of ^{235}U, 99.8% of ^{238}U. The presence of DU in the environment and its possible transfer to humans has been a matter of concern due to the potential risks to human health when ingested or inhaled. Isotope ratio by Q-ICP-MS was developed by D'Ilio *et al.* as a qualified analytical technique for this objective.[73] The LOQ for this isotopic U determination in hair was 7.21 ng g^{-1}. It was applied by these authors to the specificity studies for hair contaminated by DU with different background levels of NU. The major interest of hair analysis for U exposure was confirmed by a radiochemical technique by Israelsson *et al.* who established a strong correlation between the activity ratios measured in water and in hair ($r = 0.91$).[74] They concluded that hair may serve as a robust indicator of chronic U intake in water as they reflected a large exposure period compared to urine.

12.4.7 Reference Material and Certified Reference Material (CRM)

The need for hair CRM in element composition has rapidly increased in laboratories engaged in chemical analyses of such matrices. To meet the ever-increasing demand, international organizations and government-supported bodies are actively involved in the issuance of these certified reference materials. It is 40 years since human hair powder was first prepared and certified as a reference material, to assist in the validation of analytical procedures used in clinical and environmental laboratories. Trace-element analysis of human hair has been carried out in laboratories throughout the world for the purpose of assessing the nutritional and toxicological status of individuals. However, the reliability of data provided by laboratories engaged in hair analysis is subject to question as regards analysis reference method but also appropriate CRM. In 1982, the IAEA was the first to prepare human hair reference material for inter-laboratory study of trace and other elements.[75] Subsequently, many CRM were prepared by the NIES in Japan by using the data obtained by various analytical techniques.[70,76] Chinese CRM have also been developed. Therefore, ICP-MS is a suitable analytical technique due to its multi-element analysis capability in a single run with high sensitivity detection and the ability to measure a large range of concentrations. So, ICP-MS is the technique of choice for determining metals and metalloids in CRM. Nevertheless, NAA has been proven to be a reliable technique for ultra-trace analysis, particularly in the certification of some ultra-trace elements like REEs.[77] Hg isotopic compositions in certified reference material NIES no. 13 human hair by CV-MC-ICP-MS was determined by Yamakawa *et al.*[70] A certified reference material, NIMD-01, was developed for the analysis of Hg speciation (MeHg and total Hg) in human hair by Haraguchi *et al.*[78] Tehrani *et al.* have described the production of four reference materials (NYS RMs 18-01, 18-02, 18-03, and 18-04) based on caprine horn, a keratinized tissue similar to human hair, intended to serve as a resource for calibration, quality control, and method validation purposes.[79] The elemental content and homogeneity of these candidate reference materials were characterized for 17 elements using ICP-MS. The method was validated against three CRM based on human hair: GBW 07601, ERM DB001, and NIES no. 13. It is also mandatory to participate in a QC program.[8]

12.4.8 Element Reference Values in Hair and Interpretation

The use of hair samples in an assessment of exposure to various metals has received a great deal of attention in the literature. Major variations are common in published normal ranges.[80] This diversity reflects variation due to many factors, that have long been known:

- preparation of hair samples with the washing procedure and the digestion step being the critical steps;
- analytical methods: the development of new analytical techniques and the major improvement of sensitivity and specificity of the techniques, MS and high resolution, bioimaging techniques in particular;
- individual variation that may depend on age, sex, hair colour, smoking habits, and cosmetic treatments;
- variation in element content due to regional or local dietary habits, lifestyle, or geochemical environment.

As inter-laboratory agreement on normal values is limited, each laboratory has to establish its own hair reference ranges. Using the analytical procedure previously described,[8] we determined our own reference values with median and reference ranges in hair including 300 healthy volunteers from Normandy (France) for 23 elements as a part of a case of a control study assessing the relationship of systemic sclerosis related to exposure to heavy metals.[81] The study cohort (100) and controls (300) consisted of 22% men and 78% women with a median age of 52 years (IQR: 45–61 years). Age (±5 years), gender, and smoking habit (48% smokers) matched controls were also selected for each patient. In the control, 19% were living in rural areas and 44% in towns with more than 20 000 inhabitants. The results are reported in Table 12.2. Several reviews have published reference concentrations, but there is no consensus method, because the washing step and the preparation of the hair are highly heterogeneous.[7,82,83] In their review, Kempson and Lombi include an overview of variables attributed to altering hair concentrations, such as age, gender, melanin content, and other less reported factors.[82] Hair elemental concentrations are reviewed with regard to morbidity, with specific examples of disease related effects summarized. The application of hair analysis for epidemiology and etiologic studies is enforced. A section is dedicated specifically to the area of population studies with regards to mercury, which highlights how endogenous and exogenous incorporation relies on species dependant metabolism and metabolic products. Many of the considerations are relevant to other areas of interest in hair analysis, such as for isotopic analysis. Inclusion of a table of elemental concentrations in hair should act as a valuable ref. 82.

In another issue, Mikulewicz *et al.* have collected reference values of elements published between 2000 and 2013, among at least 120 healthy

Table 12.2 Normal hair concentrations (300 patients).[81] [a]

Element	Concentration
• Aluminium	3.83 (2.34–5.98)
• Antimony	0.009 (0.005–0.016)
• Arsenic	0.033 (0.027–0.049)
• Barium	0.95 (0.50–1.60)
• Beryllium	0.013 (0.010–0.021)
• Bismuth	0.009 (0.005–0.023)
• Boron	0.91 (0.64–1.42)
• Cadmium	0.011 (0.005–0.025)
• Chromium	0.42 (0.30–0.66)
• Cobalt	0.027 (0.012–0.076)
• Copper	18.5 (13.0–29.7)
• Germanium	0.005 (0.004–0.006)
• Lead	0.38 (0.17–0.79)
• Lithium	0.017 (0.010–0.031)
• Manganese	0.12 (0.06–0.21)
• Mercury	0.47 (0.27–0.91)
• Molybdenum	0.015 (0.011–0.022)
• Nickel	0.34 (0.16–0.72)
• Palladium	0.010 (0.006–0.018)
• Platinum	0.0004 (0.0003–0.0008)
• Rubidium	0.034 (0.016–0.087)
• Selenium	0.46 (0.33–0.58)
• Silver	0.087 (0.039–0.193)
• Strontium	2.35 (0.76–4.58)
• Thallium	0.0004 (0.0003–0.0006)
• Tellurium	0.002 (0.002–0.003)
• Tin	0.16 (0.059–0.73)
• Tungsten	0.002 (0.001–0.005)
• Uranium	0.004 (0.002–0.008)
• Vanadium	0.021 (0.014–0.034)
• Zinc	152.2 (128.8–174.1)

[a] (IQR) in scalp hair samples ($\mu g\, g^{-1}$).

individuals, with a minimum of five elements analysed.[83] Due to analytical methodology that differed, in particular sample preparation, digestion, and analysis (ICP-AES or ICP-MS), the levels of hair minerals reported as reference values varied due to these differences. For these authors, it is necessary to elaborate standard procedures and furtherly validate hair mineral analysis and deliver detailed methodology. Only then it would be possible to provide meaningful reference ranges and take advantage of the potential that lies in hair mineral analysis as a medical diagnostic technique.[83] In another review Pozebon *et al.* evaluated articles referring to elemental hair analysis published between 2007 and 2017.[7] Elemental analysis of hair is critically reviewed, with particular focus on ICP-MS employment. Standard protocols of hair washing and sample decomposition are compiled. They demonstrate the usefulness of hair elemental and/or isotopic analysis. The interest and the potential of LA-ICP-MS for time tracking elements in hair is also highlighted.[7] It is worth noting that the content and the main conclusions of this recent review are remarkably similar to those published as early as 1986 by Taylor.[18]

12.4.8.1 Interpretation and Pitfalls of Element Analysis in Hair

Metal content in hair indicates past exposure, but it is not always reflective of body burden and many pitfalls for metal interpretation in hair were reported in medical practice.[84] It is still currently being used as the only biological material by some health care practitioners and promoted by commercial laboratories as a clinical assessment tool as well as a method to identify deficiency or toxic exposures. A split hair sample taken from near the scalp of a single healthy volunteer was submitted for analysis to six commercial US laboratories.[85] The differences in highest and lowest reported mineral concentrations for the same sample exceeded 10-fold for 12 minerals. The hair washing procedures were also different. Moreover, these laboratories also provided conflicting dietary and nutritional supplement recommendations based on their results. Adults[84] and pediatric[86] cases of suspected heavy metal poisoning were reported. They most often presented non-specific multi-systemic symptoms: joint pain, muscle aches, fatigue, flu-like symptoms, constipation, loss of appetite, headache, *etc.* Hair results showed abnormal levels of many elements, including heavy metals. A diagnosis of heavy metal poisoning was concluded and chelating therapy was proposed for each patient. Some patients also had amalgam fillings removed and replaced. The original diagnosis of heavy metal poisoning was therefore not substantiated. The patients did not have any previous history of exposure to heavy metals or specific clinical characteristics of heavy metal poisoning. Moreover, blood and urine determinations were normal or within normal ranges. Additional tests such as chelating were also within normal limits or difficult to interpret. However, it should be noted that hair analysis can be useful in certain settings. Hg, and As, or Tl poisonings have been largely documented with the use of hair analysis. It is important to distinguish the use of hair metal analysis, in a research setting, from the use of a panel of hair metal measurements to make an individual diagnosis. This is particularly true with patients whose symptoms and exposure history may suggest a low likelihood of metal toxicity.

12.4.8.2 Metal Reference Values in Hair and Age, Gender, Colour of Hair, Smoking Habits, Food Habits

In our control study, for men, we observed higher hair concentration for Al, Be, B, Cd, Pb, Ru, and V (from 1.5 to 2.0 times) compared to women. The women exhibited higher hair concentrations for Ba, Bi, Co, Cu, Li, Ni, Pd, Sr, Sn (from 1.5 to 6.0 times for Sr), which was in accordance with Llorente Ballesteros for Ba, Bi, Co, Cu, and Sr, but not for Mn and Tl.[87] Wolfsperger *et al.* have measured significant hair differences for smokers when compared with non-smokers for many elements: As, Cd, Co, Cr, Pb, Ni.[88] Some differences were observed between males and females. Other studies have also pointed out element differences depending on gender, age, or smoking habits.[3,89–93] For Khalique *et al.* no appreciable change in metal hair concentration was observed as a function of age for both sexes.[89] The concentrations of Ag, Al, As, Au, B, Ba, Be, Ca, Cd, Co, Cr, Cu, Fe, Hg, K, Mg, Mn, Mo, Na, Ni, P, Pb, Pt, Sb, Se, Si, Sr, Sn, Ti, V, W, Zn, and Zr in hair samples from inhabitants of Wroclaw (Poland) were determined by ICP-MS and ICP-AES. The effect of age, sex, smoking habit, and hair colour on the element content of hair was evaluated.[94] In their survey of 5908 occupationally non-exposed adults (4384 women and 1524 men) living in Moscow and region, Skalny *et al.* measured the hair levels of Al, As, Be, Bi, Cd, Hg, Li, Ni, Pb, Sn, and Sr by ICP-MS.[95] All subjects were occupationally non-exposed adults. They concluded that sex-related differences influenced the levels of these elements in hair, with significantly higher levels of Al, As, Cd, Hg, Li, and Pb in male hair samples and, conversely, higher concentrations of Bi, Ni, Sn. and Sr in those from women, these results were in accordance with our results except for Hg and Li in men.[81] In a recent paper, Rumiantseva *et al.* pointed out the role of the environmental and social factors among the population of the Vologda region in Northwest Russia.[96] Hair Hg of people older than 44 years (0.875 $\mu g\,g^{-1}$) was three times higher than in the hair of children under 18 years of age (0.270 $\mu g\,g^{-1}$). Their data showed the importance of food habits. The main source of Hg exposure was fish, with a hair Hg concentration of 0.172 $\mu g\,g^{-1}$, for a consumption 1–2 times per month; 0.409 $\mu g\,g^{-1}$, once per week 0.555 $\mu g\,g^{-1}$; and several times per week 0.995 $\mu g\,g^{-1}$.

12.4.9 Hair and Medical Applications (Drug Monitoring, Clinical Toxicology)

Several publications reported the interest of hair analysis for drug monitoring for As, Gd, Pt, or Tl. Luo *et al.* observed the spatial distribution of

As along the hair in a hair sample from a leukaemia patient treated with arsenic trioxide (As_2O_3).[97] The actual and estimated times over which the drug entered the body were compared and discussed. As speciation in hair of acute promyelocytic leukaemia (APL) patients undergoing As_2O_3 treatment was developed by Chen *et al.*[98] Compared to As species in urine, which reflect As exposure in the past 1–3 days, As species in the hair represents As exposure in the longer term. Determination of species in hair complements the urinary As speciation analysis, contributing to the monitoring of therapeutic treatment of APL with As_2O_3. Zheng *et al.* have carried out As speciation in hair to monitor As methylation metabolism in APL patients receiving treatment with As_2O_3.[99] Exposure to Gd from Gd-containing contrast agents for MRI was evaluated by Saussereau *et al.* in a renal failure woman who underwent several such examinations.[100] Gd overdose associated with acute nephrogenic systemic fibrosis and Gd contents in blood, hair, and in fingernails were 300 to 1000 times higher compared to controls. The authors considered that the method had considerable value for monitoring patients required to undertake repeated MRI examinations. In another paper Pozebon *et al.*[101] reported that the concentration of Pt along a single hair strand from a patient treated for cancer with cisplatin clearly reflected the administered doses. Measurements were obtained by LA-ICP-MS. The LOD was 0.029 $\mu g g^{-1}$ and the RSD was in the range 15–22%. Pt concentration and Pt oxidation states were analysed in body fluids, tissue, hair, and explants from women exposed to silicone and saline breast implants by Lykissa *et al.*[102] Pt hair concentration determined by ICP-MS was 2.13 ± 2.984 ng g^{-1} ($n = 9$). Goullé *et al.* reported a very rare case of survival after accidental Cr poisoning which has an extremely poor prognosis and usually leads to rapid death.[103] Seven segments of 1 cm hair had ICP-MS concentrations ranging from 0.73 $\mu g g^{-1}$ to 1.29 $\mu g g^{-1}$, all higher than the normal range (<0.52 $\mu g g^{-1}$), consistent with regular use of potassium dichromate and probably related to external contamination which could not be removed.

12.4.10 Hair and Forensic Applications

Hair analysis has been widely applied to various minerals and in particular to inorganic As, Tl, or Hg poisonings with suicidal or homicidal intent. Kucera and Kofronova applied INNA and ICP-MS techniques in a suspicious suicide case by the ingestion of As_2O_3 due to contrary findings between the two techniques (As level of 26.4 $\mu g g^{-1}$ by ICP-MS and INNA in the range of 0.16–0.26 $\mu g g^{-1}$).[104] INNA excluded chronic poisoning, because the person died after approximately 14 h after the As_2O_3 ingestion and highlight the importance of confirmatory techniques. Many human fatalities due to Tl poisoning have been published and documented with hair analysis. Li *et al.* reported a married couple who died from Tl sulfate intended homicidal poisoning.[105] The distribution of Tl was established by ICP-MS including whole blood, urine, hair, brain, heart, lung, liver, kidney, and stomach, with hair samples showing the highest Tl concentration: 10.2 $\mu g g^{-1}$ for the woman and 13.8 μg g^{-1} for the man. Ash *et al.* have described a single hair analysis with LA-ICP-MS to investigate an historic case of Tl poisoning and reconstructed the poisoning chronology of an unsolved cold case (1995).[106] Ablation line scans of an entire length of around 7 cm hair revealed about four months of repeated exposure to Tl with increased doses and frequency toward the end, while a scan of a part of the hair of about 0.7 cm revealed approximately two weeks of constant ingestions of large doses of Tl (an exceptionally high peak of about 0.53 μg g^{-1} in the middle of this hair shaft indicated a single exposure to an unusually large dose). The endogenous origin of Tl was confirmed by the preservation of the same longitudinal distribution profile in the inner part of hair. This paper demonstrates how hair analysis with LA-ICP-MS can add very useful temporal information in poisoning cases as single hair analysis re-constituted the original poisoning chronology. Moreover, the height and frequency of Tl peaks correlated with the victim's symptoms. A case study reporting fatal poisoning of three individuals in the same family due to adulterated infusion with Tl sulfate was published by Di Candia *et al.*[107] Hair Tl content by ICP-MS at autopsy was in the range 5.72–10.38 μg g^{-1}. Single hair analysis was applied to time-resolved monitoring of heavy-metal intoxication (Hg and Pt) in forensic cases using LA-DRC-ICP-MS.[108] Stadlbauer *et al.* performed analysis from the root to the tip with a spatial resolution of 20 μm, *i.e.*, approximately 1.5 h growth of the hair. LOD for Pt was 0.5 ng g^{-1} with a standard uncertainty that was around 15%. Samples taken after treatment showed the metal concentration over time. A major increase of Hg concentration, by a factor of 50, at the time of HgO administration was shown, and variation of Pt in the hair strands could be used to monitor during the time the Pt intake by the patient. The measurements also revealed that the concentrations of Hg and Pt in the outer and inner parts of the hair varied at the same time, despite the fact that the concentration in the core of the hair was about a quarter lower than that at the surface for both elements. In another paper, Lech performed LA-ICP-MS detection of Hg in human organs and hair in a case of a homicidal poisoning of a woman exhumed six years after death.[109] Toxicological analysis by cold vapour atomic absorption spectrometry (CVAAS) of samples from the corpse, exhumed six years after death revealed high Hg concertation ranging from 0.23 μg g^{-1} in hair to 1.52 μg g^{-1} in the small intestine. Soil Hg from places near the coffin found negligible traces of Hg (<1.5 ng g^{-1}), so, contamination by Hg from the environment was excluded and supported the evidence that this woman may have been poisoned. Duncan *et al.* studied the variability in hair Gd concentrations among decedents who received Gd-based contrast agents.[110] Analysis of autopsy hair samples by LA-ICP-MS showed correlations between Gd measurements, Gd dose, and time since exposure. This gives potential for evaluating Gd exposure in subjects *in vivo*, however clinical utility may be hindered by inter-individual and intra-individual variation of the Gd peak position observed along the hair shaft. Tipple *et al.* have analysed by MC-ICP-MS the $^{87}Sr/^{86}Sr$ ratio of 98 human hairs in 16 states of the USA.[111] The ratio was related to the tap water ratio, that varied geographically, being governed by the geology of local bedrock and was used to track the origins of forensic samples. The ratio found that hair provided geographic information about an individual's recent residence. Differences in the Sr isotopes ratios of hair may be concomitant with travel and could potentially be used as a screening tool to identify recent movements. In another paper Tipple *et al.* had previously shown the interest of $^{87}Sr/^{86}Sr$ isotopes ratio of hair to track the origins of samples (see Section 12.3.1 Hair Washing Procedure).[19]

12.4.11 Hair and Workplace Exposure

One has to remember there is no known "cleaning" procedure that can be proven to be adequately effective and particularly in cases of environment or workplace exposure, metals can contribute to major contamination of the hair. So, results should be interpreted with the utmost caution. Martin *et al.* have analysed by μ-XRF a single hair from a lead smelter worker.[112] The results clearly demonstrated the impact of surface contamination on Pb concentrations determined in this matrix. The distribution of Pb shows that the environment rather than ingestion is the prime source of the bulk of the metal in this sample. It is also shown that washing procedures can have a significant effect on reducing the internal elemental signal levels in hair. In another paper Cheajesadagul *et al.* used LA-ICP-MS to measure Pb in hair strands from workers at a battery manufacturing plant in Thailand.[43] They have developed an innovative calibration strategy for hair analysis. A Pb-doped keratin film was used as a calibration standard for the determination of Pb in hair strands. Fernande *et al.* studied human hair as an indicator of Hg workplace exposure by CV-AAS.[113] The Hg levels in the hair of 19 individuals professionally exposed to Hg at odontological areas and 32 controls, finding mean concentrations (± 1 SD) of 2.05 ± 2.12 and 1.86 ± 90 $\mu g\, g^{-1}$ for exposed and control individuals, respectively. These values were not significantly different ($p > 0.05$). They concluded that human hair is a useful biological sample to determine the environmental contamination by Hg at odontological work areas. Queipo-Abad *et al.* determined the concentration of Hg species in hair, blood, and urine of individuals occupationally exposed to gaseous elemental Hg in Asturias three years after an accident.[114] Quantification of MeHg, EtHg and Hg(II) in the samples was performed by GC-ICP-MS using triple spike isotope dilution analysis. The use of three labelled analogues, each one enriched in a different Hg isotope, allows the correction of species interconversions during the analytical procedure. The concentrations of Hg species in the intoxicated individuals were not significantly different from those obtained in the control group except for MeHg in blood. In another work, Sherman *et al.* measured total Hg concentrations in hair and urine as they are often used as biomarkers of exposure to fish-derived MeHg and gaseous elemental Hg, respectively.[115] They used Hg stable isotopes to assess the validity of these biomarkers among small-scale gold mining populations in Ghana and Indonesia. MC-ICP-MS Hg isotopes indicate that Hg in urine comes from Hg mining activity and suggest that Hg in hair is a mixture of fish MeHg and inorganic Hg.

12.4.12 Hair and Geochemical Applications

Many papers have been published regarding As, Pb, Sr, Tl, REE, or U. In two papers Tipple *et al.* emphasized the interest of $^{87}Sr/^{86}Sr$ isotope ratios in hair to track the geographic origin of specimens as it is related to their ratio in tap water, which varied geographically and provides information about an individual's recent residence (see Section 12.3.1 Hair Washing Procedure).[19,111] Mandal *et al.* collected biological samples (urine, hair, fingernails) from 41 people of West Bengal, India, who were drinking As-contaminated water, whereas 25 samples were collected from a population who stopped drinking As-contaminated water two years before biological collection.[62] They reported that As in urine, fingernails, and hair was positively correlated with water As, suggesting that any of these measurements could be considered as a biomarker for As exposure in endemic areas. Steely *et al.* applied a rapid approach for assessment of geochemical As exposure by elemental analysis of a single strand of hair using LA-ICP-MS from individuals who were chronically exposed to arsenic contaminated drinking water from a village in the Atacama Desert in northern Chile.[116] Their results were satisfactorily correlated with total As concentration previously measured by HG-ICP-MS.[63] The sample throughput was high and it takes around 3 min per hair sample including mounting, focusing, and analysis. They concluded that the LA-ICP-MS method can be used for the rapid identification and screening of toxic and nutritionally important elements in hair. To measure groundwater As geographical particular contamination in Bangladesh, Chakraborti *et al.* measured As concentrations in hair, nails, urine, and skin-scales collected from the As-affected areas of Bangladesh.[117] FI-HG-AAS median hair concentration was 2.34 $\mu g\, g^{-1}$ among 4536 people living in Bangladesh and 1.32 $\mu g\, g^{-1}$ for 8665 people living in West Bengal. So, in Bangladesh As hair median concentration was nearly double that in Bengal linked to geographical factors due to a higher As level in water. Three cleaning techniques that remove external contamination of human hair were assessed by Font *et al.* to investigate the potential use of Sr and Pb isotope variations determined by ICP-MS linked to human geographic location.[118] The isotope variations in hair could be used to follow up geographical movements. Karpas *et al.* determined the ratio $^{234}U/^{238}U$ by MC-ICP-MS in drinking water, hair, nails, and urine from volunteers who were given U in drinking water.[119] Their results conclusively demonstrated that U found in the bioassays can be traced back to the drinking water, thus providing a direct link to the source of exposure. The correlation ratio between water and the hair or nails was 0.97. Hair, as it is an effective bio-concentrator, may serve as an excellent indicator of environmental or occupational exposure to U and provide information regarding its source. For the authors, hair bioassay could also be used to assess exposure to depleted U long after the subjects had left the area suspected of contamination. In another paper the point at which an individual moved from an area with high concentrations of U in drinking water to another (concentration in water changed from 2000 to 30 ng L^{-1}) was easily demonstrated by Sela *et al.*[120] as the U determined on a single hair by ICP-sector filed mass spectrometry coupled to LA (LA-ICP-SFMS) decreased from 218 to 18 ng g^{-1}. Furthermore, measurements of U isotope ratios showed a natural isotopic composition throughout the whole period in the drinking water, as well as in the hair samples. In order to demonstrate the validity of using scalp hair rare earth element (REE) content as a biomarker of human geochemical exposure, Tong *et al.* analysed by ICP-MS, 16 REEs in 60 scalp hair samples from children aged 11–15 years old, living in an area of Southern China mined for REEs.[121] The samples from the mining area were significantly higher than those from the reference area and were much higher than those published in the literature and reflected REEs in the mine and the atmosphere shrouding this area. Using ICP-MS, Campanella *et al.*[122] monitored Tl exposure of a local population in Tuscany (Italy) by measuring Tl levels in 318 hair and 150 urine samples as tap water was contaminated with Tl up to 10 μg L^{-1}. Hair concentrations, representing longer term exposure, were in the range 1–498 ng g^{-1} (values in unexposed subjects: 0.1–6 ng g^{-1}) whilst urine concentrations reflected more recent exposure. There was a clear relationship between elevated Tl levels in biological samples and residency in zones where tap water was contaminated with Tl. In the same area, Aprea *et al.* analysed by ICP-MS, 2154 urine and 254 hair samples in three different exposure areas. Hair concentrations were in the range 0.1–28.3 ng g^{-1}.[123]

12.4.13 Hair and Archaeological Applications

In two papers Tipple *et al.* have used the $^{87}Sr/^{86}Sr$ isotope ratio of hair to track the origin of archaeological specimens as it is related to their ratio in tap water, which varied geographically and provides information about an individual's recent residence (see Section 12.3.1 Hair Washing Procedure).[19,111] Matsumoto and Yoshinaga measured isotope ratios of lead $^{207}Pb/^{206}Pb$ and $^{208}Pb/^{206}Pb$ by ICP-MS in Japanese women's hair in the twentieth century.[124] They measured historic hair samples from the 1910s to 1968 and compared them to contemporary hair samples to evaluate Pb contamination of the human proximate environment in that period. The oldest hair samples between 1910 and the 1920s had the highest concentration and the most distinct isotope ratios were from those of Japanese domestic Pb indicating serious contamination of the proximate environment of people. Contemporary Pb levels were quite low (<1 μg g^{-1}); however, isotope ratios were close to those in the 1960s when hair concentration was assumed to be strongly affected by leaded gasoline. Bartkus *et al.* have explored Pb exposure in ancient Chilean mummies by LA-ICP-MS using a single strand of hair.[125] Ancient hair samples ($n = 49$) were excavated from several burial sites in Northern Chile. Samples belonged to an ancient fisher hunter–gatherer culture who occupied regions of the Atacama Desert from approximately 5000–1500 BCE. The Pb distribution in the central medulla in a 100 μm cross-section scan of hair strands demonstrated minimal exogenous contamination. Of the 49 samples, 21 (43%) showed concentrations higher (range 1.1–228.0 $\mu g\,g^{-1}$) than the average value of 5 $\mu g\,g^{-1}$ for unexposed individuals. In general, all profiles representing the Pb signal are mainly confined to the cortex and medulla regions and not to the outer cuticle. The authors suggest that the observed Pb in these examined mummy hairs is mostly endogenous and could probably be attributed to the consumption of Pb-contaminated geochemical water. Lorentz *et al.* analysed ancient hair ($n=8$) from the third millennium BCE by SR-μXRF spectrometry.[126] The samples came from a large urban site in Iran known for metal working. The results revealed differences in Cu levels and distribution with larger concentrations located in the cortex. These elevated Cu levels, together with the contextual data, indicate a biogenic rather than environmental source for Cu uptake. Arriaza *et al.* studied boron in 144 strands of ancient mummy hair from the Atacama Desert (Chile), ranging from 3000 BCE to 1500 CE using LA-ICP-MS.[127] All samples showed high B concentrations, ranging from 1.5 to 4 times above the average B concentration in contemporary hair that were <0.85 $\mu g\,g^{-1}$. The B concentration in mummy hair varied across major geographic areas because rivers in northern Chile are known to have high geogenic concentrations of B.

12.4.14 Methylmercury in Hair

A special section is devoted to Hg/MeHg in hair as the element is the only mineral for which a relationship has been established and recognized between Hg concentration in blood and Hg concentration in hair after MeHg exposure. The Hg levels in hair are generally 250–300 times higher than those observed in blood. A hair to blood mercury ratio of 250 : 1 was established as early as 1990 by the World Health Organisation (WHO).[128] According to the WHO, human hair is a biomarker of choice for MeHg exposure and is more stable than blood because Hg levels in blood change over time, peaking at between 4 h and 14 h after exposure, and then being cleared.[129] In contrast, MeHg deposited in the hair remains stable for a long time. Growing approximately 1 cm per month, hair provides a historical record of MeHg exposure, with quantification of Hg content in 1 cm corresponding to an exposure of approximately one month.[130] Interestingly, hair Hg is also proportional to Hg concentrations in other tissues. For example, MeHg in hair is approximately 250 times higher than in whole blood, and 50 times higher than in the brain (see Figure 12.3).[131]

12.5 Single Hair Analysis by LA-ICP-MS

Bulk analysis provides only very limited information about the longitudinal distribution of the minerals in the hair, without the possibility of tracing the

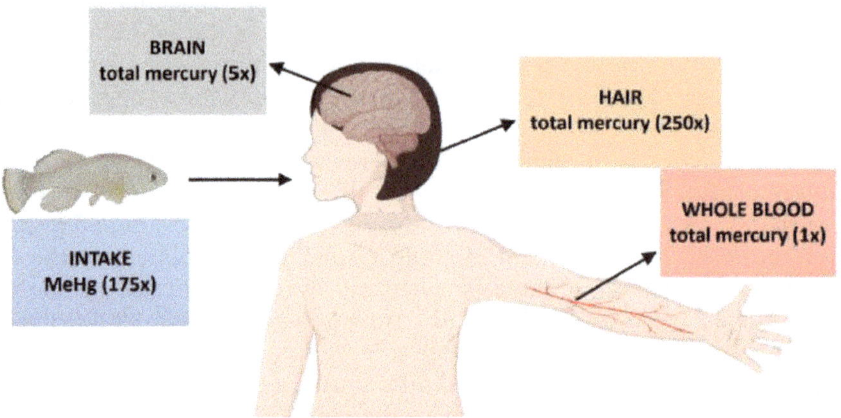

Figure 12.3 Estimated total mercury burden in different tissues according to the methylmercury (MeHg) intake for fish consumers without occupational exposure. Reproduced from ref. 130 with permission from Elsevier, Copyright 2021.

chronological intake and/or exposure to elements. Compared to bulk analysis, single hair analysis for the determination of minerals is a major analytical advance. This approach has progressed a lot since the last edition of the book in 2006. Single hair analysis uses bioimaging methods and particularly LA-ICP-MS. The potential of this technique gives the opportunity to examine hair samples with greater chronological resolution than other techniques and constitutes a major advance in the field leading to many applications. For example, element concentration along the hair can reveal periods of exposition and/or intoxication. In addition, high spatial laser resolution, as low as 10 μm, can explore a very short exposure period, corresponding to about 1 h of hair growth and a 200 μm resolution will provide sufficient resolution, corresponding to about 18 h hair growth – and will allow monitoring of acute *versus* chronic exposures. The high sensitivity of the technique is compatible with the element's quantification at the $\mu g\, g^{-1}$ or $ng\, g^{-1}$ depending on the element. The quantification with most of the routine techniques gives only monthly information by cutting the hair into 1 cm segments and analysing separately the whole segments. A commercial instrument combining combustion, gold amalgamation, and atomic absorption spectrometry (C-GA-AAS) was evaluated for the measurement of Hg in 1 cm segments from a single hair.[132] The analyser included various modules, first one for sample combustion, then one for amalgamation of the released Hg vapour with gold and finally one to measure Hg by AAS. The LOQ was established at 0.1 ng of total Hg and the between-strand variability was around 6.5%. With such performance it was possible to monitor monthly Hg exposure by analysis of just 1 cm of a single hair. Good agreement was demonstrated with hair samples and CRM materials between C-GA-AAS and conventional CV-AAS, but chronological reconstitution of element exposure along the 1 cm segment isn't possible.

The many advantages of the LA-ICP-MS are highlighted:

- a very small amount of the sample is sufficient for analysis,
- chronological reconstitution of mineral exposure is a major advantage compared to bulk analysis,
- the preparation step is considerably reduced as it is not necessary to grind and digest the hair,
- as the strands of hair can be analysed directly, the sample throughput is higher compared to other techniques,
- coupled to multi-collector, LA-MC-ICP-MS provides the possibility to determine the isotope ratio of elements that has the powerful added advantage of isotopic sensitivity which is of major interest in various applications. Santamaria-Fernandez *et al.* reported for the first time the measurement of longitudinal variations of $^{34}S/^{32}S$ isotope ratios in single hair strands using LA-MC-ICP-MS.[133] This determination has the potential to give the geographic origin of a subject and to detect their recent movements.

One of the first applications of LA-ICP-MS was devoted to the direct detection of Hg in single human hair strands by Legrand *et al.*[42] Biomonitoring of essential and toxic metals in single hair using on-line solution-based calibration was described by Dressler *et al.*[44] Kumtabtim *et al.* used this technique for biomonitoring As, toxic, and essential metals in single hair strands.[45] Workplace exposure was explored by Cheajesadagul *et al.*[43] to measure Pb in hair strands from workers at a battery manufacturing plant in Thailand; by Elish *et al.* to reveal a three-day period occupational exposure event to U.[134] LA coupled with ICP-MS demonstrates also how hair analysis can add useful information in forensic cases and give valuable temporal details in an unsolved poisoning case with Tl for exmple;[106] or as reported by Stadlbauer *et al.* document two forensic cases involving Hg and Pt using LA-DRC-ICP-MS with a high spatial resolution of 20 μm corresponding to approximately 1.5 h hair growth, whose results clearly reflected the administered doses.[108] Also, but not last, this technique is particularly relevant for drug monitoring due to its ability to track the incorporation of a drug over time as mentioned by Luo *et al.* in a leukaemia patient treated by for As_2O_3,[97] or by Pozebon *et al.* for cancer treatment with Pt.[101] The interest of the technique for geochemical and nutritional applications using LA-ICP-SFMS was clearly demonstrated by Sela *et al.*[120] Last, the interest of LA-ICP-MS in archaeology was underlined by Bartkus *et al.* who have explored Pb distribution in ancient Chilean mummies by LA-ICP-MS using a single strand of hair,[125] and by Arriaza *et al.* who demonstrated B overexposure in the hair of ancient Chilean mummies.[127]

12.6 Conclusion

Hair is recognized as an important biological material by the WHO, however, mineral analysis of hair has several obstacles as described in this chapter. The first is related to the multiple and complex modes of incorporation of these elements in the hair whose origin is endogenous or/and exogenous and which it is necessary to differentiate. The second is linked to the procedures for washing and digesting the samples, which are numerous, so an international consensus is necessary in order to harmonize these steps, as they will provide reliable data.

References

1. J. Marsh, *Edinburgh New Philos. J.*, 1836, **21**, 229.
2. G. A. Cooper, R. Kronstrand and P. Kintz, *Forensic Sci. Int.*, 2012, **218**, 20.
3. I. Rodushkin and M. D. Axelsson, *Sci. Total Environ.*, 2000, **262**, 21.
4. J. P. Goullé, L. Mahieu, J. Castermant, N. Neveu, L. Bonneau, G. Lainé, D. Bouige and C. Lacroix, *Forensic Sci. Int.*, 2005, **153**, 39.
5. I. M. Kempson and E. Lombi, *Chem. Soc. Rev.*, 2011, **40**, 3915.
6. M. Mikulewicz, K. Chojnacka, T. Gedrange and H. Górecki, *Environ. Toxicol. Pharmacol.*, 2013, **36**, 1077.
7. D. Pozebon, G. L. Scheffler and V. L. Dressler, *Anal. Chim. Acta*, 2017, **992**, 1.
8. J. P. Goullé, E. Saussereau, L. Mahieu and M. Guerbet, *Bioanalysis*, 2014, **6**, 2245.
9. V. L. Dressler, E. I. Müller and D. Pozebon, Bioimaging Metallomics, in *Metallomics. Advances in Experimental Medicine and Biology*, ed. M. Arruda, 2018, vol. 1055, DOI: 10.1007/978-3-319-90143-5_7.
10. S. Seidel, R. Kreutzer, D. Smith, S. McNeel and D. Gilliss, *JAMA*, 2001, **285**, 67.
11. M. Mikulewicz, K. Chojnacka, T. Gedrange and H. Górecki, *Environ. Toxicol. Pharmacol.*, 2013, **36**, 1077.
12. A. Israelsson, M. Eriksson and H. B. L. Pettersson, *Spectrochim. Acta*, 2015, **108**, 28.
13. H. Mikasa, Y. Suzuki, N. Fujii and K. Nishiyama, *Biol. Trace Elem. Res.*, 1988, **16**, 59.
14. V. J. Bush, T. P. Moyer, K. P. Batts and J. E. Parisi, *Clin. Chem.*, 1995 **41**, 284.
15. S. Q. Abad, P. Rodriguez-Gonzalez and J. I. G. Alonso, *J. Trace Elem. Med. Biol.*, 2016, **36**, 16.
16. E. C. Toro, J. J. M. De Goeij, J. Bacso, Y. D. Cheng, L. Kinova, J. Matsubara, S. Niese, T. Sato, G. R. Wesenberg, Y. Muramatsu and R. M. Parr, *J. Radioanal. Nucl. Chem.*, 1993, **167**, 413.
17. I. M. Kempson and W. M. Skinner, *Biol. Trace Elem. Res.*, 2012, **150**, 10.
18. A. Taylor, *Ann. Clin. Biochem.*, 1986, **23**, 364.

19. B. J. Tipple, T. Chau, L. A. Chesson, D. P. Fernandez and J. R. Ehleringer, *Anal. Chim. Acta*, 2013, **798**, 64.
20. M. Muikku and T. Heikkinen, *Health Phys.*, 2012, **102**, 699.
21. R. R. Martin, I. M. Kempson, S. J. Naftel and W. M. Skinner, *Chemosphere*, 2005, **58**, 1385.
22. G. Samanta, R. Sharma, T. Roychowdhury and D. Chakraborti, *Sci. Total Environ.*, 2004, **326**, 33.
23. R. F. Puchyr, D. A. Bass, R. Gajewski, M. Calvin, W. Marquardt, K. Urek, M. E. Druyan and D. Quig, *Biol. Trace Elem. Res.*, 1998, **62**, 167.
24. S. Grassin-Delyle, M. Martin, O. Hamzaoui, E. Lamy, C. Jayle, E. Sage, I. Etting, P. Devillier and J. C. Alvarez, *Talanta*, 2019, **199**, 228.
25. A. Monaster and D. W. Golightly, *Inductively coupled plasma in Analytical Atomic Spectroscopy*, Weinheim, New York, 1987.
26. I. B. Razagui and S. J. Haswell, *J. Anal. Toxicol.*, 1997, **21**, 149.
27. A. N. Anthemidis, G. A. Zachariadis, C. E. Michos and J. A. Stratis, *Anal. Bioanal. Chem.*, 2004, **379**, 764.
28. P. Montuori, E. Jover, R. Alzaga, S. Diez and J. M. Bayona, *J. Chromatogr. A*, 2004, **1025**, 71.
29. J. Moreda-Piñeiro, E. Alonso-Rodríguez, P. López-Mahía, S. Muniategui-Lorenzo, D. Prada-Rodríguez, A. Moreda-Piñeiro and P. Bermejo-Barrera, *Anal. Bioanal. Chem.*, 2007, **388**, 441.
30. A. Morado Piñeiro, J. Moreda-Piñeiro, E. Alonso-Rodríguez, P. López-Mahía, S. Muniategui-Lorenzo and D. Prada-Rodríguez, *Talanta*, 2013, **105**, 422.
31. F. Zheng and B. Hu, *Anal. Chim. Acta*, 2007, **605**, 1.
32. F. Zheng and B. Hu, *Talanta*, 2011, **85**, 1166.
33. G. Cheng, M. He, H. Peng and B. Hu, *Talanta*, 2012, **88**, 507.
34. S. Ma, M. He, B. Chen, W. Deng, Q. Zheng and B. Hu, *Talanta*, 2016, **146**, 93.
35. H. Wang, Z. Wu, Y. Zhang, B. Chen, M. He and B. Hu, *J. Anal. At. Spectrom.*, 2013, **28**, 1660.
36. S. Z. Chen, Y. X. Liu, J. T. Yan, C. L. Wang and D. B. Lu, *At. Spectrosc.*, 2020, **41**, 169.
37. R. Liu, M. Xu, Z. Shi, J. Zhang, Y. Gao and L. Yang, *Talanta*, 2013, **117**, 371.
38. R. Lam and E. D. Salin, *J. Anal. At. Spectrom.*, 2007, **22**, 1430.
39. C. Ilmiawati, T. Yoshida, T. Itoh, Y. Nakagi, Y. Saijo, Y. Sugioka, M. Sakamoto, A. Ikegami, M. Ogawa and F. Kayama, *Environ. Health Prev. Med.*, 2015, **20**, 18.
40. I. Martorell, M. Nadal, L. Vilavert, F. Garcia, M. Schuhmacher and J. L. Domingo, *Trace Elem. Electrolytes*, 2015, **32**, 43.
41. S. Wu, C. Hu, M. He, B. Chen and B. Hu, *Talanta*, 2013, **115**, 342.
42. M. Legrand, R. Lam, M. Jensen-Fontaine, E. D. Salin and H. M. Chan, *J. Anal. At. Spectrom.*, 2004, **19**, 1287.
43. P. Cheajesadagul, W. Wananukul, A. Siripinyanond and J. Shiowatana, *J. Anal. At. Spectrom.*, 2011, **26**, 493.

44. V. L. Dressler, D. Pozebon, M. F. Mesko, A. Matusch, U. Kumtabtim, B. Wu and J. Sabine Becker, *Talanta*, 2010, **82**, 1770.
45. U. Kumtabtim, A. Matusch, S. U. Dani, A. Siripinyanond and J. S. Becker, *Int. J. Mass Spectrom.*, 2011, **307**, 185.
46. I. M. Kempson and W. M. Skinner, *Sci. Total Environ.*, 2005, **338**, 213.
47. Z. Gajdosechova and Z. Mester, *Anal. Bioanal. Chem.*, 2019, **411**, 4277.
48. J. N. Audinot, S. Schneider, M. Yegles, P. Hallegot, R. Wennig and H. N. Migeon, *Appl. Surf. Sci.*, 2004, **231–232**, 490.
49. R. Wennig, S. Schneider, M. Yegles, P. Hallegot, P. Kintz, J. P. Goullé, J. N. Audinot and H. N. Migeon, *41th Meeting of The International Association of Forensic Toxicologists*, Melbourn, 16–20 november 2003.
50. J. Md Khudzari, H. Wagiran, I. Hossain and N. Ibrahim, *J. Environ. Radioact.*, 2013, **115**, 1.
51. R. R. Martin, I. M. Kempson, S. J. Naftel and W. M. Skinner, *Chemosphere*, 2005, **58**, 1385.
52. R. Kondo, T. Yamato, A. Munoz-Noval, S. Honda, Y. Nishiwaki, K. Komaguchi and S. Hayakawa, *J. Anal. At. Spectrom.*, 2021, **36**, 1041.
53. H. Sela, Z. Karpas, M. Zoriy, C. Pickhardt and J. S. Becker, *J. Radioanal. Nucl. Chem.*, 2008, **276**, 101.
54. J. Kučera, *J. Radioanal. Nucl. Chem.*, 2007, **273**, 273.
55. J. Kučera and K. Kofronova, *J. Radioanal. Nucl. Chem.*, 2011, **287**, 769–772.
56. P. Montuori, E. Jover, R. Alzaga, S. Diez and J. M. Bayona, *J. Chromatogr. A*, 2004, **1025**, 71.
57. M. Chen, T. Yang and J. Wang, *Anal. Chim. Acta*, 2009, **631**, 74.
58. M. T. Liu, T. P. Liu, J. X. Liu, X. F. Mao, X. Na, L. Ding, G. Y. Chen and Y. Z. Qian, *J. Anal. At. Spectrom.*, 2019, **34**, 526.
59. X. Deng, R. Li and S. Deng, *J. Fluoresc.*, 2020, **30**, 949.
60. M. L. Astolfia, C. Protano, E. Marconi, D. Piamonti, L. Massimi, M. Brunori, M. Vitali and S. Canepari, *Microchem. J.*, 2019, **150**, 104186.
61. R. Clough, C. F. Harrington, S. J. Hill, Y. Madrid and J. F. Tyson, *J. Anal. At. Spectrom.*, 2022, **37**, 1387.
62. B. K. Mandal, Y. Ogra, K. Anzai and K. T. Suzuki, *Toxicol. Appl. Pharmacol.*, 2004, **198**, 307.
63. J. Yanez, V. Fierro, H. Mansilla, L. Figueroa, L. Cornejo and R. M. Barnes, *J. Environ. Monit.*, 2005, **7**, 1335.
64. E. Sanz, R. Munoz-Olivas, C. Dietz, J. Sanz and C. Camara, *J. Anal. At. Spectrom.*, 2007, **22**, 131.
65. E. Sanz, R. Munoz-Olivas, C. Camara, M. Kumar-Sengupta and S. Ahamed, *J. Environ. Sci. Health*, 2007, **42**, 1695.
66. Y. Gao, S. De Galan, A. De Brauwere, W. Baeyens and M. Leermakers, *Talanta*, 2010, **82**, 1919.
67. S. Ma, M. He, B. Chen, W. Deng, Q. Zheng and B. Hu, *Talanta*, 2016, **146**, 93.
68. S. Queipo Abad, P. Rodríguez-González, W. C. Davis and J. L. García Alonso, *Anal. Chem.*, 2017, **89**, 6731.

69. I. Petry-Podgórska, V. Schrenková, M. Migašová, T. Matoušek and J. Kratzer, *Microchem. J.*, 2021, **170**, 106606, ISSN: 0026-265X.
70. A. Yamakawa, A. Takeuchi, Y. Shibata, S. Berail and O. F. X. Donard, *Accredit. Qual. Assur.*, 2016, **21**, 197.
71. L. S. Sherman, J. D. Blum, A. Franzblau and N. Basu, *Environ. Sci. Technol.*, 2013, **47**, 3403.
72. T. Ohno, A. Shinohara, M. Chiba and T. Hirata, *Anal. Sci.*, 2005, **21**, 425.
73. S. D'Ilio, N. Violante, O. Senofonte, C. Majorani and F. Petrucci, *Anal. Methods*, 2010, **2**, 1184.
74. A. Israelsson and H. Pettersson, *Health Phys.*, 2014, **107**, 143.
75. S. B. M'Baku and R. M. Parr, *J. Radioanal. Chem.*, 1982, **69**, 171.
76. K. Okamoto, M. Morita, H. Quan, T. Uehiro and K. Fuwa, *Clin. Chem.*, 1985, **31**, 1592.
77. L. Cao, W. Tian, B. Ni, P. Wang and Y. Zhang, *Anal. Bioanal. Chem.*, 2002, **372**, 397.
78. K. Haraguchi, M. Sakamoto, A. Matsuyama, M. Yamamoto, D. T. Hung, H. Nagasaka, K. Uchida, Y. Ito, H. Kodamatani, M. Horvat, H. M. Chan, M. Rand, C. M. Cirtiu, B. G. Kim, F. Nielsen, A. Yamakawa, N. Mashyanov, N. Panichev, E. Panova, T. Watanabe, N. Kaneko, J. Yoshinaga, R. F. Herwati, A. E. Suoth and H. Akagi, *Anal. Sci.*, 2020, **36**, 561.
79. M. W. Tehrani, K. X. Yang and P. J. Parsons, *Anal. Bioanal. Chem.*, 2020, **412**, 1847.
80. J. P. Goullé, in *Analytical and Practical Aspects of Drug Testing in Hair*, ed. P. Kintz, Taylor and Francis, Boca Raton, FL, 2006, vol. 16, pp. 343–369.
81. I. Marie, J. F. Gehanno, M. Bubenheim, A. B. Duval-Modeste, P. Joly, S. Dominique, P. Bravard, D. Noël, A. F. Cailleux, J. Benichou, H. Levesque and J. P. Goullé, *Autoimmun. Rev.*, 2017, **16**, 223.
82. I. M. Kempson and E. Lombi, *Chem. Soc. Rev.*, 2011, **40**, 3915.
83. M. Mikulewicz, K. Chojnacka, T. Gedrange and H. Górecki, *Environ. Toxicol. Pharmacol.*, 2013, **36**, 1077.
84. M. Frisch and B. S. Schwartz, *Environ. Health Perspect.*, 2002, **110**, 433.
85. S. Seidel, R. Kreutzer, D. Smith, S. McNeel and D. Gilliss, *JAMA*, 2001, **285**, 67.
86. W. T. Poon, S. C. Ling, A. Y. W. Chan and T. W. L. Mak, *Hong Kong Med. J.*, 2004, **10**, 197.
87. M. T. Llorente Ballesteros, I. Navarro Serrano and S. Izquierdo Álvarez, *J. Trace Elem. Med. Biol.*, 2017, **43**, 113.
88. M. Wolfsperger, G. Hauser, W. Gössler and C Schlagenhaufen, *Sci. Total Environ.*, 1994, **156**, 235.
89. A. Khalique, S. Ahmad, T. Anjum, M. Jaffar, M. H. Shah, N. Shaheen, R. S. Tariq, S. and Manzoor, *Environ. Monit. Assess.*, 2005, **104**, 45.
90. B. Benes, J. Sladká, V. Spevácková and J. Smid, *Cent. Eur. J. Public Health*, 2003, **11**, 184.
91. J. Chłopicka, Z. Zachwieja, P. Zagrodzki, J. Frydrych, P. Słota and M. Krośniak, *Biol. Trace Elem. Res.*, 1998, **62**, 229.

92. Z. Meng, *Biol. Trace Elem. Res.*, 1998, **61**, 79.
93. W. Ashraf, M. Jaffar and D. Mohammad, *Sci. Total Environ.*, 1994, **151**, 227.
94. K. Chojnacka, H. Gorecka and H. Gorecki, *Environ. Toxicol. Pharmacol.*, 2006, **22**, 52.
95. A. V. Skalny, M. G. Skalnaya, A. A. Tinkov, E. P. Serebryansky, V. A. Demidov, Y. N. Lobanova, A. R. Grabeklis, E. S. Berezkina, I. V. Gryazeva, A. A. Skalny and A. A. Nikonorov, *Environ. Toxicol. Pharmacol.*, 2016, **40**, 18.
96. O. Rumiantseva, E. Ivanova and V. Komov, *Int. Arch. Occup. Environ. Health*, 2022, **95**, 1027.
97. R. Luo, X. Su, W. Xu, S. Zhang, X. Zhuo and D. Ma, *Sci. Rep.*, 2017, **7**, 3426.
98. B. Chen, F. Cao, X. Lu, S. Shen, J. Zhou and X. C. Le, *Talanta*, 2018, **184**, 446.
99. Y. Zheng, Y. F. Mao, H. J. Zhao, L. Chen, L. N. Wang, Y. X. Zhang, J. Hu, J. M. Li, X. Y. Li and H. M. Zhu, *Exp. Hematol. Oncol.*, 2021, **10**, 10.
100. E. Saussereau, C. Lacroix, A. Cattaneo, L. Mahieu and J. P. Goullé, *Forensic Sci. Int.*, 2008, **176**, 54.
101. D. Pozebon, V. L. Dressler, A. Matusch and J. S. Becker, *Int. J. Mass Spectrom.*, 2008, **272**, 57.
102. E. D. Lykissa and S. V. M. Maharaj, *Anal. Chem.*, 2006, **78**, 2925.
103. J. P. Goullé, E. Saussereau, J. Grosjean, C. Doche, L. Mahieu, J. M. Thouret, M. Guerbet and C. Lacroix, *Forensic Sci. Int.*, 2012, **217**, e8.
104. J. Kucera and K. Kofronova, *J. Radioanal. Nucl. Chem.*, 2011, **287**, 769–772.
105. S. Li, W. Huang, Y. Duan, J. Xing and Y. Zhou, *J. Forensic Sci.*, 2015, **60**, 247.
106. R. D. Ash and M. He, *Forensic Sci. Int.*, 2018, **292**, 224.
107. D. Di Candia, E. Muccino, A. Battistini, M. Boracchi, G. Gentile and R. Zoja, *Leg. Med.*, 2020, **42**, 101661.
108. C. Stadlbauer, T. Prohaska, C. Reiter, A. Knaus and G. Stingeder, *Anal. Bioanal. Chem.*, 2005, **383**, 500.
109. T. Lech, *Am. J. Forensic Med. Pathol.*, 2015, **36**, 227.
110. B. R. Duncan, M. Hasegawa, D. A. Marshall, L. F. Gonzalez-Cuyar, M. Paulsen, M. Kobayashi, K. R. Maravilla and C. D. Simpson, *Anal. Bioanal. Chem.*, 2021, **413**, 1571.
111. B. J. Tipple, L. O. Valenzuela, T. H. Chau, L. Hu, C. P. Bataille, L. A. Chesson and J. R. Ehleringer, *Rapid Commun. Mass Spectrom.*, 2019, **33**, 461.
112. R. R. Martin, I. M. Kempson, S. J. Naftel and W. M. Skinner, *Chemosphere*, 2005, **58**, 1385.
113. D. R. Fernande, A. D. Vasquez, A. M. Ocando, S. C. Bello, M. C. Rodriguez and V. A. Granadillo, *At. Spectrosc.*, 2008, **29**, 224.
114. S. Queipo-Abad, P. R. González, E. Martínez-Morillo, W. C. Davis and J. I. García Alonso, *Sci. Total Environ.*, 2019, **672**, 314.

115. L. S. Sherman, J. D. Blum, N. Basu, M. Rajaee, D. C. Evers, D. G. Buck, J. Petrlik and J. DiGangi, *Environ. Res.*, 2015, **137**, 226.
116. S. Steely, D. Amarasiriwardena, J. Jones and J. Yanez, *Microchem. J.*, 2007, **86**, 235.
117. D. Chakraborti, M. M. Rahman, A. Mukherjee, M. Alauddin, M. Hassan, R. N. Dutta, S. Pati, S. C. Mukherjee, S. Roy, Q. Quamruzzman, M. Rahman, S. Morshed, T. Islam, S. Sorif, M. Selim, M. R. Islam and M. M. Hossain, *J. Trace Elem. Med. Biol.*, 2015, **31**, 237.
118. L. Font, G. van der Peijl, I. van Wetten, P. Vroon, B. van der Wagt and G. Davies, *J. Anal. At. Spectrom.*, 2012, **27**, 719.
119. Z. Karpas, A. Lorber, H. Sela, O. Paz-Tal, Y. Hagag, P. Kurttio and L. Salonen, *Health Phys.*, 2005, **89**, 315.
120. H. Sela, Z. Karpas, M. Zoriy, C. Pickhardt and J. S. Becker, *Int. J. Mass Spectrom.*, 2007, **261**, 199.
121. S. L. Tong, W. Z. Zhu, Z. H. Gao, Y. X. Meng, R. L. Peng and G. C. Lu, *J. Environ. Sci. Health, Part A*, 2004, **39**, 2517.
122. B. Campanella, M. Onor, A. D'Ulivo, R. Giannecchini, M. D'Orazio, R. Petrini and E. Bramanti, *Sci. Total Environ.*, 2016, **548**, 33.
123. M. C. Aprea, D. Nuvolone, D. Petri, F. Voller, S. Bertelloni and I. Aragona, *PLoS One*, 2020, **15**, e0241223.
124. M. Matsumoto and J. Yoshinaga, *Environ. Sci. Pollut. Res. Int.*, 2010, **17**, 643.
125. L. Bartkus, D. Amarasiriwardena, B. Arriaza, D. Bellis and J. Yañez, *Microchem. J.*, 2011, **98**, 267.
126. K. O. Lorentz, W. de Nolf, M. Cotte, G. Ioannou, F. Foruzanfar, M. R. Zaruri and S. M. S. Sajjadi, *J. Archaeol. Sci.*, 2020, **120**, 105193.
127. B. Arriaza, D. Blumenstiel, D. Amarasiriwardena, V. G. Standen and A. Vizcarra, *Am. J. Phys. Anthropol.*, 2021, **174**, 254.
128. World Health Organization, *Environmental Health Criteria 101: Methylmercury*, World Health Organization, Geneva, 1990.
129. World Health Organization, *Guidance for Identifying Populations at Risk from Mercury Exposure*, 2008, accessed 10th October 2022, https://www.who.int/publications/m/item/guidance-for-identifying-populations-at-risk-from-mercury-exposure
130. M. E. Crespo-Lopez, M. Augusto-Oliveira, A. Lopes-Araújo, L. Santos-Sacramento, P. Yuki Takeda, B. M. Macchi, J. L. M. do Nascimento, C. S. F. Maia, R. R. Lima, G. P. Arrifano, *Environ. Int.*, 2021, **146**, 106223.
131. T. W. Clarkson, J. B. Vyas and N. Ballatori, *Am. J. Ind. Med.*, 2007, **50**, 757.
132. M. Legrand, C. J. Passos, D. Mergler and H. M. Chan, *Environ. Sci. Technol.*, 2005, **39**, 4594.
133. R. Santamaria-Fernandez, J. Giner Martínez-Sierra, J. M. Marchante-Gayón, J. I. García-Alonso and R. Hearn, *Anal. Bioanal. Chem.*, 2009, **394**, 225.
134. E. Elish, Z. Karpas and A. Lorber, *J. Anal. At. Spectrom.*, 2007, **22**, 540.

CHAPTER 13

Controversies in Hair Analysis... or 30 Years of Challenges and Issues

PASCAL KINTZ

X-Pertise Consulting, 42 rue principale, 67206 Mittelhausbergen, France
Email: pascal.kintz@wanadoo.fr

13.1 Introduction

Since the late 1980s, the most crucial issue facing hair analysis is the avoidance of false-positive results caused by passive exposure to the drug. This is the reason why it was suggested by a group of active scientists that a decontamination step should be added to the analytical procedure.[1] In addition, some years later, in 2004, the same group suggested that positive cut-offs should be used to report the presence of a specific drug.[2] However, this was not enough and other scientists, mostly from the United States and those very actively engaged in the business of urine testing, continued to challenge the validity of hair tests due to potential external contamination.[3,4] As a consequence, new studies were published to propose alternative approaches to the issue of external contamination.[5–7] However, these scenarios were also challenged... and so on. The debate will probably never end. This situation is an example of the difficulties in implementing a new technology in science where various scientific, but also economic parameters have to be taken into consideration. The late Professor Manfred Moeller taught me that it is easier for two toxicologists to share their toothbrush than their analytical procedures.

Issues in Toxicology No. 47
Perspectives and Challenges of Hair Analysis
Edited by Pascal Kintz, Alberto Salomone and Marco Vincenti

Published by the Royal Society of Chemistry, www.rsc.org

Other pitfalls have been encountered. It has been discussed by those challenging the validity of hair tests that a negative result could be due to the absence of drug consumption, but also the consequence of a normal washout during standard hygiene practices (shower, shampoo...). If a 5–10% drug loss (washout) per month has been reported for head hair, they claimed during some trials that the use of deodorant for axillae should prevent a complete washout in 2–3 months, although this was not supported by any literature reference. Those against hair tests have also strongly indicated that pubic hair can be contaminated by urine and this should prevent the use of this body hair. They simply missed that pubic hair is the best specimen to demonstrate abstinence to alcohol as this matrix is very sensitive to ethyl glucuronide.

Although it has been reported that for each step of hair tests for drugs some problems do exist, obviously major controversies are in the interpretation of the data. It has also been advocated that a hair test is not able to determine how and when a drug was used. This is true and accepted, but is exactly the same limitation observed with urine. A dose will never be estimated from a urine concentration.

The aim of this chapter is to review some issues that are still under debate, a debate that will probably continue over the years.

From my experience of more than 30 years[8] (my first work on drug detection in hair was published in 1992), the following points are still under strong discussion:

- possible external contamination
- use of pulverized hair *versus* cut hair
- differences between a child's hair and an adult's hair
- single hair *versus* a strand of hair
- minimal detectable drug dose in hair and significance of a negative result
- should FAEE (ethyl palmitate) be always tested in addition to EtG to demonstrate excessive alcohol consumption?
- is GHB testing in hair reliable to document a single exposure?
- reliability of doping control using hair in case of contamination
- influence of drug stability in case of cosmetic treatment

13.2 Possible External Contamination

The major issue facing hair analysis is the avoidance of technical and evidentiary false-positives. Technical false-positives are caused by errors in the collection, processing, and analysis of specimens, while evidentiary false-positives are caused by passive exposure to the drug. Contamination of hair through environmental exposure also has the potential to cause a false positive result. There are various approaches to avoid evidentiary and technical false-positives. The use of a proper washing process will not only remove surface contamination but also can provide useful information on the source of drug uptake in hair, *i.e.*, from growing hair follicles or from surface bodily fluids.

There is no consensus or uniformity in the washing procedures, however most have similar characteristics. The agents used in washing are detergents such as shampoo, surgical scrubbing solutions, surfactants such as 0.1% sodium dodecylsulfate, phosphate buffer, or organic solvents such as acetone, diethyl ether, methanol, ethanol, dichloromethane, hexane, or pentane of various volumes and various contact times. A single washing step is the most common procedure, although a second identical wash is sometimes performed. If external contamination is found by analyzing the wash solution, the washout kinetics of repeated washing can demonstrate that external contamination is rapidly removed. It has been proposed that the concentration of drug in the hair after washing should exceed the concentration in the last wash by at least a factor of ten.

It must be understood that washing (rinsing hair) can remove a drug from the interior as well as from the exterior surface of hair during a decontamination procedure.

However, it is not always possible, during experimental studies, to distinguish a drug-contaminated subject from an active user. Thus, while a negative result excludes both chronic use and contact with drugs, a positive result cannot be interpreted as a sure sign of drug addiction, particularly if not supported by other investigations.[9,10]

The guidelines of the Society of Hair Testing (SoHT)[11] have suggested using specific metabolites (norcocaine, cocaethylene, or THC-COOH) to discriminate external contamination from systemic passing of the drug. Even the detection of drug metabolite(s) in hair, whose presence could not be explained by hydrolysis or environmental exposure, has been challenged as a proof of ingestion, *e.g.*, for cocaine or cannabis.[12,13]

The presence of similar concentrations of drug in consecutive segments can be considered as indicative of potential contamination from an individual's body fluids or tissues or other forms of contamination unless it can be shown that the person was in treatment with this substance over the period represented by the segments.[14]

Various approaches to document external contamination have been recently suggested in the literature, including micro-segmental analysis,[15] testing for artefacts (Δ9-THCA-A) as indirect markers[16] or the use of very sophisticated instruments, such as matrix-assisted laser desorption/ionization mass spectrometry (MALDI-MS) with longitudinal sectioning of single hairs and different decontamination protocols.[17]

Whatever the result, a proper interpretation of hair test results is critical and should be done ideally with other information available, *e.g.*, medical history, witness statements, and the available circumstances of the matter.

Results from a single segment of hair cannot be used to discriminate long-term exposure to a drug from a single exposure. This means that segmental testing is needed if a history of drug exposure is required.

13.3 Use of Pulverized Hair *Versus* Cut Hair

In the more recent version of the consensus of the SoHT[18] for the use of alcohol markers in hair for supporting the assessment of chronic alcohol consumption, it is indicated that powdering hair for the extraction of ethyl glucuronide (EtG) is best practice. This is associated with a positive cut-off at 30 pg mg^{-1}. In addition, if non-pulverized samples are used for extraction, the laboratory must demonstrate that comparable results are achieved through proficiency testing. This is not surprising as it has been demonstrated on several occasions that pulverized hair will favour EtG extraction when compared to cut hair.[19,20] Therefore, when using positive cut-offs, irrespective of the nature of the drug, it is of importance to indicate which preparation has been used, in particular when the same hair specimen is tested in two different laboratories for comparison. As a general rule, it is important for each laboratory to build its own ranges of concentrations for specific populations and to avoid comparing single measures from lab to lab, as the final result is highly dependent on the whole analytical procedure.

13.4 Differences Between a Child's Hair and an Adult's Hair

Whereas the detection of drugs in a child's hair unambiguously shows drug handling in the environment of the child, it is difficult to distinguish between systemic incorporation into hair after ingestion or inhalation and external deposition into hair from smoke, dust, or contaminated surfaces. However, the interpretation of hair results with respect to systemic or only external exposure is particularly important in the case of children for a realistic assessment of the toxic health risk.

Hair from children is finer and more porous in comparison with that from adults (risk of higher contamination by sweat *versus* adults). Apart from the lower amount of biological material in children *versus* adults, there is no specific analytical problem when processing samples from children. Obviously, the same procedure can be used. The issue is the interpretation of the findings with respect to the different pharmacological parameters. Children are not "little adults" but rather immature individuals whose bodies and organ functions are in a continuing state of development. It is not surprising, therefore, that the pharmacokinetics and toxicity of most drugs vary considerably throughout the paediatric age range and may differ profoundly from findings in adults.

In utero, the activation of the hair follicles of the fetus starts at about the fourth month of pregnancy to produce the lanugo, which is very thin, curly, and non-pigmented. The lanugo will be replaced by a down, short, non-pigmented, thin (between 5 and 40 μm), with a slow growing rate and located between 0.5 and 1.5 mm under the skin. Starting in the eighth month of pregnancy, the final hair will appear. It is long, pigmented, thick (more than 80 μm), with a rapid growing rate, and located between

2.5 and 5 mm under the skin. After delivery, to avoid becoming bald, hair growing is asynchronous (variation in the anagen/catagen phases) during the first 3–4 months. Hair loss can be observed during the first six months after delivery, followed by a slow growing rate during the next six months. After one year, the normal rate of hair growth (1 cm per month as in adults) starts. The growth rate is first 0.2 mm per day, and then it increases to 0.3–0.5 mm per day, to finally stabilize at about 0.35 mm per day. As a consequence, it is very difficult to put any window of detection when testing for drugs in young children.[21]

It is in fact even more complicated as it has been demonstrated that drugs can be incorporated during pregnancy in the hair of the fetus, which will contribute to positive findings after delivery.[22,23] The trans-placental transfer of drugs with accumulation in fetal hair has been demonstrated. Hair measurement for drugs of abuse or pharmaceuticals in neonates is a useful screening method to detect intra-uterine exposure to the drug. It seems that there is a good correlation between the amount of drug that was consumed by the mother during pregnancy and the hair findings of the new-born at the time of delivery. Several weeks or months after delivery, identification of a drug in hair can indicate: (1) *in utero* exposure, or (2) exposure after delivery, or (3) a mix of both situations. There have been no publications about the disappearance of drug from hair after discontinuation of use or exposure (in case of *in utero* exposure). When the ratio of the concentration of the proximal segment to the concentration of the distal segment is lower than 0.5, it has been proposed by Kintz[23] that for the interpretation of the results to consider it 100% *in utero* origin. This can be applied only when the child is under one year old and the hair shaft length is at least 4 cm (to achieve suitable segmentation). These restrictive criteria (age of the child, length of the hair strand) were included in his statement as they apply to the non-homogenous growing phases in the hair and variability of growth rates in children aged less than one year. A minimal hair length (4 cm) is needed to achieve segmental analyses with good representation of the various periods of growth.

In recent years, several papers have demonstrated that the final interpretation can remain inconclusive due to possible environmental contamination.[24–27]

The final interpretation of drug findings in children's hair is very complicated as this can result from drug oral administration by the parents or relatives, close contact to drug users (hands, bedding, dishes), or self-consumption after touching contaminated items. Infants are at a higher risk for environmental exposure to drugs from caregivers who use drugs. Higher concentrations of drugs have been found in infants and toddlers than in older children.[24] Younger children spend most of their time with their parents and, therefore, the chance of passive exposure to medications from sweat or sebum from parents is greater. In a young child, the identification of an active compound in hair does not always mean administration. It can be the result of administration or accidental intake, but also of environmental contamination. The potential contamination by external sources of drugs remains the critical point when dealing with children. The concentration in hair is not always a measure of the degree of danger and a child with lower hair values may be more endangered than another with higher values. Due to a lack of controlled studies, it is not possible to establish the frequency and the dosages of exposure.

Therefore, hair results should not be used in isolation but should be seen as a part of the whole picture and be corroborated by further investigations before decisions are made regarding the child.

13.5 Single Hair *Versus* a Strand of Hair

Due to the recent development of imaging instruments and their availability in toxicological laboratories, it has been proposed to use these tools to test for drugs using a single hair. Matrix-assisted laser desorption/ionization (MALDI) combined with imaging is an innovative and powerful tool used for a few years, especially in forensic research. Some authors have proposed innovative methods to monitor drug consumption through direct mapping of the compounds with a high spatial distribution in human hair samples by MALDI-MS.[28,29] In addition to higher temporal resolution, this approach can be used to document potential external contamination.

However, mass spectrometry imaging techniques are very expensive and are unlikely to be useful for routine analysis during the next few years. For 12 years, there have been publications that claim that drug(s) can be detected in a single hair using highly sensitive conventional LC–MS/MS or LC–HRMS. Alibe *et al.*[30] have reviewed such possibilities. It seems that the use of a single segmented hair in comparison with a complete lock of hair may be more informative to obtain a retrospective timeline of drug exposure. To exclude the risk of using hair fibres in different growth cycles (anagen *vs.* catagen + telogen), one has to test several single hairs. The variations due to hair thickness between individual hairs can be limited by expressing the results in pg per length unit, such as mm. Segmental analysis of a single hair can overcome the issues of poor alignment and irregular hair growth which are sometimes observed after segmentation of a complete hair lock.

13.6 Minimal Detectable Drug Dose in Hair and Significance of a Negative Result

When using hair analysis as a matrix during investigative analysis, the important question is whether the analytical procedure was sensitive enough to identify traces of drugs; this is particularly important when the urine sample(s) of the subject was positive and the hair sample(s) was negative.

It has been accepted in the forensic community that a negative hair result cannot exclude the administration of a particular drug, or one of its precursors and the negative findings in hair should not overrule a positive urine result. Nevertheless, a negative hair test can, on occasion, cast doubt on the positive urine analysis, resulting in substantial legal debate and various consequences for the subject. A negative hair result is also a result. However, this can be interpreted in two different ways: (1) the owner of the hair did not take or was not exposed to the specific drug, or (2) the procedure is not sensitive enough to detect the drug.

Until laboratories have sensitive enough methodologies to detect drugs following a single use, care should be taken to compare urine and hair findings.[31]

The concept of minimal detectable dosage in hair is of interest to document negative findings, but limited data is currently available in the scientific literature. Such data includes cocaine, codeine, ecstasy, ketamine, some benzodiazepines, and some unusual compounds tested for particular cases, mostly criminal or during anti-doping controls. There is obviously a lack of data in the scientific literature. As an example, the minimum detectable dosage in hair has not been established for common drugs of abuse, such as cannabis, heroin, morphine, methadone, or amphetamine.

Several variables can influence the detection of a drug in hair and Table 13.1 lists reasons that can account for the absence of analytical response in hair after drug administration. The drug may not be incorporated in hair: that is the case for large bio-molecules, such as hormones, which cannot be transferred from the blood capillaries to growing cells of hair. Cosmetic treatments (perming, colouring, bleaching) or environmental aggressions (UV radiation, thermal application) will always reduce the concentrations. It is also possible that the administered parent compound is not the target compound in hair. This is, for example, the case with ethanol and ethyl glucuronide. Given that drugs are incorporated into hair according to several parameters such as melanin affinity, lipophilicity, and membrane permeability, some are well incorporated while others are poorly incorporated.

A recent publication[32] has reviewed the factors that can affect drug detection in hair and Table 13.1 lists the 10 possible scenarios, based on the experience of the author.

Table 13.1 Factors that can affect the detection of a drug in hair.[32]

Item	Factors that can affect the detection of a drug in hair
1	The subject does not take the drug
2	The drug is not incorporated into hair
3	The drug is not the correct target in hair (chemical marker)
4	The drug is fully extracted during hair decontamination
5	The drug is hydrolyzed/destroyed during hair preparation
6	The drug is unstable in hair once incorporated
7	The analytical method is not sensitive enough
8	Cosmetic treatment(s) of the hair have extracted/destroyed the drug
9	Environmental degradation of hair occurred (UV light, water...)
10	The subject has used detox products

Knowing the expected concentration after a single exposure is of great interest in two major situations, including drug-facilitated crimes and doping control. Indeed, in the first case, the victim is generally exposed to a specific drug on a single occasion. Due to the progress of the analytical instruments, it is now possible to identify in hair a single exposure to most sedatives or hypnotics.[31] Concentrations, after a single exposure to these drugs are in the range 1–10 pg mg^{-1}. It is therefore easy to compare the concentration measured in a specific case to what is expected after a controlled single administration and this will help in interpreting any data, including a negative hair test result. In the second situation, *i.e.*, doping control, it is accepted by active scientists that a single IM injection or oral administration of an anabolic steroid will not show up in hair due to limited incorporation. This has resulted in numerous challenging hearings with sport authorities that claimed that a negative hair test for steroids has no interest as doping violation can be established on a single administration. However, when it is possible to demonstrate that a single exposure of a drug, even at therapeutic or recreational dose, will produce a significant concentration in hair, it is much easier to document a scenario of contamination during intimate moments or the use of poor quality supplements. For example, a single exposure to cocaine (32 mg) will produce a response in hair close to 100 pg mg^{-1}[31]. Therefore, measuring something lower than 10–20 pg can be consistent with drinking coca tea from bags containing small amounts (3–5 mg) of cocaine. Such a situation has also been described for diuretics or letrozole in the scientific literature.

13.7 Should FAEE (Ethyl Palmitate) Always be Tested for in Addition to EtG to Demonstrate Excessive Alcohol Consumption?

Considering the large scale of alcohol-associated problems, the diagnosis of excessive alcohol consumption is an important task from a medical and legal point of view. Major markers of ethanol consumption in hair are ethyl glucuronide (EtG) and fatty acid ethyl esters (FAEEs). Detection of EtG in hair can be associated with excessive alcohol consumption, whereas a negative result does not unambiguously exclude alcohol intake. In particular, acute intoxication or binge drinking cannot be diagnosed through hair analysis.

FAEEs can also be used to monitor excessive alcohol consumption. FAEEs are formed in the presence of ethanol and free fatty acids, triglycerides, lipoproteins, or phospholipids by a FAEE synthase found in the liver and also in hair roots. Measurement of FAEEs has been suggested as they appear responsive to alcohol-induced organ damage. Initially, four FAEEs (ethyl myristate, ethyl palmitate, ethyl oleate, and ethyl stearate) were proposed as the most suitable markers for the detection of heavy alcohol consumption. Today, the single measure of ethyl palmitate is considered as the standard.[18]

Pragst *et al.*[33] have extensively encouraged that EtG and FAEEs are tested for simultaneously with several promising epidemiological studies and numerous graphs. Other scientists, mostly from private laboratories, were of the same opinion. One cannot exclude that financial income was a major reason for this situation. Indeed, when testing hair specimens simultaneously for both EtG and FAEE, a large number of discrepancies have been observed. For example, Kintz *et al.*[34] reported that the hair of 97 subjects was analyzed for EtG and FAEE according to the Society of Hair Testing guidelines to examine the role of both tests in documenting chronic excessive alcohol drinking. There were 27 (27.8%) results that were EtG negative and FAEE positive, when applying the SoHT cut-offs, probably due to the use of alcohol-containing hair products. There were four cases (4.1%) which were EtG positive and FAEE negative that were attributed to the use of herbal lotions containing EtG. To date, the use of herbal lotions has been described as the unique source of false positive EtG results.[35]

For the author, the best approach to document excessive alcohol testing is the combination of measuring EtG in hair and carbohydrate-deficient transferrin (CDT) or phosphatidylethanol (PEth) in blood.

13.8 Is GHB Testing in Hair Reliable to Document a Single Exposure?

Since GHB is present in the hair of the general population under physiological concentrations, toxicologists must be able to discriminate between endogenous levels and a concentration resulting from exposure. The implementation of a cut-off concentration must be done cautiously, due to the wide distribution of endogenous concentrations, from 0.5 to 12.0 $ng\,mg^{-1}$ for example in the pioneer study of Kintz *et al.*[36] The solution is to use each subject as his own control. From the demonstration that physiological concentrations are stable along the hair shaft, except at the root due to sweat contamination, one can suppose that exposure will lead to a peak concentration that can be detected. Use of tandem mass spectrometry is mandatory because of the low amount of hair that needs to be tested as a consequence of the short 3 mm segments that need to be analyzed. To avoid difficulties in interpretation due to potential contamination by sweat, it is necessary to wait for 4–5 weeks before hair collection. This will permit the migration of the GHB spot along the length of the hair shaft.

Since then, several papers have been published in the literature, sometimes with very conflicting statements.[37–43] Indeed, for some scientists it is simply not possible to document a single exposure to GHB during a drug-facilitated event using the hair matrix. In a very recent paper,[43] the authors have clearly demonstrated that the initial approach of Kintz *et al.*[36] is not applicable. In my opinion, the presented study raised numerous concerns, which need to be carefully discussed to avoid providing a wrong scientific message. The protocol proposed in 2003[36] must be strictly applied, including hair collection within 4–5 weeks after possible exposure (and not after several months), segmentation with short segments, *i.e.*, 3 mm (and not with 10 mm segments that will produce drug dilution) and use of sophisticated analytical equipment of gas chromatography coupled to tandem mass spectrometry (and not a single quadrupole). In order to avoid wrong scientific interpretation and approaches like the ones adopted by these authors, there are some pre-requisites that must be followed as GHB is a very complicated substance.

13.9 Reliability of Doping Control Using Hair in Cases of Contamination

As it is the case for the general population, amateur and professional athletes can be exposed to a specific drug by environmental factors. This has been suggested by Thevis *et al.*,[44] under the concept of athlete exposome. Part of this situation is due to the development of very sensitive instruments that can be coupled to ultra specific detectors, such as high-resolution mass spectrometry. Indeed, laboratories are now able to measure drugs in urine at the pg mL^{-1} level with suitable accuracy. As a consequence, an increasing number of adverse analytical findings (AAF) have been recently observed during doping controls, with estimated drug concentration close to or even much lower than 1 ng mL^{-1}. These AAF are generally vigorously challenged by the athletes, claiming that the findings correspond to a scenario of contamination and not to an anti-doping rules violation (ADRV).

A low or a very low urine concentration can be interpreted in two different ways: (1) it can be the tail end of a drug used to enhance performance; or (2) it is the direct consequence of a contamination. By contamination, one can include laced supplements either voluntarily or by lack of hygiene measures, contaminated meat by growth promoters, poor quality pharmaceuticals with chemical residues and finally drug transfer during intimate moments. This last type of contamination is presented by journalists as the "kissing defence" or the "sex defence".

In all these cases, after the ADRV had been notified, a ban (period of ineligibility where the athlete is not allowed to compete) was imposed on the athlete. As the athletes denied cheating and taking prohibited substances, they appointed a legal team to look after their defence. In the World anti-doping code, it is indicated that if someone establishes in an individual case that the athlete bears no fault or negligence, the otherwise applicable period of ineligibility shall be eliminated. The task of the legal representative, as per the rules of the Code to satisfy the no fault or negligence, was to address the following points: (1) the athlete must present verified circumstances of contamination; (2) the source of contamination must be clearly identified; and (3) the athlete must demonstrate that the violation was not intentional (the athlete did not knowingly take the controlled substance).

Acceptance by sport federations or CAS of third-party contamination to account for an AAF has some pre-requisites, which include, at least, the following points.[45]

(1) To support incidental exposure, only minute amounts of prohibited drug or metabolite(s) must be present in the urine collected for control. Concentrations which are observed after recreational abuse or for doping purposes cannot account for contamination. The detection of the parent drug can be of interest to document the moment of the last exposure and will allow estimating, for some drugs (*e.g.*, cocaine), its dose. This dose has to be too small to have any recreational or performance-enhancing effect.

(2) The athlete hair test, which can support either (1) a negative result (with a suitable limit of detection) to confirm incidental exposure or (2) a result that can support external or environmental contamination, must be presented. This hair test must cover the period of time where the specimen was collected for doping control (one can use the standard head hair growth rate of 1 cm per month – alternatively, body hair or nails have to fit to the requested window of detection). A negative hair test that does not have the required sensitivity, does not exclude intentional doping and should not be presented as such. In addition, only fully validated methods should be applied and not testing developed on the spot.

(3) The athlete's legal team must present verified circumstances of contamination or evidence of drug possession (contaminated environment, claims of drug use, proof of drug, supplement or ingredient purchase, invoices, receipts, *etc.*). The source of contamination must be established, as, on a balance of probability, the indicted athlete bears the burden of persuading the judging body that the occurrence of the circumstances on which they rely is more probable than their non-occurrence or more probable than other possible explanations of the doping offence.

(4) There must be verified claims of the athlete about the fact that they did not knowingly take the prohibited substance, *i.e.*, that the violation was not intentional, although no test exists that can positively and definitively exclude the intentional use of a drug prior to the collection of a specimen for doping control.

(5) An experienced pharmacologist or toxicologist has to perform a pharmacological verification of the athlete's claims. This means that pharmacokinetic and pharmacodynamics parameters of the specific drug must fit with the submitted scenario and the athlete must establish how the prohibited substance entered their system.

(6) A third-party positive hair test to support the claims of drug use by the partner and therefore being the source of contamination during intimate moments, must be presented.

13.10 Influence of Drug Stability in Cases of Cosmetic Treatment

An important issue of concern for drug analysis in hair is the change in the drug concentration induced by cosmetic treatment of hair. Hair is continuously subjected to natural factors, such as sunlight, weather, water, pollution, *etc.*, which affect and damage the cuticle but hair cosmetic treatments enhance that damage. Particular attention has been focused on the effects of repeated shampooing, perming, relaxing, and dyeing of hair. Repeated shampooing was found to have no significant action on the drug content of hair. After cosmetic treatments, drug concentrations decline dramatically by decreasing from 50–80% the original concentration. The products used for cosmetic treatments, like bleaching, permanent waving, dyeing, or relaxing, are strong bases. They will cause hair damage and affect drug content (by loss) or affect directly drug stability.

Cosmetic treatments (bleaching, coloring) can be responsible for

- axial diffusion of drugs along the hair fiber[46]
- loss of some amounts of drug as coloring will open the hair cuticle (the outside protein matrix). This damage to the ultrafine structure of the hair cuticle (surface, endocuticle, and cell membrane complex) and cortex (cell membrane complex) is a situation prone to drug evasion.[47]

The author administered to a volunteer a 6 mg tablet of bromazepam. Hair was collected 32 days after and tested for the drug by LC–MS/MS. Immediately after collection, a strand of hair was bleached, collected, and also analysed by LC–MS/MS. Segmental analyses are presented in the Table 13.2, clearly showing axial diffusion and drug loss.

In some cases of thermal straightening, an increase of drug concentration was observed that may be explained by denaturation of the hair matrix by thermal treatment possibly causing a better extraction of the drugs.[48]

Cosmetic treatment, such as coloring, will increase hair porosity. In this situation, hair swelling and water absorption capacity increase, producing a displacement of the spot of exposure by radial migration of the drug. In addition, thermal smoothing has also been reported to produce drug migration.[49]

It has been reported that radial swelling of the hair fiber with radial diffusion was the first and main process to appear when hair was exposed to water. The authors have indicated that this phenomenon was particularly observed in damaged hair.[50]

Table 13.2 Segmental analysis of a volunteer receiving 6 mg bromazepam.

Event	Segment 0–2 cm	Segment 2–4 cm	Segment 4–6 cm
Before bleaching	11 pg mg^{-1}	Not detected	Not detected
After bleaching	4 pg mg^{-1}	3 pg mg^{-1}	1 pg mg^{-1}

It has also been reported[51] that for quetiapine a significant dose–concentration relationship was observed in segmented hair from individuals with natural hair colour but not for individuals with dyed/bleached hair due to a higher variation in consecutive segmental concentrations. Finally, in another study,[52] it has been published that after a single controlled administration of diazepam, in hair collected one month later, the drug was present in segment 0–2 cm (which was expected), but also in segment 2–4 cm, and even in segment 4–6 cm for some subjects (which was not expected). The authors indicated that this axial migration could be due to sweat and sebum.

Indeed, it has been known since 1996[53] that a single administration of drug (here cocaine) can produce several positive consecutive segments.

13.11 Conclusion

Hair is a unique and challenging matrix for toxicological testing in that detailed information on historical use is recorded over time as compared to traditional blood and urine matrices. In order to make the best use of hair testing in forensic casework, a thorough understanding of the advantages and limitations of this matrix is essential. The value of hair analysis for the identification of drug exposure in a person is steadily gaining recognition. This can be seen from its growing use in various forensic and clinical applications.

Hair analysis may be a useful adjunct to conventional (urine, blood) drug testing in toxicology. Specimens can be more easily obtained with less difficulty, and hair can provide a more accurate history of drug use. GC and LC–MS are the methods of choice for hair analysis including use of tandem MS for detection of low dose compounds. There are still controversies on how to best interpret the results and it must be accepted that they can be challenged during tribunal hearings. This demonstrates that the debate is open and more data are needed to firmly establish the reliability of hair testing for drugs.

References

1. Society of Hair Testing, *Forensic Sci. Int.*, 1997, **84**, 3.
2. Society of Hair Testing, *Forensic Sci. Int.*, 2004, **145**, 83.
3. M. A. Lebeau and M. A. Montgomery, *J. Anal. Toxicol.*, 2009, **33**, 343.
4. D. Mantinieks, P. Wright, M. Di Rago and D. Gerostamoulos, *Drug Test. Anal.*, 2019, **11**, 1542.
5. V. A. Hill, M. I. Schaffer, R. B. Paulsen and G. N. Stowe, *Drug Test. Anal.*, 2022, **14**, 1557.
6. M. D. M. Ramirez Fernandez, S. M. R. Wille, M. Yegles and N. Samyn, *Drug Test. Anal.*, 2022, **14**, 1155.
7. F. Pragst, H. Sachs and P. Kintz, *J. Anal. Toxicol.*, 2010, **34**, 354.
8. P. Kintz, A. Tracqui and P. Mangin, *Int. J. Leg. Med.*, 1992, **105**, 1.

9. P. R. Stout, J. D. Ropero-Miller, M. R. Baylor and J. M. Mitchell, *J. Anal. Toxicol.*, 2006, **30**, 490.
10. G. Romano, N. Barbera and I. Lombardo, *Forensic Sci. Int.*, 2001, **123**, 119.
11. G. A. A. Cooper, R. Kronstrand and P. Kintz, *Forensic Sci. Int.*, 2012, **218**, 20.
12. M. Schaffer, C. C. Cheng, O. Chao, V. Hill and P. Matsui, *Anal. Bioanal. Chem.*, 2016, **408**, 2043.
13. B. Moosmann, N. Roth and V. Auwärter, *Sci. Rep.*, 2015, **5**, 14906.
14. P. Kintz, *Forensic Sci. Int.*, 2012, **215**, 73.
15. K. Kuwayama, H. Miyaguchi, T. Kanamori, K. Tsujikawa, T. Yamamuro, H. Segawa, Y. Okada and Y. T. Iwata, *Forensic Toxicol.*, 2022, **40**, 64.
16. C. Scholz, M. M. Madry, T. Kraemer and M. R. Baumgartner, *J. Anal. Toxicol.*, 2022, **46**, 504.
17. R. Erne, L. Bernard, A. E. Steuer, M. R. Baumgartner and T. Kraemer, *Anal. Chem.*, 2019, **91**, 4132.
18. http://www.soht.org/images/pdf/Revision_2019_Alcoholmarkers.pdf, accessed 14 February 2023.
19. M. E. Albermann, F. Musshoff, L. Aengenheister and B. Madea, *Anal. Bioanal. Chem.*, 2012, **403**, 769.
20. N. Kummer, S. M. Wille, V. Di Fazio, M. Ramirez Fernadez, M. Yegles, W. E. Lambert and N. Samyn, *J. Anal. Toxicol.*, 2015, **39**, 17.
21. P. Kintz, A. Farrugia, A. Ameline, A. Eibel and J. S. Raul, *Forensic Sci. Int.*, 2017, **280**, 176.
22. N. Alibe, H. Eysseric-Guérin, P. Kintz, M. Bartoli, C. Bost-Bru, F. Grenier, V. Scolan and F. Stanke-Labesque, *Forensic Sci. Int.*, 2015, **249**, 53.
23. P. Kintz, *Forensic Sci. Int.*, 2015, **249**, 314.
24. F. Pragst, F. Krumbiegel, D. Thurmann, L. Westendorf, M. Methling, A. Niebel and S. Hartwig, *Forensic Sci. Int.*, 2019, **297**, 161.
25. J. C. Alvarez, L. Lasne, I. Etting, G. Chéron, V. Abadie, N. Fabresse and I. A. Larabi, *Int. J. Leg. Med.*, 2018, **132**, 165.
26. P. Kintz, A. Ameline, A. Eibel, L. Gheddar, E. Feisthaurer, A. Geraut, L. Berthelon, A. Farrugia and J. S. Raul, *Curr. Pharm. Biotechnol.*, 2017, **18**, 791.
27. P. Kintz, A. Ameline and J. S. Raul, *J. Anal. Toxicol.*, 2020, **44**, 747.
28. A. Kernalléguen, C. Enjalbal, J. C. Alvarez, O. Belgacem, G. Léonetti, D. Laffite and A. L. Pélissier-Alicot, *Anal. Chim. Acta*, 2018, **1041**, 87.
29. R. Erne, L. Bernard, A. E. Steuer, M. R. Baumgartner and T. Kraemer, *Anal. Chem.*, 2019, **91**, 4132.
30. N. Alibe, P. Kintz, A. faure, F. Paysant, A. P. Michard-Lenoir, F. Stanke-Labesque, V. Scolan and H. Eysseric-Guérin, *Curr. Pharm. Des.*, 2017, **23**, 5502.
31. P. Kintz, *Forensic Sci. Int.*, 2012, **218**, 28.
32. P. Kintz, *Clin. Chem. Lab. Med.*, 2020, **59**, 267.
33. F. Pragst, M. Rothe, B. Moench, M. Hastedt, S. Herre and D. Simmert, *Forensic Sci. Int.*, 2010, **196**, 101.

34. P. Kintz and D. Nicholson, *Forensic Sci. Int.*, 2014, **243**, 44.
35. P. Kintz and D. Nicholson, *Forensic Toxicol.*, 2014, **32**, 176.
36. P. Kintz, V. Cirimele, C. jamey and B. Ludes, *J. Forensic Sci.*, 2003, **48**, 195.
37. J. P. Goullé, M. Chèze and G. Pépin, *J. Anal. Toxicol.*, 2003, **27**, 574.
38. E. Bertol, F. Mari, F. Vaiano, G. Romano, S. Zaami, G. Baglio and F. P. Busardo, *Drug Test. Anal.*, 2015, **7**, 376.
39. P. Kintz, *J. Anal. Toxicol.*, 2016, **40**, 563.
40. F. P. Busardo, F. Vaiano, G. Mannocchi, E. Bertol, S. Zaami and E. Marinelli, *Drug Test. Anal.*, 2017, **9**, 953.
41. F. P. Busardo, S. Pichini, S. Zaami, R. Pacifici and P. Kintz, *Clin. Chem. Lab. Med.*, 2018, **56**, 198.
42. E. C. Strickland, J. L. Thomas, E. W. Lloyd, M. A. Smith, M. A. Lebeau, M. A. Montgomery, R. P. Karas, E. M. Peters and M. L. Miller, *J. Anal. Toxicol.*, 2020, **44**, 637.
43. S. Mestria, S. Odoardi, G. Biosa and S. Strano Rossi, *Drug Test. Anal.*, 2022, DOI: 10.1002/dta.3423.
44. M. Thevis, T. Kuuranne, M. Fedoruk and H. Geyer, *Drug Test. Anal.*, 2021, **13**, 1814.
45. P. Kintz, *J. Anal. Toxicol.*, 2021, **45**, e3.
46. P. Kintz, *Ther. Drug Monit.*, 2013, **35**, 408.
47. G. Zhang, R. L. McMullen and L. Kulcsar, *J. Cosmet. Sci.*, 2016, **67**, 1.
48. J. Ettlinger and M. Yegles, *Forensic Sci. Int.*, 2016, **265**, 13.
49. M. D. M. Ramirez Fernandez, S. M. R. Wille, V. Di Fazio and N. Samyn, *Forensic Sci. Int.*, 2016, **297**, 277.
50. G. Skopp, L. Pötsch and R. Aderjan, *J. Forensic Sci.*, 1996, **41**, 117.
51. K. N. Günther, S. S. Johansen, M. K. K. Nielsen, P. Wicktor, J. Banner and K. Linnet, *Forensic Sci. Int.*, 2018, **285**, 58.
52. X. Wang, S. S. Johansen, Y. Zhang, J. Jia, Y. Rao, F. Jiang and K. Linnet, *Int. J. Leg. Med.*, 2017, **131**, 131.
53. G. Henderson, M. R. Harkey, C. Zhou, R. T. Jones and P. Jacob 3rd, *J. Anal. Toxicol.*, 1996, **20**, 1.

Subject Index